Hilbert C^*-Modules

Contents

Preface

Hilbert C^*-modules provide a natural generalization of Hilbert spaces arising when the field of scalars $\mathbf{C}$ is replaced by an arbitrary C^*-algebra. This generalization, in the case of commutative C^*-algebras, appeared in the paper [**59**] of I. Kaplansky; however the noncommutative case seemed too complicated at that time. The general theory of Hilbert C^*-modules (i.e., for an arbitrary C^*-algebra serving as 'scalars') appeared 25 years ago in the pioneering papers of W. Paschke [**100**] and M. Rieffel [**108**]. This theory has proved to be a very convenient tool in the theory of operator algebras, allowing one to obtain information about C^*-algebras by studying Hilbert C^*-modules over them. In particular, a series of results about some classes of C^*-algebras (like AW^*-algebras and monotone complete C^*-algebras) was obtained in this way [**38**]. An important notion of Morita equivalence for C^*-algebras was also formulated in terms of Hilbert C^*-modules [**109, 18**]. This notion also has applications in the theory of group representations. It turned out to be possible to extract information on group actions from Hilbert C^*-modules arising from these actions [**102, 110**]. Some results about conditional expectations of finite index [**5, 133**] and about completely positive maps of C^*-algebras [**2**] were also obtained using Hilbert C^*-modules.

The theory of Hilbert C^*-modules may also be considered as a noncommutative generalization of the theory of vector bundles [**33, 69**]. This was the reason Hilbert modules became a tool in topological applications — namely in index theory of elliptic operators, in K- and KK-theory [**92, 90, 62, 63, 64, 65, 128**] and in noncommutative geometry as a whole [**24, 29**].

Among other applications, one should emphasize the theory of quantum groups [**135, 136**], unbounded operators as a tool for Kasparov's KK-theory [**3, 4**] (also Section 8.4) and some physical applications [**72, 80**].

Alongside these applications, the theory of Hilbert C^*-modules itself has been developed as well. A number of results about the structure of Hilbert modules and about operators on them have been obtained [**74, 42, 88, 77, 81, 127**]. Besides these results, an axiomatic approach in the theory of Hilbert modules based on the theory of operator spaces and tensor products was developed [**12, 11**].

A detailed bibliography of the theory of Hilbert C^*-modules can be found in [**43**].

Some results presented here were only announced in the literature or the proofs were discussed rather briefly. We have tried to fill such gaps. We could not discuss all the aspects of the theory of Hilbert C^*-modules here, but we tried to explain in detail the basic notions and theorems of this theory, a number of important examples, and also some results related to the authors' interest.

A major part of this book formed the content of the lecture course presented by the authors at the Department of Mechanics and Mathematics of Moscow State University in 1997.

We are grateful to A. S. Mishchenko for introducing the theory of Hilbert C^*-modules to us. Together with Yu. P. Solovyov, he has acquainted us with the circle of problems related to its applications in topology.

While working in this field and in the process of writing the present text, a significant influence on us was made by our friend and co-author M. Frank.

We have discussed a number of problems of the Hilbert C^*-module theory with L. Brown, A. A. Irmatov, G. G. Kasparov, R. Nest, G. K. Pedersen and W. Paschke. Some applications were considered also with J. Cuntz, A. Ya. Helemskii, J. Kaminker, V. Nistor, J. Rosenberg, K. Thomsen, B. L. Tsygan and others.

Our research was partially supported by a series of subsequent Russian Foundation of Basic Research grants.

CHAPTER 1

Basic Definitions

1.1. C^*-algebras

For basic information about C^*-algebras we refer to the books [**31, 104, 123, 58**]. We present here only some results on C^*-algebras, which will be necessary for our purpose.

Recall that an involutive Banach algebra A is called a *C^*-algebra* if the equality

$$\|a^*a\| = \|a\|^2$$

holds for each element $a \in A$. Any C^*-algebra can be realized as a norm-closed subalgebra of the algebra of bounded operators $\mathcal{B}(H)$ on a Hilbert space H. We do not assume existence of the unit element in C^*-algebras. By A^+ we denote the C^*-algebra obtained from the C^*-algebra A by *unitalization*. As a linear space with involution, A^+ coincides with $A \oplus \mathbf{C}$ and multiplication in A^+ is given by the formula $(a, z)(b, w) = (ab + zb + aw, zw)$, $a, b \in A$, $z, w \in \mathbf{C}$. Any pair $(a, z) \in A^+$ defines an operator $A \to A$ by $b \mapsto ab + zb$ and the norm of (a, z) is the norm of this operator.

The *spectrum* of an element a of a unital C^*-algebra is the set $\mathrm{Sp}(a)$ of complex numbers z such that $a - z \cdot 1$ is not invertible. If a C^*-algebra A has no unit, then the spectrum of an element $a \in A$ is defined as its spectrum in the C^*-algebra $A^+ \supset A$. The spectrum is a compact subset of $\mathbf{C}$. An element $a \in A$ is called *positive* (we write $a \geq 0$) if it is *Hermitian*, i.e., if it satisfies the condition $a^* = a$, and if one of the following equivalent [**31**, 1.6.1] conditions holds:

(i) $\mathrm{Sp}(a) \subset [0, \infty)$;
(ii) $a = b^*b$ for some $b \in A$;
(iii) $a = h^2$ for some Hermitian $h \in A$.

The set of all positive elements $P^+(A)$ forms a closed convex cone in A and $P^+(A) \cap (-P^+(A)) = 0$. Among the Hermitian elements h defined in (iii) there exists a unique positive element, which is called the *positive square root* of a (we write $h = a^{1/2}$).

A linear functional $\varphi : A \longrightarrow \mathbf{C}$ is called *positive* if $\varphi(a) \geq 0$ for any positive element $a \in P^+(A)$. A positive linear functional is called a *state* if $\|\varphi\| = 1$. We have $\|a\| = \sup \varphi(a)$, where $a \geq 0$ and the supremum is taken over all states.

A C^*-homomorphism of an algebra A into the C^*-algebra $\mathcal{B}(H)$ of all bounded operators on a Hilbert space H is called a *representation*. A vector $\xi \in H$ is called *cyclic* for the representation $\pi : A \longrightarrow \mathcal{B}(H)$ if the set of all vectors of the form $\pi(a)\xi$, $a \in A$, is dense in H. The vector $\xi \in H$ is called *separating* for the representation $\pi : A \longrightarrow \mathcal{B}(H)$ if the equality $\pi(a)\xi = 0$ implies $a = 0$.

To each positive linear functional ω on a C^*-algebra A we can associate a unique (up to the unitary equivalence) representation π_ω of the algebra A on some

Hilbert space H_ω and a vector $\xi_\omega \in H_\omega$ such that $\omega(a) = (\pi_\omega(a)\xi_\omega, \xi_\omega)$ for all $a \in A$ and the vector ξ_ω is cyclic. The construction of such a representation is called the *GNS-construction.*

An *approximate unit* of a C^*-algebra A is an increasing net $e_\alpha \in A$, $\alpha \in \mathcal{A}$, such that $\|e_\alpha\| \le 1$ and $\lim \|a - ae_\alpha\| = 0$ for any $a \in A$. Each C^*-algebra has an approximate unit e_α such that $e_\alpha \ge 0$ and $e_\alpha \ge e_\beta$ for $\alpha \ge \beta$ [**31**].

DEFINITION 1.1.1. A C^*-algebra possessing a countable approximate unit is called *σ-unital.*

DEFINITION 1.1.2. An element $h \in A$ is called *strictly positive* if for any positive nonzero linear functional φ (or, equally, for any state) one has $\varphi(h) > 0$.

REMARK 1.1.3. Existence of a strictly positive element is equivalent to existence of a countable approximate unit. One can assume that $e_i \ge 0$. Then $h := \sum_i e_i/2^i$ is strictly positive. Conversely, $e_i := h^{1/i}$ is a countable approximate unit. Any separable C^*-algebra is σ-unital. The details can be found in [**104**].

We will often use the following statements.

LEMMA 1.1.4 ([**104**, Lemma 1.4.4]). *Let x, y and a be elements of a C^*-algebra A such that $a \ge 0$ and*

$$x^*x \le a^\alpha, \qquad yy^* \le a^\beta, \qquad \alpha + \beta > 1.$$

Then the sequence $u_n = x[(1/n) + a]^{-1/2} y$ is norm-convergent in A to an element u such that $\|u\| \le \|a^{(\alpha+\beta-1)/2}\|$.

PROOF. Put $d_{nm} := [(1/n) + a]^{-1/2} - [(1/m) + a]^{-1/2}$. Then

$$\begin{aligned}
\|u_n - u_m\|^2 &= \|x d_{nm} y\|^2 = \|y^* d_{nm} x^* x d_{nm} y\| \\
&\le \|y^* d_{nm} a^\alpha d_{nm} y\| = \|a^{\alpha/2} d_{nm} y\|^2 \\
&= \|a^{\alpha/2} d_{nm} y y^* d_{nm} a^{\alpha/2}\| \le \|a^{\alpha/2} d_{nm} a^\beta d_{nm} a^{\alpha/2}\| = \|d_{nm} a^{(\alpha+\beta)/2}\|^2.
\end{aligned}$$

Using, for example, the Dini theorem, we can see that the sequence of functions

$$[(1/n) + t]^{-1/2} t^{(\alpha+\beta)/2}, \qquad t \in \mathrm{Sp}(a),$$

is uniformly convergent to $t^{(\alpha+\beta-1)/2}$ on the spectrum of a, hence, $\|d_{nm} a^{(\alpha+\beta)/2}\| \to 0$. Therefore, by the Cauchy criterion, $\{u_n\}$ is norm-convergent to an element $u \in A$. Then, reasoning as above, we obtain

$$\|u_n\| = \|x[(1/n) + a]^{-1/2} y\| \le \|a^{\alpha/2}[(1/n) + a]^{-1/2} a^{\beta/2}\| \le \|a^{(\alpha+\beta-1)/2}\|.$$

Hence $\|u\| \le \|a^{(\alpha+\beta-1)/2}\|$. □

PROPOSITION 1.1.5 ([**104**, Prop. 1.4.5]). *Let x and a be elements of a C^*-algebra A such that $a \ge 0$ and $x^*x \le a$. For any $0 < \alpha < \frac{1}{2}$ there exists an element $u \in A$ such that $\|u\| \le \|a^{\frac{1}{2}-\alpha}\|$ and $x = ua^\alpha$.*

PROOF. Put $u_n := x[(1/n) + a]^{-\frac{1}{2}} a^{\frac{1}{2}-\alpha}$. By Lemma 1.1.4, $\{u_n\}$ is norm-convergent to an element $u \in A$ such that

$$\|u\| \le \|a^{\frac{1}{2}(1+1-2\alpha-1)}\| = \|a^{\frac{1}{2}-\alpha}\|.$$

Then

$$\|x - u_n a^\alpha\|^2 = \|x(1 - [(1/n) + a]^{-1/2} a^{1/2})\|^2$$
$$\leq \|a^{1/2}(1 - [(1/n) + a]^{-1/2} a^{1/2})\|^2 \longrightarrow 0$$

as $n \longrightarrow \infty$, by the Dini theorem applied to the corresponding functions on the spectrum. Thus, $x = ua^\alpha$. □

1.2. Pre-Hilbert modules

Let $\mathcal{M}$ be a module over a C^*-algebra A. An action of an element $a \in A$ on $\mathcal{M}$ is denoted by $x \cdot a$, where $x \in \mathcal{M}$.

DEFINITION 1.2.1. A *pre-Hilbert A-module* is a (right) A-module $\mathcal{M}$ equipped with a sesquilinear form $\langle \cdot, \cdot \rangle : \mathcal{M} \times \mathcal{M} \longrightarrow A$ with the following properties:

(i) $\langle x, x \rangle \geq 0$ for any $x \in \mathcal{M}$;
(ii) $\langle x, x \rangle = 0$ implies that $x = 0$;
(iii) $\langle y, x \rangle = \langle x, y \rangle^*$ for any $x, y \in \mathcal{M}$;
(iv) $\langle x, y \cdot a \rangle = \langle x, y \rangle a$ for any $x, y \in \mathcal{M}$ and any $a \in A$.

The map $\langle \cdot, \cdot \rangle$ is called an *A-valued inner product.*

Here are some examples.

EXAMPLE 1.2.2. Let $J \subset A$ be a right ideal. Then J can be equipped with the structure of a pre-Hilbert A-module with the inner product of elements $x, y \in J$ defined by $\langle x, y \rangle := x^* y$.

EXAMPLE 1.2.3. Let $\{J_i\}$ be a countable set of right ideals of a C^*-algebra A and let $\mathcal{M}$ be the linear space of all sequences (x_i), $x_i \in J_i$, satisfying the condition $\sum_i \|x_i\|^2 < \infty$. Then $\mathcal{M}$ becomes a right A-module if the action of A is defined by $(x_i) \cdot a := (x_i a)$ for $(x_i) \in \mathcal{M}$, $a \in A$, and becomes a pre-Hilbert A-module if the inner product of elements $(x_i), (y_i) \in \mathcal{M}$ is defined by $\langle (x_i), (y_i) \rangle := \sum_i x_i^* y_i$.

Let $\mathcal{K}$ be a right A-module equipped with a sesquilinear map $[\cdot, \cdot] : \mathcal{K} \times \mathcal{K} \longrightarrow A$ satisfying all properties of Definition 1.2.1 except (ii). Put

$$N := \{x \in \mathcal{K} : [x, x] = 0\}.$$

For each positive linear functional φ on the C^*-algebra A the map $(x, y) \mapsto \varphi([x, y])$ is a (degenerate) inner product on $\mathcal{K}$, hence the set $N_\varphi = \{x \in \mathcal{K} : \varphi([x, x]) = 0\}$ is a linear subspace in $\mathcal{K}$. By taking the intersection of all such subspaces we see that $N = \bigcap_\varphi N_\varphi$ is also a linear subspace in $\mathcal{K}$. It follows from properties (iii) and (iv) of Definition 1.2.1 that $N \cdot A \subset N$. Therefore N is a submodule in $\mathcal{K}$. The quotient module $\mathcal{M} = \mathcal{K}/N$ is equipped with the obvious structure of a pre-Hilbert A-module with the inner product $\langle x + N, y + N \rangle := [x, y]$.

Let $\mathcal{M}$ be a pre-Hilbert A-module, $x \in \mathcal{M}$. Put $\|x\|_{\mathcal{M}} := \|\langle x, x \rangle\|^{1/2}$. We usually skip the subscript $\mathcal{M}$ when it does not lead to confusion of norms.

PROPOSITION 1.2.4 ([**100**]). *The function $\|\cdot\|_{\mathcal{M}}$ is a norm on $\mathcal{M}$ and satisfies the following properties:*

(i) $\|x \cdot a\|_{\mathcal{M}} \leq \|x\|_{\mathcal{M}} \cdot \|a\|$ *for any* $x \in \mathcal{M}$, $a \in A$;
(ii) $\langle x, y \rangle \langle y, x \rangle \leq \|y\|_{\mathcal{M}}^2 \langle x, x \rangle$ *for any* $x, y \in \mathcal{M}$;
(iii) $\|\langle x, y \rangle\| \leq \|x\|_{\mathcal{M}} \|y\|_{\mathcal{M}}$ *for any* $x, y \in \mathcal{M}$.

PROOF. For any positive linear functional φ on A, the function $x \mapsto \varphi(\langle x,x\rangle)^{1/2}$ defines a seminorm on $\mathcal{M}$. For each $x \in \mathcal{M}$,

$$\|x\|_{\mathcal{M}} = \|\langle x,x\rangle\|^{1/2} = \sup\{\varphi(\langle x,x\rangle)^{1/2}\},$$

where the supremum is taken over all states φ on A. Therefore $\|\cdot\|_{\mathcal{M}}$ is a seminorm and, by property (ii) of Definition 1.2.1, $\|\cdot\|_{\mathcal{M}}$ is a norm on $\mathcal{M}$. Statement (i) follows from the equality

$$\|x\cdot a\|_{\mathcal{M}}^2 = \|\langle x\cdot a, x\cdot a\rangle\| = \|a^*\langle x,x\rangle a\| \le \|a\|^2\,\|\langle x,x\rangle\| = \|x\|_{\mathcal{M}}^2\,\|a\|^2 .$$

To prove (ii) we take $x, y \in \mathcal{M}$ and a positive linear functional φ on A. Applying the Cauchy–Bunyakovskii inequality for the (degenerate) inner product $\varphi(\langle\cdot,\cdot\rangle)$ on $\mathcal{M}$ we obtain

$$\begin{aligned}\varphi(\langle x,y\rangle\langle y,x\rangle) &= \varphi(\langle x, y\cdot\langle y,x\rangle\rangle)\\ &\le \varphi(\langle x,x\rangle)^{1/2}\cdot\varphi(\langle y\cdot\langle y,x\rangle, y\cdot\langle y,x\rangle\rangle)^{1/2}\\ &= \varphi(\langle x,x\rangle)^{1/2}\cdot\varphi(\langle x,y\rangle\langle y,y\rangle\langle y,x\rangle)^{1/2}\\ &\le \varphi(\langle x,x\rangle)^{1/2}\cdot\|\langle y,y\rangle\|^{1/2}\cdot\varphi(\langle x,y\rangle\langle y,x\rangle)^{1/2}.\end{aligned}$$

Thus, for any positive linear functional φ, we have $\varphi(\langle y,x\rangle\langle x,y\rangle) \le \|y\|_{\mathcal{M}}^2\cdot\varphi(\langle x,x\rangle)$, so statement (ii) is proved. This evidently implies statement (iii). □

We call inequality (ii) (and also its consequence, inequality (iii)) of Proposition 1.2.4 the *Cauchy–Bunyakovskii inequality* for Hilbert C^*-modules.

REMARK 1.2.5. For any C^*-pre-Hilbert module, or more precisely, for any sesquilinear form $\langle\cdot,\cdot\rangle$, the *polarization equality*

$$4\langle y,x\rangle = \sum_{k=0}^{3} i^k\langle x+i^k y, x+i^k y\rangle$$

is obviously satisfied for all x, $y \in \mathcal{M}$.

1.3. Hilbert C^*-modules

DEFINITION 1.3.1. An A-module $\mathcal{M}$ that is at the same time a Banach space with a norm $\|\cdot\|$ satisfying the inequality $\|x\cdot a\| \le \|x\|\,\|a\|$, $x \in \mathcal{M}$, $a \in A$, is called a *Banach* A-module.

DEFINITION 1.3.2. A pre-Hilbert A-module $\mathcal{M}$ is called a *Hilbert C^*-module* if it is complete with respect to the norm $\|\cdot\|_{\mathcal{M}}$.

If $\mathcal{M}$ is a pre-Hilbert A-module, then the action of the C^*-algebra A and the A-valued inner product on $\mathcal{M}$ can be extended to the completion $\widetilde{\mathcal{M}}$, which thus becomes a Hilbert C^*-module. Consider some examples.

EXAMPLE 1.3.3. If $J \subset A$ is a right ideal, then the pre-Hilbert module J is complete with respect to the norm $\|\cdot\|_J = \|\cdot\|$. In particular, the C^*-algebra A itself is a *free* Hilbert A-module with one generator.

EXAMPLE 1.3.4. If $\{\mathcal{M}_i\}$ is a finite set of Hilbert A-modules, then one can define the direct sum $\oplus\mathcal{M}_i$. The inner product on $\oplus\mathcal{M}_i$ is given by the formula $\langle x,y\rangle := \sum_i\langle x_i,y_i\rangle$, where $x=(x_i), y=(y_i) \in \oplus\mathcal{M}_i$. We denote the direct sum of n copies of a Hilbert module $\mathcal{M}$ by $\mathcal{M}^n$ or $L_n(\mathcal{M})$.

Here and further we denote by $\oplus$ the direct sum of orthogonal submodules and by $\widetilde{\oplus}$ the direct sum of Banach subspaces without orthogonality.

EXAMPLE 1.3.5. If $\{\mathcal{M}_i\}$, $i \in \mathbf{N}$, is a countable set of Hilbert A-modules, then one can define their direct sum $\oplus\mathcal{M}_i$. On the A-module $\oplus\mathcal{M}_i$ of all sequences $x = (x_i) : x_i \in \mathcal{M}_i$, such that the series $\sum_i \langle x_i, x_i\rangle$ is norm-convergent in the C^*-algebra A, we define the inner product by

$$\langle x, y\rangle := \sum\nolimits_i \langle x_i, y_i\rangle \quad \text{for } x, y \in \oplus\mathcal{M}_i.$$

Let us check that this series converges. Since the series $\sum_i \langle x_i, x_i\rangle$ and $\sum_i \langle y_i, y_i\rangle$ are convergent, for any $\varepsilon > 0$ there exists a number N such that for all $n > 0$ we have

$$\left\| \sum_{i=N}^{N+n} \langle x_i, x_i\rangle \right\| < \varepsilon, \qquad \left\| \sum_{i=N}^{N+n} \langle y_i, y_i\rangle \right\| < \varepsilon.$$

Then

$$\left\| \sum_{i=N}^{N+n} \langle x_i, y_i\rangle \right\| \le \left\| \sum_{i=N}^{N+n} \langle x_i, x_i\rangle \right\| \cdot \left\| \sum_{i=N}^{N+n} \langle y_i, y_i\rangle \right\| < \varepsilon^2.$$

This proves that the inner product is well defined.

Let us verify completeness of the module $\oplus\mathcal{M}_i$. Let $x^{(n)} = (x_i^{(n)}) \in \oplus\mathcal{M}_i$ be a Cauchy sequence. Then for any $\varepsilon > 0$ there exists a number N such that

$$\left\| \sum_i \langle x_i^{(n)} - x_i^{(m)}, x_i^{(n)} - x_i^{(m)}\rangle \right\| < \varepsilon \tag{1.1}$$

for all $n, m \ge N$. Since all summands in (1.1) are positive, the inequality

$$\left\| \langle x_i^{(n)} - x_i^{(m)}, x_i^{(n)} - x_i^{(m)}\rangle \right\| < \varepsilon$$

holds for each number i separately. But then the sequences $x_i^{(n)} \in \mathcal{M}_i$ are Cauchy sequences, hence they converge to the limits $x_i = \lim x_i^{(n)} \in \mathcal{M}_i$. Let us verify that the series $\sum_i \langle x_i, x_i\rangle$ is norm-convergent in A. Let us fix $\varepsilon > 0$. There exists a number $n > N$ such that the estimate (1.1) holds. Let us choose a number K such that

$$\left\| \sum_{i=K}^{\infty} \langle x_i^{(n)}, x_i^{(n)}\rangle \right\| < \varepsilon.$$

Then, for any $k > 0$, we have

$$\begin{aligned} &\left\| \sum_{i=K}^{K+k} \Big(\langle x_i^{(m)}, x_i^{(m)}\rangle + \langle x_i^{(n)} - x_i^{(m)}, x_i^{(m)}\rangle + \langle x_i^{(m)}, x_i^{(n)} - x_i^{(m)}\rangle + \langle x_i^{(n)}, x_i^{(n)}\rangle \Big) \right\| \\ &\quad = \left\| \sum_{i=K}^{K+k} \langle x_i^{(n)} - x_i^{(m)}, x_i^{(n)} - x_i^{(m)}\rangle \right\| \le \left\| \sum_{i=1}^{\infty} \langle x_i^{(n)} - x_i^{(m)}, x_i^{(n)} - x_i^{(m)}\rangle \right\| < \varepsilon. \end{aligned}$$

Therefore

$$\left\|\sum_{i=K}^{K+k}\langle x_i^{(m)},x_i^{(m)}\rangle\right\| < 2\varepsilon+\left\|\sum_{i=K}^{K+k}\langle x_i^{(n)}-x_i^{(m)},x_i^{(m)}\rangle\right\|+\left\|\sum_{i=K}^{K+k}\langle x_i^{(m)},x_i^{(n)}-x_i^{(m)}\rangle\right\|$$
$$\leq 2\varepsilon+2\left\|\sum_{i=K}^{K+k}\langle x_i^{(n)}-x_i^{(m)},x_i^{(n)}-x_i^{(m)}\rangle\right\|^{1/2}\left\|\langle x_i^{(m)},x_i^{(m)}\rangle\right\|^{1/2}$$
$$\leq 2\varepsilon+2\varepsilon^{1/2}\left\|\langle x_i^{(m)},x_i^{(m)}\rangle\right\|^{1/2}.$$

Now, by solving the quadratic inequality, we obtain that

$$\left\|\sum_{i=K}^{K+k}\langle x_i^{(m)},x_i^{(m)}\rangle\right\| < (1+\sqrt{3})^2\varepsilon < 8\varepsilon. \tag{1.2}$$

Passing to the limit $m\to\infty$ in the inequality (1.2), we obtain that

$$\left\|\sum_{i=K}^{K+k}\langle x_i,x_i\rangle\right\| < 8\varepsilon.$$

This proves that the series $\sum_i\langle x_i,x_i\rangle$ is norm-convergent.

The direct sum of a countable number of copies of a Hilbert module $\mathcal{M}$ is denoted by $l_2(\mathcal{M})$ or $H_{\mathcal{M}}$. The Hilbert C^*-module $l_2(A)$ (another notation is H_A) is called the *standard* Hilbert C^*-module over A. If the C^*-algebra is unital, then the Hilbert module H_A possesses the standard basis $\{e_i\}$, $i\in\mathbf{N}$, where $e_i=(0,\dots,0,1,0,\dots,0,\dots)$ with the unit being the i-th entry.

EXAMPLE 1.3.6. Let $B\subset A$ be a C^*-subalgebra of a C^*-algebra A. Let A and B be unital with the common unit. Assume that there exists a linear map $E:A\longrightarrow B$ that is a projection (i.e., $E^2=E$) of norm ≤ 1. Such a map is called a *conditional expectation* from A to B. Conditional expectation is a positive map, i.e., $E(a^*a)\geq 0$ for all $a\in A$, and it satisfies the equality

$$E(b_1ab_2)=b_1E(a)b_2 \quad \text{for } a\in A, b_1,b_2\in B$$

(see [**123**]). A conditional expectation is called *faithful* if, for any positive element $a\in P^+(A)$, the equality $E(a)=0$ implies $a=0$. When the conditional expectation is faithful, one can introduce the structure of a pre-Hilbert B-module on the C^*-algebra A by

$$\langle x,y\rangle=E(x^*y), \quad x,y\in A.$$

We will give a condition for this module to be a Hilbert C^*-module (i.e., to be complete) in Section 4.5.

Let $\mathcal{N}\subset\mathcal{M}$ be a closed submodule of a Hilbert C^*-module $\mathcal{M}$. We define the *orthogonal complement* $\mathcal{N}^\perp$ by the formula

$$\mathcal{N}^\perp=\{y\in\mathcal{M}:\langle x,y\rangle=0 \quad \text{for all } x\in\mathcal{N}\}.$$

Then $\mathcal{N}^\perp$ is a closed submodule of the Hilbert C^*-module $\mathcal{M}$ too. However, the equality $\mathcal{M}=\mathcal{N}\oplus\mathcal{N}^\perp$ does not always hold, as the following example shows.

EXAMPLE 1.3.7. Let $A=C[0,1]$ be the C^*-algebra of all continuous functions on the segment $[0,1]$. Consider, in the Hilbert A-module $\mathcal{M}=A$, the submodule $\mathcal{N}=C_0(0,1)$ of functions that vanish at the end points of the segment. Then, obviously, $\mathcal{N}^\perp=0$.

If $\mathcal{M}$ is a Hilbert A-module, then we denote by $\mathcal{M} \cdot A$ the closure in $\mathcal{M}$ of the linear span of all the elements of the form $x \cdot a$, $x \in \mathcal{M}$, $a \in A$.

LEMMA 1.3.8. *$\mathcal{M} \cdot A = \mathcal{M}$.*

PROOF. Let $e_\alpha \in A$ be an approximate unit. Then, for any $x \in \mathcal{M}$,

$$\begin{aligned} \|x - x \cdot e_\alpha\|^2 &= \|\langle x - x \cdot e_\alpha, x - x \cdot e_\alpha \rangle\| \\ &\leq (1 + \|e_\alpha^*\|)\, \|\langle x, x\rangle - \langle x, x\rangle e_\alpha\| \to 0, \end{aligned}$$

hence the elements of the form $x \cdot e_\alpha$ are dense in $\mathcal{M}$. □

We will often use the following statement.

LEMMA 1.3.9. *For any $x \in \mathcal{M}$*

$$x = \lim_{\varepsilon \to 0} x\langle x, x\rangle (\langle x, x\rangle + \varepsilon)^{-1}.$$

PROOF. Let $\langle x, x\rangle = a$. Then

$$\begin{aligned} &\|x\langle x, x\rangle(\langle x, x\rangle + \varepsilon)^{-1} - x\|^2 \\ &\quad = \|\langle x(\langle x, x\rangle(\langle x, x\rangle + \varepsilon)^{-1} - 1), x(\langle x, x\rangle(\langle x, x\rangle + \varepsilon)^{-1} - 1)\rangle\| \\ &\quad = \|a(a^2(a+\varepsilon)^{-2} - 2a(a+\varepsilon)^{-1} + 1)\| \\ &\quad = \|a^3(a+\varepsilon)^{-2} - 2a^2(a+\varepsilon)^{-1} + a\| \longrightarrow 0, \end{aligned}$$

since the following inequalities hold under the condition $t \geq 0$:

$$\begin{aligned} |t^3(t+\varepsilon)^{-2} - t| &= \left| t\left(\left(\frac{t}{t+\varepsilon}\right)^2 - 1\right)\right| = \left| t\left(\frac{-\varepsilon^2 - 2\varepsilon t}{(t+\varepsilon)^2}\right)\right| \\ &= \varepsilon \left| \frac{\varepsilon t + 2t^2}{(t+\varepsilon)^2}\right| \leq \varepsilon\left(\frac{1}{2} + 2\right) = \frac{3}{2}\varepsilon \end{aligned}$$

and

$$|t^2(t+\varepsilon)^{-1} - t| = \left|\frac{t\varepsilon}{t+\varepsilon}\right| < \varepsilon.$$

□

The following statement is an analog of the polar decomposition for Hilbert C^*-modules. We will see below that, similarly to the case of C^*-algebras, the exact polar decomposition exists only in the case of Hilbert C^*-modules over W^*-algebras.

PROPOSITION 1.3.10 ([**71**]). *Let $\mathcal{M}$ be a Hilbert A-module, $x \in \mathcal{M}$, and $0 < \alpha < 1/2$. Then there exists an element $z \in \mathcal{M}$ such that $x = z \cdot \langle x, x\rangle^\alpha$.*

PROOF. For $n \in \mathbf{N}$ put

$$g_n(\lambda) = \begin{cases} n^{-\alpha/2}, & \text{if } \lambda \leq 1/n, \\ \lambda^{\alpha/2}, & \text{if } \lambda > 1/n. \end{cases}$$

Then, by the spectral theorem,

$$\begin{aligned} \|x \cdot (g_n(\langle x, x\rangle) - g_m(\langle x, x\rangle))\| &= \left\|\langle x, x\rangle (g_n(\langle x, x\rangle) - g_m(\langle x, x\rangle))^2\right\|^{1/2} \\ &= \sup\{|\lambda(g_n(\lambda) - g_m(\lambda))| : \lambda \in \mathrm{Sp}(\langle x, x\rangle)\}. \end{aligned}$$

Therefore the sequence $x \cdot g_n(\langle x, x\rangle)$ is a Cauchy sequence, so it has a limit $z \in \mathcal{M}$. Then

$$\begin{aligned}\|z\langle x,x\rangle^\alpha - x\| &= \lim_{n\to\infty} \|x \cdot g_n(\langle x,x\rangle)\langle x,x\rangle^\alpha - x\| \\ &= \lim_{n\to\infty} \|x(g_n(\langle x,x\rangle)\langle x,x\rangle^\alpha - 1)\| \\ &= \lim_{n\to\infty} \sup\{|\lambda^{1/2}(g_n(\lambda)\lambda^\alpha - 1)| : \lambda \in \operatorname{Sp}\langle x,x\rangle\} = 0.\end{aligned}$$

This completes the proof. □

1.4. The standard Hilbert module H_A

DEFINITION 1.4.1. A Hilbert C^*-module $\mathcal{M}$ is called *finitely generated* if there exists a finite set $\{x_i\} \subset \mathcal{M}$ such that $\mathcal{M}$ equals the linear span (over $\mathbf{C}$ and A) of this set. A Hilbert C^*-module $\mathcal{M}$ is called *countably generated* if there exists a countable set $\{x_i\} \subset \mathcal{M}$ such that $\mathcal{M}$ equals the norm-closure of the linear span (over $\mathbf{C}$ and A) of this set.

THEOREM 1.4.2 (Kasparov stabilization theorem, [**63**]). *Let A be a C^*-algebra and $\mathcal{M}$ a countably generated Hilbert A-module. Then $\mathcal{M} \oplus H_A \cong H_A$.*

PROOF. We start by proving the theorem for the case where A is unital. It is convenient here to use the procedure of almost orthogonalization [**36**]. An element x of the Hilbert C^*-module $\mathcal{N}$ is called *nonsingular* if the element $\langle x, x\rangle \in A$ is invertible. The set $\{x_i\} \in \mathcal{N}$ is called *orthonormal* if $\langle x_i, x_j\rangle = \delta_{ij}$. It is called a *basis* of the module $\mathcal{N}$ if finite sums of the form $\sum_i x_i \cdot a_i$, $a_i \in A$, are dense in $\mathcal{N}$.

LEMMA 1.4.3 ([**36**]). *Let $\mathcal{N}$ be a Hilbert A-module that contains the orthonormal elements $e_1, \dots, e_n$, $x \in \mathcal{N}$, $\varepsilon > 0$. If an element $y \in \mathcal{N}$ satisfies $\langle y, y\rangle = 1$ and $y \perp \{x, e_1, \dots, e_n\}$, then there exists an element $e_{n+1} \in \mathcal{N}$ such that*

(i) *the elements $e_1, \dots, e_n, e_{n+1}$ are orthonormal,*
(ii) $e_{n+1} \in \operatorname{Span}_A(e_1, \dots, e_n, x, y)$,
(iii) $\operatorname{dist}(x, \operatorname{Span}_A(e_1, \dots, e_{n+1})) \le \varepsilon$.

PROOF. Let

$$x' = x - \sum_{i=1}^n e_i\langle e_i, x\rangle, \qquad x'' = x' + \varepsilon y.$$

Then

$$\langle x'', x''\rangle = \langle x', x'\rangle + \varepsilon^2 \ge \varepsilon^2 > 0.$$

Therefore the element x'' is nonsingular. Put $e_{n+1} = x'' \cdot \langle x'', x''\rangle^{-1/2}$. Then

$$e_{n+1} \in \operatorname{Span}_A(x', y) \perp \{e_1, \dots, e_n\}.$$

Therefore the elements $e_1, \dots, e_n, e_{n+1}$ are orthonormal. Since we have taken $x' \in \operatorname{Span}_A(x, e_1, \dots, e_n)$ and $e_{n+1} \in \operatorname{Span}_A(x', y)$, we obtain (ii). Finally, put

$$w = e_{n+1}\langle x'', x''\rangle^{1/2} + \sum_{i=1}^n e_i\langle e_i, x\rangle \in \operatorname{Span}_A(e_1, \dots, e_{n+1}).$$

Then the equality $\|w - x\| = \|x'' - x'\| = \|\varepsilon y\| = \varepsilon$ proves (iii). □

We return now to the proof of Theorem 1.4.2. Let $\{y_n\}$ be the sequence of all generators of module $\mathcal{M}$. By $\{e_n\}$ we denote the standard basis of the module H_A. Let $\{x_n\} \subset \{e_n\} \cup \{y_n\}$ be a sequence, in which one meets each element e_n and each element y_n *infinitely* many times. Then the set $\{x_n\}$ is generating for the module $\mathcal{M} \oplus H_A$. We will prove the theorem by induction. Let us assume that the orthonormal elements $\overline{e}_1, \ldots, \overline{e}_n \in \mathcal{M} \oplus H_A$ and the number $m(n) \geq n$ are already constructed in such a way that

(i) $\{\overline{e}_1, \ldots, \overline{e}_n\} \subset \operatorname{Span}_A(x_1, \ldots, x_n, e_1, \ldots, e_{m(n)})$,
(ii) $\operatorname{dist}(x_k, \operatorname{Span}_A(\overline{e}_1, \ldots, \overline{e}_k)) \leq \frac{1}{k}$, $1 \leq k \leq n$.

Since each element x_i is equal to e_j or y_k, one can find a number $m' > m(n)$ such that $e_{m'} \perp \{x_1, \ldots, x_{n+1}\}$. Since $e_{m'} \perp \{e_1, \ldots, e_{m(n)}\}$, it follows from (i) that

$$e_{m'} \perp \{x_{n+1}, \overline{e}_1, \ldots, \overline{e}_n\}.$$

By Lemma 1.4.3, there exists an element

$$\overline{e}_{n+1} \in \operatorname{Span}_A(\overline{e}_1, \ldots, \overline{e}_n, x_{n+1}, e_{m'}) \tag{1.3}$$

such that the elements $\overline{e}_1, \ldots, \overline{e}_n, \overline{e}_{n+1}$ are orthonormal and

$$\operatorname{dist}(x_{n+1}, \operatorname{Span}_A(\overline{e}_1, \ldots, \overline{e}_{n+1})) \leq \frac{1}{n+1}.$$

It follows from (1.3) and from condition (i) that

$$\{\overline{e}_1, \ldots, \overline{e}_{n+1}\} \subset \operatorname{Span}_A(x_1, \ldots, x_{n+1}, e_1, \ldots, e_{m'}).$$

By setting $m(n+1) = m'$, we complete the induction step. Thus, an orthonormal sequence $\overline{e}_n$ satisfying properties (i) and (ii) has been constructed. But property (ii) means that this sequence generates the whole module $\mathcal{M} \oplus H_A$, so $\mathcal{M} \oplus H_A \cong H_A$.

Thus, Theorem 1.4.2 is proved for unital C^*-algebras. Let A be a nonunital C^*-algebra and let A^+ be its unitalization. Defining the action of A^+ on the Hilbert A-module $\mathcal{M}$ by the formula $x \cdot (a, \lambda) := x \cdot a + x\lambda$, $x \in \mathcal{M}$, $(a, \lambda) \in A^+$, $\lambda \in \mathbf{C}$, we equip $\mathcal{M}$ with the structure of a Hilbert A^+-module. Consider the A^+-module H_{A^+} and denote by $H_{A^+}A$ the closure of the linear span of all the elements of the form $x \cdot a$, $x \in H_{A^+}$, $a \in A$, in H_{A^+}. It is easy to see that $H_{A^+}A = H_A$. The isomorphism $\mathcal{M} \oplus H_{A^+} \cong H_{A^+}$ implies the isomorphism

$$\mathcal{M} \oplus H_A = \mathcal{M}A \oplus H_{A^+}A = (\mathcal{M} \oplus H_{A^+})A \cong H_{A^+}A = H_A.$$

□

DEFINITION 1.4.4. A Hilbert A-module $\mathcal{M}$ is called a *finitely generated projective* A-module if there exists a Hilbert A-module $\mathcal{N}$ such that $\mathcal{M} \oplus \mathcal{N} \cong L_n(A)$ for some n.

The following two theorems of Dupré and Fillmore show that finite-dimensional projective submodules lie in Hilbert C^*-modules in the simplest way.

THEOREM 1.4.5 (Dupré – Fillmore, [**36**]). *Let A be a unital C^*-algebra and let $\mathcal{M}$ be a finite-dimensional projective A-submodule in the standard Hilbert A-module H_A. Then*

(i) *the nonsingular elements of the module $\mathcal{M}^\perp$ are dense in $\mathcal{M}^\perp$;*
(ii) *$H_A = \mathcal{M} \oplus \mathcal{M}^\perp$;*
(iii) *$\mathcal{M}^\perp \cong H_A$.*

PROOF. We begin the proof of the theorem with the case where $\mathcal{M} \cong L_n(A)$. Let $g_1, \ldots, g_n$ be an orthonormal basis in $\mathcal{M}$. Fix $\varepsilon > 0$. For each m put

$$e'_m = e_m - \sum_{i=1}^{n} g_i \langle g_i, e_m \rangle.$$

Then $e'_m \in \mathcal{M}^\perp$ and

$$\langle e'_m, e'_m \rangle = 1 - \sum_{i=1}^{n} \langle e_m, g_i \rangle \langle g_i, e_m \rangle.$$

Since $\langle x, e_m \rangle \to 0$ for each $x \in H_A$, we conclude that $\langle e'_m, e'_m \rangle \to 1$. Therefore there exists a number m_0 such that for any $m \geq m_0$, the element e'_m is nonsingular. Then one can define

$$e''_m = e'_m \langle e'_m, e'_m \rangle^{-1/2}$$

with $\langle e''_m, e''_m \rangle = 1$. Let $x \in \mathcal{M}^\perp$. Then

$$\langle e''_m, x \rangle = \langle e'_m, e'_m \rangle^{-1/2} \langle e'_m, x \rangle = \langle e'_m, e'_m \rangle^{-1/2} \langle e_m, x \rangle \to 0.$$

Choose a number $m \geq m_0$ such that $\|\langle e''_m, x \rangle\| < \varepsilon$ and set

$$x' = x + \varepsilon\, e''_m.$$

It is easy to see that

$$\|x' - x\| = \varepsilon. \tag{1.4}$$

Let us check that the element x' is nonsingular. Put

$$u = x - e''_m \langle e''_m, x \rangle, \qquad v = e''_m (\langle e''_m, x \rangle + \varepsilon\, 1).$$

Then $u \perp v$ (since $u \perp e''_m$) and $x' = u + v$. Therefore

$$\langle x', x' \rangle = \langle u, u \rangle + \langle v, v \rangle = \langle u, u \rangle + (\langle e''_m, x \rangle + \varepsilon\, 1)^* (\langle e''_m, x \rangle + \varepsilon\, 1), \tag{1.5}$$

and the right-hand side of equality (1.5) is invertible since $\|\langle e''_m, x \rangle\| < \varepsilon$. Therefore $\langle x', x' \rangle$ is invertible too. Together with estimate (1.4), this proves statement (i).

Let $\{x_n\}$ be a sequence in which each element e_m is repeated *infinitely* many times. Put $x = x_1 - \sum_{i=1}^{n} g_i \langle g_i, x_1 \rangle$. Then (taking $\varepsilon = 1$) one can find an element $g_{n+1} \in \mathcal{M}^\perp$ such that $\langle g_{n+1}, g_{n+1} \rangle = 1$, $\operatorname{dist}(x, g_{n+1} A) \leq 1$. Therefore $\operatorname{dist}(x_1, \operatorname{Span}_A(g_1, \ldots, g_{n+1})) \leq 1$. At the next step we replace the module $\mathcal{M}$ by $\operatorname{Span}_A(g_1, \ldots, g_{n+1})$, x_1 by x_2, and $\varepsilon = 1$ by $\varepsilon = 1/2$. Going on with this procedure, we obtain an orthonormal basis $\{g_k\}$, $k \in \mathbf{N}$, extending the basis $g_1, \ldots, g_n$ of submodule $\mathcal{M}$, and the remaining part $\{g_k : k > n\}$ is a basis of the module $\mathcal{M}^\perp$. This proves statements (ii) and (iii).

We pass now to the case of an arbitrary finitely generated projective module $\mathcal{M}$. Let $\mathcal{M} \oplus \mathcal{N} \cong L_n(A)$. By Theorem 1.4.2, $\mathcal{N} \oplus H_A \cong H_A$, hence

$$L_n(A) \cong \mathcal{N} \oplus \mathcal{M} \subset \mathcal{N} \oplus H_A \cong H_A.$$

Therefore, if $\mathcal{K}$ is the orthogonal complement to the submodule $\mathcal{N} \oplus \mathcal{M}$ in the module $\mathcal{N} \oplus H_A$, then $\mathcal{K} \cong H_A$ and $\mathcal{N} \oplus \mathcal{M} \oplus \mathcal{K} = \mathcal{N} \oplus H_A$. But $\mathcal{K} = \mathcal{M}^\perp$ is obviously the orthogonal complement to the submodule $\mathcal{M}$ in the module H_A. □

THEOREM 1.4.6 ([**36**]). *Let A be a unital C^*-algebra and let $\mathcal{M}$ be a finitely generated projective Hilbert submodule in an arbitrary Hilbert A-module $\mathcal{N}$. Then $\mathcal{N} = \mathcal{M} \oplus \mathcal{M}^\perp$.*

PROOF. As in the previous theorem, the proof can be reduced to the case where $\mathcal{M}$ is a free module, $\mathcal{M} = L_n(A)$. If $\{g_1, \ldots, g_n\}$ is the standard basis of $\mathcal{M}$, then put $x' = x - \sum_{i=1}^{n} g_i \langle g_i, x \rangle$ for $x \in \mathcal{N}$. Then $x' \in \mathcal{M}$ and $x - x' \in \mathcal{M}^\perp$, hence $\mathcal{N} = \mathcal{M} \oplus \mathcal{M}^\perp$. □

For a Hilbert A-module $\mathcal{M}$ denote by $\langle \mathcal{M}, \mathcal{M} \rangle \subset A$ the closure of the linear span of all $\langle x, x \rangle$, $x \in \mathcal{M}$. The set $\langle \mathcal{M}, \mathcal{M} \rangle$ is obviously a closed two-sided involutive ideal in the C^*-algebra A.

DEFINITION 1.4.7. A Hilbert A-module is called *full* if $\langle \mathcal{M}, \mathcal{M} \rangle = A$.

One can always consider any Hilbert module as a full Hilbert module over the C^*-algebra $\langle \mathcal{M}, \mathcal{M} \rangle$.

The standard Hilbert C^*-module is obviously full.

1.5. Hilbert C^*-bimodules and strong Morita equivalence

In this section we briefly discuss the case where a module has two C^*-module structures over two C^*-algebras. Such C^*-bimodules were studied by Rieffel in [**109**] and we follow these papers here.

Let A and B be two C^*-algebras. Let $\mathcal{M}$ be a right pre-Hilbert C^*-module over A with the inner product $\langle \cdot, \cdot \rangle_A$ and a left pre-Hilbert C^*-module over B with the inner product $\langle \cdot, \cdot \rangle_B$. The latter means that the sesquilinear form $\langle \cdot, \cdot \rangle_B$ is conjugate linear in the first variable.

DEFINITION 1.5.1. The module $\mathcal{M}$ is called a *pre-Hilbert A-B-bimodule* if the following conditions hold:

(i) $\langle x, y \rangle_B z = x \langle y, z \rangle_A$ for any $x, y, z \in \mathcal{M}$;
(ii) $\langle bx, bx \rangle_A \leq \|b\|^2 \langle x, x \rangle_A$ and $\langle xa, xa \rangle_B \leq \|a\|^2 \langle x, x \rangle_B$.

LEMMA 1.5.2. *One has* $\|\langle x, x \rangle_A\| = \|\langle x, x \rangle_B\|$ *for any* $x \in \mathcal{M}$.

PROOF. Denote $\langle x, x \rangle_A = a$, $\langle x, x \rangle_B = b$. Then $bx = xa$ for any $x \in \mathcal{M}$. Since

$$a^3 = \langle xa, xa \rangle_A = \langle bx, bx \rangle_A \leq \|b\|^2 a,$$

$\|a\|^3 \leq \|b\| \|a\|^2$, hence $\|a\| \leq \|b\|$. Similarly one obtains $\|b\| \leq \|a\|$. □

Thus we see that $\|x\| = \|\langle x, x \rangle_A\|^{1/2} = \|\langle x, x \rangle_B\|^{1/2}$ defines a norm on $\mathcal{M}$.

DEFINITION 1.5.3. A pre-Hilbert A-B-bimodule $\mathcal{M}$ is called *an A-B-equivalence bimodule* if it is complete with respect to the norm and if it is full both as a right and as a left Hilbert C^*-module.

LEMMA 1.5.4. *One has* $\langle bx, y \rangle_A = \langle x, b^* y \rangle_A$ *and* $\langle xa, y \rangle_B = \langle x, ya^* \rangle_B$ *for any* $x, y \in \mathcal{M}$, $a \in A$, $b \in B$.

PROOF. It suffices to prove the first statement since the second is similar. Since the bimodule $\mathcal{M}$ is full as a B-module, we can assume without loss of generality that $b = \langle z, t \rangle_B$ for some $z, t \in \mathcal{M}$. Then

$$\begin{aligned} \langle bx, y \rangle_A &= \langle z \langle t, x \rangle_A, y \rangle_A = \langle t, x \rangle_A^* \langle z, y \rangle_A \\ &= \langle x, t \rangle_A \langle z, y \rangle_A. \end{aligned}$$

On the other hand,

$$\begin{aligned}\langle x, b^*y\rangle_A &= \langle x, \langle t, z\rangle_B y\rangle_A = \langle x, t\langle z, y\rangle_A\rangle_A \\ &= \langle x, t\rangle_A\langle z, y\rangle_A.\end{aligned}$$

□

DEFINITION 1.5.5. Two C^*-algebras A and B are called *strongly Morita equivalent* if there exists an A-B-equivalence bimodule.

LEMMA 1.5.6. *Strong Morita equivalence is an equivalence relation.*

PROOF. Reflexivity of this equivalence relation is clear, symmetry follows from considering the conjugate module $\widetilde{X}$, which consists of the same elements, but the bimodule structure is given by $a\widetilde{x} = xa^*$ and $\widetilde{x}b = b^*x$, $x \in \mathcal{M}$, $a \in A$, $b \in B$. So we have to check transitivity. Let $\mathcal{M}$ be an A-B-equivalence bimodule and let $\mathcal{N}$ be a B-C-equivalence bimodule, where C is one more C^*-algebra. Then $\mathcal{N} \otimes_B \mathcal{M}$ is a left C-module and a right A-module. For finite sums $\sum_i n_i \otimes m_i$, $\sum_j n'_j \otimes m'_j$ in $\mathcal{N} \otimes_B \mathcal{M}$ define an A-valued inner product by

$$\langle \sum_i n_i \otimes m_i, \sum_j n'_j \otimes m'_j\rangle_A = \sum_{ij} \langle\langle n_i, n'_j\rangle_B m_i, m'_j\rangle_A$$

and similarly define a C-valued inner product. It is easy to check that these inner products satisfy all the necessary properties, except, possibly, nondegeneracy. Let $\mathcal{L} \subset \mathcal{N} \otimes_B \mathcal{M}$ be the (maybe empty) set of all finite sums $l = \sum_i n_i \otimes m_i \in \mathcal{N} \otimes_B \mathcal{M}$ with $\langle l, l\rangle_A = 0$. Then the completion of $\mathcal{N} \otimes_B \mathcal{M}/\mathcal{L}$ with respect to the norm obtained from any of the two inner products gives us an A-C-equivalence bimodule (for example, fullness follows from Lemma 1.3.8). □

EXAMPLE 1.5.7. Consider the standard right Hilbert A-module H_A and endow it with the left module structure over the C^*-algebra $A \otimes \mathcal{K}$, where $\mathcal{K}$ denotes the C^*-algebra of compact operators on a separable Hilbert space. Elements of $A \otimes \mathcal{K}$ can be written as infinite matrices of the form (a_{ij}), $i, j \in \mathbf{N}$, with entries in A. The left action of such a matrix (a_{ij}) on a sequence $(x_i) \in H_A$ results in the sequence $(\sum_j a_{ij}x_j)$. Define the $A \otimes \mathcal{K}$-valued inner product on H_A by

$$\langle (x_i), (y_j)\rangle_{A\otimes\mathcal{K}} = (x_i y_j^*) \in A \otimes \mathcal{K}$$

(it is easy to check that the matrix $(x_i y_j^*)$ defines an element from $A \otimes \mathcal{K}$ because both sequences (x_i) and (y_j) vanish at infinity). All properties of an A-$A \otimes \mathcal{K}$-equivalence bimodule can be easily checked and thus we see that A is strongly Morita equivalent to $A \otimes \mathcal{K}$. Similarly, taking A^n instead of H_A, we obtain the strong Morita equivalence between A and the C^*-algebra $M_n(A)$ of $n \times n$-matrices with entries from A.

This example and Lemma 1.5.6 imply the following statement.

COROLLARY 1.5.8. *Let A and B be C^*-algebras. If the C^*-algebras $A \otimes \mathcal{K}$ and $B \otimes \mathcal{K}$ are isomorphic, then A and B are strongly Morita equivalent.*

A criterion for the strong Morita equivalence was obtained in [**15, 18**]. Since the method of proof is far removed from the Hilbert C^*-module technique, we skip the proof.

THEOREM 1.5.9 ([**15, 18**]). *Let A and B be strongly Morita equivalent C^*-algebras. If they both have countable approximate identities, then the C^*-algebras $A \otimes \mathcal{K}$ and $B \otimes \mathcal{K}$ are isomorphic.*

CHAPTER 2

Operators on Hilbert Modules

2.1. Bounded and adjointable operators

Let $\mathcal{M}, \mathcal{N}$ be Hilbert C^*-modules over a C^*-algebra A. A bounded $\mathbf{C}$-linear A-homomorphism from $\mathcal{M}$ to $\mathcal{N}$ is called an *operator* from $\mathcal{M}$ to $\mathcal{N}$. Let $\mathrm{Hom}_A(\mathcal{M}, \mathcal{N})$ denote the set of all operators from $\mathcal{M}$ to $\mathcal{N}$. If $\mathcal{N} = \mathcal{M}$, then $\mathrm{End}_A(\mathcal{M}) = \mathrm{Hom}_A(\mathcal{M}, \mathcal{M})$ is obviously a Banach algebra. However, we shall soon see that there is no natural involution on this algebra. Let $T \in \mathrm{Hom}_A(\mathcal{M}, \mathcal{N})$. We say that T is *adjointable* if there exists an operator $T^* \in \mathrm{Hom}_A(\mathcal{N}, \mathcal{M})$ such that

$$\langle Tx, y\rangle = \langle x, T^*y\rangle$$

for all $x \in \mathcal{M}$, $y \in \mathcal{N}$.

LEMMA 2.1.1. *Let $\mathcal{M}$ be a Hilbert A-module and let $T : \mathcal{M} \to \mathcal{M}$ and $T^* : \mathcal{M} \to \mathcal{M}$ be maps such that*

$$\langle x, Ty\rangle = \langle T^*x, y\rangle$$

for all $x, y \in \mathcal{M}$. Then T is a bounded $\mathbf{C}$-linear A-homomorphism (and T^* is as well).

PROOF. For any $x, y, z \in \mathcal{M}$, $w \in \mathbf{C}$ and $a \in A$ one has

$$\begin{aligned}
\langle z, T(x+y)\rangle &= \langle T^*z, x+y\rangle = \langle T^*z, x\rangle + \langle T^*z, y\rangle \\
&= \langle z, Tx\rangle + \langle z, Ty\rangle = \langle z, Tx + Ty\rangle, \\
\langle z, Twx\rangle &= \langle T^*z, x\rangle w = \langle z, Tx\rangle w = \langle z, wTx\rangle, \\
\langle z, T(xa)\rangle &= \langle T^*z, xa\rangle = \langle T^*z, x\rangle a = \langle z, Tx\rangle a = \langle z, (Tx)a\rangle.
\end{aligned}$$

Since z is an arbitrary element, it follows that

$$T(x+y) = Tx + Ty, \qquad T(wx) = wTx, \qquad T(xa) = (Tx)a,$$

and linearity properties hold.

To prove the continuity of T we should verify that its graph is closed. Let $x_\alpha \to x$, $T(x_\alpha) \to y$ in $\mathcal{M}$, and let $z \in \mathcal{M}$ be an arbitrary element. Then

$$\begin{aligned}
0 &= \langle T^*(y - Tx), x_\alpha\rangle - \langle T^*(y - Tx), x_\alpha\rangle \\
&= \langle y - Tx, T(x_\alpha)\rangle - \langle T^*(y - Tx), x_\alpha\rangle \\
&\longrightarrow \langle y - Tx, y\rangle - \langle T^*(y - Tx), x\rangle = \langle y - Tx, y - Tx\rangle.
\end{aligned}$$

□

We show now that there exist nonadjointable operators.

EXAMPLE 2.1.2. Let A be a unital C^*-algebra. As above, the standard basis of the Hilbert module H_A consists of the elements $e_i = (0, \dots, 0, 1, 0, \dots)$, where 1 is the i-th entry. To each operator $T \in \operatorname{End}_A(H_A)$ one can associate an infinite matrix with respect to this basis,

$$\|t_{ij}\|, \qquad t_{ij} = \langle e_i, Te_j \rangle.$$

Then the adjoint operator has the matrix $\|t_{ji}^*\|$.

Let $A = C([0,1])$ and let the functions $\varphi_i \in A$, $i = 1, 2, \dots$, be defined by the formula

$$\varphi_i = \begin{cases} 0 & \text{on } [0, \frac{1}{i+1}] \text{ and } [\frac{1}{i}, 1], \\ 1 & \text{at the point } x_i = \frac{1}{2}\left(\frac{1}{i} + \frac{1}{i+1}\right), \\ \text{is linear} & \text{on } [\frac{1}{i+1}, x_i] \text{ and } [x_i, \frac{1}{i}]. \end{cases}$$

Let an operator $T \in \operatorname{End}_A(H_A)$ have the matrix

$$\begin{pmatrix} \varphi_1 & \varphi_2 & \varphi_3 & \dots \\ 0 & 0 & 0 & \dots \\ 0 & 0 & 0 & \dots \\ \dots & \dots & \dots & \dots \end{pmatrix}$$

(actually it is an operator from the module H_A to A, i.e., an A-functional). It is easy to verify that T is bounded. But the operator T^* is not well defined since it should have the matrix

$$\begin{pmatrix} \varphi_1^* & 0 & 0 & \dots \\ \varphi_2^* & 0 & 0 & \dots \\ \varphi_3^* & 0 & 0 & \dots \\ \dots & \dots & \dots & \dots \end{pmatrix}$$

and the image of the basis element e_1 should be an element of H_A having the first column as its coordinates and it has to be an element of H_A, which is impossible since the series $\sum \varphi_i \varphi_i^*$ is not norm-convergent in the C^*-algebra A.

Denote by $\operatorname{Hom}_A^*(\mathcal{M}, \mathcal{N})$ the set of all adjointable operators from $\mathcal{M}$ to $\mathcal{N}$. The algebra $\operatorname{End}_A^*(\mathcal{M}) = \operatorname{Hom}_A^*(\mathcal{M}, \mathcal{M})$ is an involutive Banach algebra. Moreover, it is a C^*-algebra; this follows from the estimate

$$\|T^*T\| \geq \sup_{x \in B_1(\mathcal{M})} \{\langle T^*Tx, x \rangle\} = \sup_{x \in B_1(\mathcal{M})} \{\langle Tx, Tx \rangle\} = \|T\|^2,$$

where $B_1(\mathcal{M})$ denotes the unit ball of the module $\mathcal{M}$.

We will use the following statement frequently without special reference.

PROPOSITION 2.1.3. *For an operator $T : \mathcal{M} \to \mathcal{M}$, the following conditions are equivalent:*

(i) *T is a positive element of the C^*-algebra* $\operatorname{End}^*(\mathcal{M})$*;*
(ii) *for any $x \in \mathcal{M}$ the inequality $\langle Tx, x \rangle \geq 0$ is fulfilled, i.e., is positive in the C^*-algebra A.*

PROOF. The first condition is equivalent to the equality $T = S^*S$ for some $S \in \operatorname{End}^*(\mathcal{M})$. Therefore

$$\langle Tx, x \rangle = \langle Sx, Sx \rangle \geq 0 \qquad \text{for any } x \in \mathcal{M}.$$

Now let $\langle Tx, x \rangle \geq 0$ for all $x \in \mathcal{M}$. Then

$$\langle Tx, x \rangle = \langle Tx, x \rangle^* = \langle x, Tx \rangle \qquad \text{for all } x \in \mathcal{M}.$$

The map $(x, y) \mapsto \langle Tx, y\rangle$ defines a sesquilinear form on $\mathcal{M}$. Therefore, by the polarization equality 1.2.5, $\langle Tx, y\rangle = \langle x, Ty\rangle$ for all $x, y \in \mathcal{M}$. By Lemma 2.1.1, this means that $T \in \mathrm{End}^*(\mathcal{M})$ and $T = T^*$. So T is a selfadjoint element of the C^*-algebra $\mathrm{End}^*(\mathcal{M})$, so (see [**31**, 1.6.5]) it can be represented as the difference $T = T_+ - T_-$ of two elements of $\mathrm{End}^*(\mathcal{M})$, where $T_+ \geq 0$, $T_- \geq 0$ and $T_+T_- = T_-T_+ = 0$. Then $\langle T_-y, y\rangle \leq \langle T_+y, y\rangle$ for any $y \in \mathcal{M}$. In particular,

$$\langle T_-^3 x, x\rangle = \langle T_-^2 x, T_- x\rangle \leq \langle T_+T_-x, T_-x\rangle = 0.$$

On the other hand, $T_- \geq 0$ and $T_-^3 \geq 0$, hence $\langle T_-^3 x, x\rangle \geq 0$ (because the statement in this direction is already proved). So the only possibility left is $\langle T_-^3 x, x\rangle = 0$ for any $x \in \mathcal{M}$. By the polarization equality, this implies $\langle T_-^3 x, y\rangle = 0$ for all $x, y \in \mathcal{M}$, hence $T_-^3 = 0$, $T_- = 0$. Thus, $T = T_+ \geq 0$. □

THEOREM 2.1.4 ([**100**]). *Let $\mathcal{M}$ and $\mathcal{N}$ be Hilbert A-modules and let $T : \mathcal{M} \to \mathcal{N}$ be a linear map. Then the following conditions are equivalent:*

(i) *the operator T is bounded and A-linear, i.e., $T(x \cdot a) = Tx \cdot a$ for all $x \in \mathcal{M}$, $a \in A$;*
(ii) *there exists a constant $K \geq 0$ such that the inequality $\langle Tx, Tx\rangle \leq K\langle x, x\rangle$ holds in A for all $x \in \mathcal{M}$.*

PROOF. To obtain the second statement from the first, let us assume that $T(x \cdot a) = Tx \cdot a$ and that $\|T\| \leq 1$. If the C^*-algebra A is not unital, then we consider $\mathcal{M}$ and $\mathcal{N}$ as modules over the C^*-algebra A^+ obtained from A by unitalizing. For $x \in \mathcal{M}$ and $n \in \mathbf{N}$ put

$$a_n = \left(\langle x, x\rangle + \frac{1}{n}\right)^{-1/2}, \qquad x_n = x \cdot a_n.$$

Then $\langle x_n, x_n\rangle = a_n^*\langle x, x\rangle a_n = \langle x, x\rangle\left(\langle x, x\rangle + \frac{1}{n}\right)^{-1} \leq 1$. Therefore $\|x_n\| \leq 1$, hence $\|Tx_n\| \leq 1$. Then the inequality $\langle Tx_n, Tx_n\rangle \leq 1$ holds in A (or in A^+) for all $n \in \mathbf{N}$. But

$$\langle Tx, Tx\rangle = a_n^{-1}\langle Tx_n, Tx_n\rangle a_n^{-1} \leq a_n^{-2} = \langle x, x\rangle + \frac{1}{n}. \tag{2.1}$$

Passing to the limit as $n \to \infty$ in the inequality (2.1), we obtain $\langle Tx, Tx\rangle \leq \langle x, x\rangle$.

To derive the first statement from the second, assume that the inequality $\langle Tx, Tx\rangle \leq \langle x, x\rangle$ holds for all $x \in \mathcal{M}$. It clearly follows that T is bounded, $\|T\| \leq 1$. Let $x \in \mathcal{M}$, $y \in \mathcal{N}$. Define the map $r : A^+ \longrightarrow A^+$ by the formula

$$r(a) = \langle y, T(x \cdot a)\rangle.$$

Then

$$\begin{aligned} r(a)^*r(a) &= \langle T(x \cdot a), y\rangle\langle y, T(x \cdot a)\rangle \leq \|y\|^2 \langle T(x \cdot a), T(x \cdot a)\rangle \\ &\leq \|y\|^2 \langle x \cdot a, x \cdot a\rangle = \|y\|^2 a^*\langle x, x\rangle a \\ &\leq \|y\|^2 \|x\|^2 a^*a. \end{aligned}$$

To complete the proof we use the following statement.

LEMMA 2.1.5 ([**56, 100**]). *Let A be a unital C^*-algebra and let $r : A \longrightarrow A$ be a linear map such that for some constant $K \geq 0$, the inequality $r(a)^*r(a) \leq Ka^*a$ holds for all $a \in A$. Then $r(a) = r(1)a$ for all $a \in A$.*

Thus $r(a) = r(1)a$, i.e.,

$$\langle y, T(x \cdot a)\rangle = \langle y, Tx\rangle a = \langle y, Tx \cdot a\rangle$$

for all $y \in \mathcal{N}$, $x \in \mathcal{M}$. This implies the first statement of the theorem. □

COROLLARY 2.1.6. *Let $\mathcal{M}, \mathcal{N}$ be Hilbert A-modules, $T \in \mathrm{End}_A(\mathcal{M}, \mathcal{N})$. Then*

$$\|T\| = \inf\{K^{1/2} : \langle Tx, Tx\rangle \le K\langle x, x\rangle \quad \forall x \in \mathcal{M}\}.$$

EXAMPLE 2.1.7. Let $\mathcal{M} = \mathcal{N} \oplus \mathcal{L}$ be a decomposition into an orthogonal direct sum of Hilbert modules. We define the operator $P : \mathcal{M} \longrightarrow \mathcal{M}$ to be the projection onto the submodule $\mathcal{N}$ along the submodule $\mathcal{L}$. Then P is bounded, $\|P\| = 1$, and $P^* = P$, hence $P \in \mathrm{End}_A^*(\mathcal{M})$.

2.2. Compact operators in Hilbert modules

Let $\mathcal{M}, \mathcal{N}$ be Hilbert A-modules, $x \in \mathcal{N}$, $y \in \mathcal{M}$. Define the operator $\theta_{x,y} : \mathcal{M} \longrightarrow \mathcal{N}$ by its action on an element $z \in \mathcal{M}$ by the formula

$$\theta_{x,y}(z) := x\langle y, z\rangle. \tag{2.2}$$

Operators of the form (2.2) are called *elementary* operators. They clearly satisfy the equalities

(i) $(\theta_{x,y})^* = \theta_{y,x}$;
(ii) $\theta_{x,y}\theta_{u,v} = \theta_{x\langle y,u\rangle,v} = \theta_{x,v\langle u,y\rangle}$ for $u \in \mathcal{M}$, $v \in \mathcal{N}$;
(iii) $T\theta_{x,y} = \theta_{Tx,y}$ for $T \in \mathrm{Hom}_A(\mathcal{N}, \mathcal{L})$;
(iv) $\theta_{x,y}S = \theta_{x,S^*y}$ for $S \in \mathrm{Hom}_A^*(\mathcal{L}, \mathcal{M})$.

We denote the closed linear span of the set of all elementary operators by $\mathcal{K}(\mathcal{M}, \mathcal{N})$. The elements of $\mathcal{K}(\mathcal{M}, \mathcal{N})$ are called *compact* operators. In the case $\mathcal{N} = \mathcal{M}$ the equalities (i)–(iv) mean that the algebra $\mathcal{K}(\mathcal{M}) = \mathcal{K}(\mathcal{M}, \mathcal{M})$ is a closed two-sided ideal in the C^*-algebra $\mathrm{End}_A^*(\mathcal{M})$. Compact operators acting on Hilbert modules are not compact operators in the usual sense, when one considers them as operators from one Banach space to another. However, they are a natural generalization of compact operators on a Hilbert space.

PROPOSITION 2.2.1. *Let H_A be the standard Hilbert module over a unital C^*-algebra A and let $L_n(A) \subset H_A$ be the free submodule generated by the first n elements of the standard basis. An operator $K \in \mathrm{End}_A(H_A)$ is compact if and only if the norms of restrictions of K onto the orthogonal complements $L_n(A)^\perp$ of the submodules $L_n(A)$ vanish as $n \to \infty$.*

PROOF. Denote by p_n the projection in H_A onto the submodule $L_n(A)^\perp$. Then, for any $z \perp L_n(A)$, one has

$$\begin{aligned}\|\theta_{x,y}(z)\|^2 = \|\langle\theta_{x,y}(z), \theta_{x,y}(z)\rangle\| &= \left\|\langle y, z\rangle^*\langle x, x\rangle\langle y, z\rangle\right\| \\ \le \|x\|^2\,\|\langle y, z\rangle\|^2 &= \|x\|^2\,\|\langle p_n y, z\rangle\|^2 \\ \le \|x\|^2 \cdot \|p_n y\|^2 \cdot \|z\|^2 &.\end{aligned}$$

Since $\|p_n y\|$ tends to zero, the same is true for the norm of the restriction of the operator $\theta_{x,y}$ to the submodule $L_n(A)^\perp$, hence, for the norm of any compact operator. Let us assume now that for some operator K, one has $\left\|K|_{L_n(A)^\perp}\right\| \to 0$.

Then, since $\sum_{m=1}^{n} Ke_m\langle e_m, z\rangle = 0$ for any $z \perp L_n(A)$, for $\|z\| \le 1$ and $z \perp L_n(A)$ one has

$$\sup_z \left\| Kz - \sum_{m=1}^{n} Ke_m\langle e_m, z\rangle \right\| = \sup_z \|Kz\| \longrightarrow 0 \tag{2.3}$$

as $n \to \infty$. If $z \in L_n(A)$, then $Kz = \sum_{m=1}^{n} Ke_m\langle e_m, z\rangle$. This means that (2.3) still holds if the supremum is taken over the unit ball of the whole module H_A. Therefore the operator K is the norm limit of the operators $K_n = \sum_{m=1}^{n} \theta_{Ke_m, e_m}$. □

Note that, in the case of modules over nonunital C^*-algebras, the statement of 2.2.1 is not valid.

Let $\mathcal{K}$ denote the C^*-algebra of compact operators on a separable Hilbert space H. Since the algebra $\mathcal{K}$ is nuclear [**70**], there is a unique C^*-seminorm on the algebraic tensor product of $\mathcal{K}$ by any C^*-algebra A and we denote its completion with respect to this seminorm by $\mathcal{K} \otimes A$. Denote by $M_n(A) = M_n \otimes A$ the C^*-algebra of all $n\times n$-matrices with entries from A, where M_n is the algebra of complex $n\times n$-matrices.

PROPOSITION 2.2.2. *There exist natural isometric isomorphisms:*

(i) $\mathcal{K}(A) \cong A$;
(ii) $\mathcal{K}(L_n(A)) \cong M_n(A)$;
(iii) $\mathcal{K}(H_A) \cong \mathcal{K} \otimes A$.

PROOF. If a C^*-algebra is unital, then statement (i) is clear. In the general case consider the map $\varphi : \mathrm{Span}_{\mathbf{C}}(\theta_{a,b} : a, b \in A) \longrightarrow A$ defined by the formula

$$\varphi\left(\sum_{i=1}^{n} \lambda_i \theta_{a_i,b_i}\right) = \sum_{i=1}^{\infty} \lambda_i a_i b_i^*.$$

Let us verify that this map is well defined. If $\sum_i \lambda_i \theta_{a_i,b_i} = \sum_j \mu_j \theta_{c_j,d_j}$, then $\sum_i \lambda_i a_i b_i^* x = \sum_j \mu_j c_j d_j^* x$ for any $x \in A$. Therefore $\sum_i \lambda_i a_i b_i^* = \sum_j \mu_j c_j d_j^*$. The map φ is multiplicative and involutive,

$$\varphi(\theta_{a,b}\, \theta_{c,d}) = \varphi(\theta_{ab^*,dc^*}) = \varphi(\theta_{a,b})\varphi(\theta_{c,d}); \qquad \varphi(\theta_{a,b}^*) = \varphi(\theta_{b,a}) = \varphi(\theta_{a,b})^*.$$

Surjectivity of φ follows from the possibility of the factorization $a = u(a^*a)^{1/4}$ for any $a \in A$ (see 1.1.5). If (u_α), $\alpha \in \mathcal{A}$, is an approximate unit of the C^*-algebra A, then

$$\lim_\alpha \left\| \sum_{i=1}^{n} \lambda_i \theta_{a_i,b_i}(u_\alpha) \right\| = \left\| \sum_{i=1}^{n} \lambda_i a_i b_i^* \right\|.$$

Therefore $\|\varphi(k)\| \le \|k\|$ for $k = \sum_{i=1}^{n} \lambda_i \theta_{a_i,b_i}$. This means that the map φ can be extended by continuity up to a map defined on the whole algebra $\mathcal{K}(A)$. The estimate

$$\left\| \sum_{i=1}^{\infty} \lambda_i \theta_{a_i,b_i} \right\| = \sup_{\|x\|\le 1} \left\| \sum_{i=1}^{\infty} \lambda_i a_i b_i^* x \right\| \le \left\| \sum_{i=1}^{\infty} \lambda_i a_i b_i^* \right\| = \left\| \varphi\left(\sum_{i=1}^{\infty} \lambda_i \theta_{a_i,b_i} \right) \right\|$$

shows that the map φ is an isometry, so statement (i) is proved. Statement (ii) can be proved in a similar way using the map

$$\varphi_n : \theta_{a_1\oplus\cdots\oplus a_n, b_1\oplus\cdots\oplus b_n} \longmapsto \begin{pmatrix} a_1 b_1^* & \dots & a_1 b_n^* \\ \vdots & & \vdots \\ a_n b_1^* & \dots & a_n b_n^* \end{pmatrix}.$$

Finally, since there exists an isometric map from the linear space $\bigcup_n \mathcal{K}(L_n(A))$ to the linear space $\bigcup_n M_n(A)$ and since these spaces are dense in C^*-algebras $\mathcal{K}(H_A)$ and $\mathcal{K}\otimes A$, respectively, statement (iii) follows. □

LEMMA 2.2.3. *For any $x \in \mathcal{M}$ there exists $z \in \mathcal{M}$ and $k = \theta_{u,v} \in \mathcal{K}(\mathcal{M})$ such that $x = kz$.*

PROOF. Put

$$u := v := z := \lim_{\varepsilon\to 0} x\,(\varepsilon + \langle x, x\rangle^{1/3})^{-1}.$$

Since $s^2(\varepsilon + s)^{-1}$ is uniformly convergent to s on bounded sets, in order to prove that u is well defined, note that for $t = \langle x, x\rangle$, one has

$$\begin{aligned}\langle x\,(\varepsilon+&\langle x, x\rangle^{1/3})^{-1} - x\,(\mu + \langle x, x\rangle^{1/3})^{-1}, x\,(\varepsilon + \langle x, x\rangle^{1/3})^{-1} - x\,(\mu + \langle x, x\rangle^{1/3})^{-1}\rangle \\ &= [(\varepsilon + t^{1/3})^{-1} - (\mu + t^{1/3})^{-1}]t[(\varepsilon + t^{1/3})^{-1} - (\mu + t^{1/3})^{-1}] \\ &= [(\varepsilon + t^{1/3})^{-1} - (\mu + t^{1/3})^{-1}]^2 (t^{1/3})^4.\end{aligned}$$

The same arguments show that $x = kz$. □

Note that we have also proved that $\mathcal{M}\langle\mathcal{M}, \mathcal{M}\rangle = \mathcal{M}$.

THEOREM 2.2.4. *A Hilbert A-module $\mathcal{M}$ is countably generated if and only if the C^*-algebra $\mathcal{K}(\mathcal{M})$ is σ-unital.*

PROOF. Let $\mathcal{K}(\mathcal{M})$ be σ-unital and let $æ_n$ be a countable approximate unit for it. Then

$$x = \lim_{n\to\infty} æ_n x \quad \text{for any } x \in \mathcal{M}. \tag{2.4}$$

Indeed, by Lemma 2.2.3, $x = kz$ holds for some $k \in \mathcal{K}(\mathcal{M})$, $z \in \mathcal{M}$. Since $æ_n k \longrightarrow k$ with respect to the norm, one has $æ_n x = æ_n kz \longrightarrow kz = x$.

By definition, any compact operator can be approximated by a linear combination of elementary ones. Hence, for each $æ_n$ there exist elements x_i^n and y_i^n in $\mathcal{M}$ such that

$$\left\| \sum_{i=1}^{s(n)} \theta_{x_i^n, y_i^n} - æ_n \right\| < \frac{1}{n}, \qquad n = 1, 2, \ldots.$$

Let us show that the countable set x_i^n, $i = 1, \ldots, s(n)$, $n = 1, 2, \ldots$, generates the module $\mathcal{M}$. Consider an arbitrary element $x \in \mathcal{M}$ and a number $\varepsilon > 0$. By (2.4) one can find a sufficiently large n such that

$$\|x - æ_n x\| < \frac{\varepsilon}{2} \qquad \text{and} \qquad \frac{1}{n} < \frac{\varepsilon}{2}.$$

Then

$$\left\| x - \sum_{i=1}^{s(n)} x_i^n \cdot \langle y_i^n, x\rangle \right\| \le \|x - æ_n(x)\| + \left\| æ_n(x) - \sum_{i=1}^{s(n)} \theta_{x_i^n, y_i^n}(x) \right\|$$
$$< \frac{\varepsilon}{2} + \frac{\varepsilon}{2} = \varepsilon.$$

Now let the module $\mathcal{M}$ be countably generated. It can be viewed as a module over the algebra A^+ obtained by unitalizing A (if it was not unital) with respect to the action $x \cdot (a, \mu) := x \cdot a + \mu x$, $x \in \mathcal{M}$, $a \in A$, $\mu \in \mathbf{C}$. If it were countably generated over A, then it should be countably generated over A^+ as well. Since in the definition of elementary compact operators only the A-valued inner product is involved, one has $\mathcal{K}_A(\mathcal{M}) = \mathcal{K}_{A^+}(\mathcal{M})$. Thus we can restrict ourselves to the case where A is unital.

So let $\mathcal{M}$ be a countably generated Hilbert module over a unital algebra A. By the Kasparov stabilization theorem, $\mathcal{M} \oplus H_A \cong H_A$. Let $\iota : \mathcal{M} \to H_A$ be the corresponding inclusion and let $\pi : H_A \to \mathcal{M}$ be the corresponding selfadjoint projection. Let $\{e_i\}$ denote the standard basis of H_A. Recall that for a C^*-algebra, the property of being σ-unital is equivalent to that of having a strictly positive element. Consider

$$æ := \sum_{n=1}^{\infty} \frac{\theta_{e_n, e_n}}{n},$$

or, in matrix form,

$$æ := \operatorname{diag}\,(1, \frac{1}{2}, \frac{1}{3}, \dots).$$

Then æ is a strictly positive element in $\mathcal{K}(H_A)$. Indeed, on one hand, by Proposition 2.2.1, we have $æ \in \mathcal{K}(H_A)$. On the other hand, if $\rho : \mathcal{K}(H_A) \to \mathbf{C}$ is a state such that $\rho(æ) = 0$, then $\rho(\theta_{e_n,e_n}) = 0$ for any n, since all $\theta_{e_n,e_n} \ge 0$. Then, for any $x = (x_1, x_2, \dots) \in H_A$, one has

$$\begin{aligned}
\rho(\theta_{e_n,x}\theta_{x,e_n}) &= \rho\left(\left(\sum_{j=1}^{\infty} \theta_{e_n, e_j x_j} \right) \left(\sum_{j=1}^{\infty} \theta_{e_n, e_j x_j} \right)^* \right) \\
&= \rho\left(\sum_{j=1}^{\infty} \theta_{e_n, e_j x_j} \theta^*_{e_n, e_j x_j} \right) = \rho\left(\sum_{j=1}^{\infty} \theta_{e_n, e_j x_j} \theta_{e_j x_j, e_n} \right) \\
&= \rho\left(\sum_{j=1}^{\infty} \theta_{e_n \cdot \langle e_j x_j, e_j x_j\rangle, e_n} \right) \le \|x\|^2 \sum_{j=1}^{\infty} \rho(\theta_{e_n,e_n}) = 0,
\end{aligned}$$

where the last inequality follows from

$$\begin{aligned}
\langle \theta_{e_n \cdot \langle e_j x_j, e_j x_j\rangle, e_n}(z), z\rangle &= \langle e_n \cdot \langle e_j x_j, e_j x_j\rangle \langle e_n, z\rangle, z\rangle \\
&= \langle z, e_n\rangle x_j^* x_j \langle e_n, z\rangle \le \|x\|^2 \langle \theta_{e_n,e_n} z, z\rangle.
\end{aligned}$$

Thus, for any $x, y, z \in \mathcal{M}$,

$$\theta_{x,y}(z) = x \cdot \langle y, z\rangle = \theta_{x,e_n}\theta_{e_n,y}(z)$$

and

$$|\rho(\theta_{x,y})| = |\rho(\theta_{x,e_n}\theta_{e_n,y})| \le \rho^{1/2}(\theta_{x,e_n}\theta_{e_n,x})\, \rho^{1/2}(\theta_{e_n,y}\theta_{y,e_n}) = 0,$$

due to the vanishing of the second factor. So ρ vanishes on a dense subset, hence, everywhere on $\mathcal{K}(H_A)$. We have shown that æ is a strictly positive element of $\mathcal{K}(H_A)$. Then $æ_n := æ^{1/n}$ is a countable approximate unit in $\mathcal{K}(H_A)$ and $\pi æ_n \iota$ is a countable approximate unit in $\mathcal{K}(\mathcal{M})$. Indeed, if $k \in \mathcal{K}(\mathcal{M})$, $\pi\iota k = k$, then $\iota k \pi \in \mathcal{K}(H_A)$ and

$$\|k - \pi æ_n \iota\| = \|\pi(\iota k \pi - æ_n)\iota\| = \|\iota k \pi - æ_n\| \longrightarrow 0 \qquad (n \longrightarrow \infty).$$

□

2.3. Complementable submodules and projections in Hilbert C^*-modules

Recall that a closed submodule $\mathcal{N}$ in a Hilbert C^*-module $\mathcal{M}$ is called orthogonally complementable if $\mathcal{M} = \mathcal{N} \oplus \mathcal{N}^\perp$. As we have already seen, a closed submodule in a Hilbert C^*-module is not necessarily orthogonally complementable.

DEFINITION 2.3.1. A closed submodule $\mathcal{N}$ in a Hilbert C^*-module $\mathcal{M}$ is called *(topologically) complementable* if there exists a closed submodule $\mathcal{L}$ in $\mathcal{M}$ such that $\mathcal{N} + \mathcal{L} = \mathcal{M}$, $\mathcal{N} \cap \mathcal{L} = \{0\}$.

The following example shows that there exist topologically complementable submodules that are not orthogonally complementable.

EXAMPLE 2.3.2. Let $J \subset A$ be a closed ideal such that the equality $Ja = 0$, $a \in A$, implies that $a = 0$. Put $\mathcal{M} := A \oplus J$,

$$\mathcal{N} := \{(b, b) : b \in J\}.$$

Then

$$\mathcal{N}^\perp = \{(c, -c) : c \in J\}.$$

Therefore $\mathcal{N} \oplus \mathcal{N}^\perp = J \oplus J \neq \mathcal{M}$. However, the submodule $\mathcal{L} = \{(a, 0) : a \in A\} \subset \mathcal{M}$ is a topological complement to $\mathcal{N}$ in $\mathcal{M}$.

Recall that we denote the nonorthogonal direct sum of Hilbert C^*-modules by $\mathcal{N} \widetilde{\oplus} \mathcal{L}$. A decomposition into a direct sum $\mathcal{M} = \mathcal{N} \widetilde{\oplus} \mathcal{L}$ allows us to define the projection P onto $\mathcal{N}$ along $\mathcal{L}$. The operator P is A-linear and, by the closed graph theorem, is bounded, hence $P \in \mathrm{End}_A(\mathcal{M})$. However, as is clear from Example 2.3.2, the projection P can be nonadjointable. But if $\mathcal{M} = \mathcal{N} \oplus \mathcal{L}$, then the corresponding projection is selfadjoint, $P \in \mathrm{End}^*_A(\mathcal{M})$. Since it is more convenient to work with orthogonal decompositions, we would like to describe situations where such a decomposition exists.

THEOREM 2.3.3 (**[90]**). *Let $\mathcal{M}$, $\mathcal{N}$ be Hilbert A-modules and $T \in \mathrm{Hom}^*_A(\mathcal{M}, \mathcal{N})$ an operator with closed image. Then*

(i) $\mathrm{Ker}\, T$ *is an orthogonally complementable submodule in* $\mathcal{M}$,
(ii) $\mathrm{Im}\, T$ *is an orthogonally complementable submodule in* $\mathcal{N}$.

PROOF. (i) Let $\mathrm{Im}\, T = \mathcal{N}_0$ and let $T_0 : \mathcal{M} \longrightarrow \mathcal{N}_0$ be an operator such that its action coincides with the action of T. By the open mapping theorem, the image of the unit ball $T_0(B_1(\mathcal{M}))$ contains some ball of radius $\delta > 0$ in $\mathcal{N}_0$. Therefore for each $y \in \mathcal{N}_0$, one can find some $x \in \mathcal{M}$ such that $T_0 x = y$ and $\|x\| \leq \delta^{-1}\|y\|$. One has

$$\|T_0^* y\|^2 = \|\langle y, T_0 T_0^* y\rangle\| \leq \|y\| \cdot \|T_0 T_0^* y\|,$$

and hence,

$$\|y\|^2 = \|\langle T_0 x, y\rangle\| = \|\langle x, T_0^* y\rangle\| \le \|x\| \cdot \|T_0^* y\| \le \delta^{-1} \|y\| \cdot \|y\|^{1/2} \|T_0 T_0^* y\|^{1/2}.$$

We obtain that for any $y \in \mathcal{N}_0$,

$$\|y\| \le \delta^{-2} \|T_0 T_0^* y\|.$$

Let us show that the spectrum of the operator $T_0 T_0^*$ does not contain the origin. Suppose the opposite, i.e., that $0 \in \mathrm{Sp}(T_0 T_0^*)$. Let f be a continuous function on $\mathbf{R}$ such that

$$f(0) = 1 = \|f\|, \qquad f(t) = 0 \text{ if } |t| \ge \frac{1}{2}\delta^{-2}.$$

Using functional calculus in the C^*-algebra $\mathrm{End}_A^*(\mathcal{M})$, we define the operator $S \in \mathrm{End}_A^*(\mathcal{M})$ by the formula $S = f(T_0 T_0^*)$. Then $\|S\| = 1$ and $\|T_0 T_0^* S\| \le \frac{1}{2}\delta^{-2}$. We can choose an element $x \in \mathcal{M}$ such that $\|x\| = 1$, $\|Sx\| > \frac{1}{2}$. Then the inequality

$$\|T_0 T_0^* Sx\| \le \frac{1}{2}\delta^{-2} < \delta^{-2} \|Sx\|$$

contradicts the assumption (with $y = Sx$). So $0 \notin \mathrm{Sp}(T_0 T_0^*)$. Therefore the operator $T_0 T_0^*$ is invertible and, in particular, surjective. For any $z \in \mathcal{M}$ one can find an element $w \in \mathcal{N}_0$ such that $T_0 z = T_0 T_0^* w$. Then $z - T_0^* w \in \mathrm{Ker}\, T$ and

$$z = (z - T_0^* w) + T_0^* w \in \mathrm{Ker}\, T + \mathrm{Im}\, T_0^*.$$

Since the module $\mathrm{Im}\, T_0^*$ is obviously orthogonal to $\mathrm{Ker}\, T$, it is a complement of $\mathrm{Ker}\, T$. This completes the proof of (i).

(ii) Since $\mathcal{M} = \mathrm{Ker}\, T \oplus \mathrm{Im}\, T_0^*$, the submodule $\mathrm{Im}\, T_0^*$ is closed. Note that $\mathrm{Im}\, T_0^* = \mathrm{Im}\, T^*$, so one can apply the proof of (i) to the case of the operator T^* instead of T and it gives the orthogonal decomposition $\mathcal{N} = \mathrm{Ker}\, T^* \oplus \mathrm{Im}\, T$. □

COROLLARY 2.3.4. *If $P \in \mathrm{End}_A^*(\mathcal{M})$ is an idempotent, then its image* $\mathrm{Im}\, P$ *is an orthogonally complementable submodule in $\mathcal{M}$.*

COROLLARY 2.3.5. *Let $\mathcal{M}, \mathcal{N}$ be Hilbert A-modules and let $F : \mathcal{M} \to \mathcal{N}$ be a topologically injective (i.e., $\|Fx\| \ge \delta \|x\|$ for some $\delta > 0$ and for all $x \in M$) adjointable A-homomorphism. Then $F(\mathcal{M}) \oplus F(\mathcal{M})^\perp = \mathcal{N}$.*

COROLLARY 2.3.6. *Let $\mathcal{M}$ be a Hilbert A-module and let $J : \mathcal{M} \to \mathcal{M}$ be a selfadjoint topologically injective A-homomorphism. Then J is an isomorphism.*

LEMMA 2.3.7 ([**90**]). *Let $\mathcal{M}$ be a finitely generated Hilbert submodule in a Hilbert module $\mathcal{N}$ over a unital C^*-algebra. Then $\mathcal{M}$ is an orthogonal direct summand in $\mathcal{N}$.*

PROOF. Let $x_1, \dots, x_n \in \mathcal{M}$ be a finite set of generators. Define the operator $F : L_n(A) \longrightarrow \mathcal{N}$ by the formula $F(e_i) = x_i$, where $e_i \in L_n(A)$, $i = 1, \dots, n$, is the standard basis. It is easy to see that F is adjointable with the adjoint $F^* : \mathcal{N} \longrightarrow L_n(A)$ acting by the formula $F^*(x) = (\langle x_1, x\rangle, \dots, \langle x_n, x\rangle)$, where $x \in \mathcal{N}$. By Theorem 2.3.3 the module $\mathrm{Im}\, F = \mathcal{M}$ is an orthogonal direct summand. □

LEMMA 2.3.8. *Let A be a unital C^*-algebra, $H_A = \mathcal{M} \widetilde{\oplus} \mathcal{N}$, $p : H_A \to \mathcal{M}$ a projection and $\mathcal{N}$ a projective module. Then $H_A = \mathcal{M} \oplus \mathcal{M}^\perp$ if and only if p is adjointable.*

PROOF. If p^* exists, then $(1-p)^* = 1-p^*$ exists as well. Therefore, by Theorem 2.3.3, $\operatorname{Ker}(1-p) = \mathcal{M}$ is the image of a selfadjoint projection.

To prove the converse, let us verify first that $H_A = \mathcal{N}^\perp + \mathcal{M}^\perp$. By the Kasparov stabilization theorem, one can assume, without loss of generality, that $\mathcal{N} = \operatorname{span}_A\langle e_1, \dots, e_n\rangle$, $\mathcal{N}^\perp = \operatorname{span}_A\langle e_{n+1}, e_{n+2}, \dots\rangle$. Let g_i be the image of e_i under the projection $\mathcal{N}$ onto $\mathcal{M}^\perp$:

$$e_1 = f_1 + g_1, \dots, e_n = f_n + g_n, \quad f_i \in \mathcal{M},\ g_i \in \mathcal{M}^\perp.$$

Since the projection induces an isomorphism of A-modules $\mathcal{N} \cong \mathcal{M}^\perp$, the elements $g_1, \dots, g_n$ are free generators and $\langle g_k, g_k\rangle > 0_A$ (i.e., the spectrum of this positive operator is separated from 0 and hence it is invertible). So, if

$$f_k = \sum_{i=1}^{\infty} f_k^i e_i, \quad \text{then} \quad e_k - f_k^k e_k = \sum_{i\neq k} f_k^i e_i + g_k.$$

On the other hand,

$$1 = \langle e_k, e_k\rangle = \langle f_k, f_k\rangle + \langle g_k, g_k\rangle, \quad 1 - (f_k^k)(f_k^k)^* \geq \langle g_k, g_k\rangle > 0,$$

i.e., the spectrum of the element $1 - f_k^k$ is separated from the origin, hence this element is invertible in A,

$$e_k = \frac{1}{1-f_k^k}\left(\sum_{i\neq k} f_k^i e_i + g_k\right) \in \mathcal{N}^\perp + \mathcal{M}^\perp \quad (k = 1, \dots, n).$$

Thus $\mathcal{N}^\perp + \mathcal{M}^\perp = H_A$. Let $x \in \mathcal{N}^\perp \cap \mathcal{M}^\perp$. Since any element $y \in H_A = \mathcal{M}\widetilde{\oplus}\mathcal{N}$ has the form $y = m+n$, one has $\langle x, y\rangle = \langle x, m\rangle + \langle x, n\rangle = 0$; in particular, $\langle x, x\rangle = 0$, thus $x = 0$. Therefore $H_A = \mathcal{N}^\perp\widetilde{\oplus}\mathcal{M}^\perp$. Consider the map $q = \begin{cases} 1 & \text{on } \mathcal{N}^\perp \\ 0 & \text{on } \mathcal{M}^\perp \end{cases}$, which is a bounded projection since $H_A = \mathcal{N}^\perp\widetilde{\oplus}\mathcal{M}^\perp$. Let $x+y \in \mathcal{M}\widetilde{\oplus}\mathcal{N}$, $x_1 + y_1 \in \mathcal{N}^\perp\widetilde{\oplus}\mathcal{M}^\perp$. Then

$$\begin{aligned} \langle p(x+y), x_1+y_1\rangle &= \langle x, x_1+y_1\rangle = \langle x, x_1\rangle, \\ \langle x+y, q(x_1+y_1)\rangle &= \langle x+y, x_1\rangle = \langle x, x_1\rangle. \end{aligned}$$

Therefore p^* exists and equals q. □

2.4. Full Hilbert C^*-modules

Let $\mathcal{M}$ be a Hilbert A-module. Recall that it is full if the closure $\langle \mathcal{M}, \mathcal{M}\rangle \subset A$ of the linear span of all $\langle x, x\rangle$, $x \in \mathcal{M}$, coincides with A. One can always consider any Hilbert module as a full Hilbert module over the C^*-algebra $\langle \mathcal{M}, \mathcal{M}\rangle$. The following statement shows that full modules are easier to handle.

THEOREM 2.4.1 ([**71, 36**]). *Let A be a σ-unital C^*-algebra and let $\mathcal{M}$ be a full Hilbert A-module. Then:*

(i) *There exists a Hilbert A-module $\mathcal{N}$ such that $l_2(\mathcal{M}) \cong H_A \oplus \mathcal{N}$. If A is unital, then there exist a number n and a Hilbert A-module $\mathcal{N}'$ such that $\mathcal{M} \oplus \cdots \oplus \mathcal{M} = \mathcal{M}^n \cong A \oplus \mathcal{N}'$.*

(ii) *If the module $\mathcal{M}$ is countably generated, then $l_2(\mathcal{M}) \cong H_A$.*

PROOF. Consider the set

$$S = \left\{ c \in A : \|c\| \leq 1, c = \sum_{i=1}^{k} \langle x_i, x_i \rangle, k \in \mathbf{N}, x_i \in \mathcal{M} \right\}.$$

To prove the theorem, two following lemmas are required.

LEMMA 2.4.2 ([**16**]). *For any $a \in A$, $a \geq 0$ and any $\varepsilon > 0$, there exists $c \in S$ such that $\|(1-c)a\| < \varepsilon$.*

PROOF. Since the module $\mathcal{M}$ is full, one can find a finite set of elements $y_i \in \mathcal{M}$ such that

$$\left\| a - \sum_{i=1}^{k} \langle y_i, y_i \rangle \right\| < \varepsilon/2. \tag{2.5}$$

Put

$$x_i = y_i\Big(\varepsilon^2 + \sum_{j=1}^{k} \langle y_j, y_j \rangle\Big)^{-1/2}, \quad i = 1, \dots, k;$$

$$c = \sum_{i=1}^{k} \langle x_i, x_i \rangle, \qquad b = \sum_{i=1}^{k} \langle y_i, y_i \rangle.$$

Then

$$\|c\| = \left\| (\varepsilon^2 + b)^{-1/2} b (\varepsilon^2 + b)^{-1/2} \right\| \leq 1,$$

hence $c \in S$. Let $f(t) := \varepsilon^4 t^2 (\varepsilon^2 + t)^{-2}$. Applying this function to the element b we obtain the estimate

$$\begin{aligned} \|f(b)\| &= \|\varepsilon^4 b^2 (\varepsilon^2 + b)^{-2}\| = \|\varepsilon^2(\varepsilon^2 + b)^{-1} b^2 \varepsilon^2 (\varepsilon^2 + b)^{-1}\| \\ &= \left\|(1-c) b^2 (1-c)\right\| \leq \varepsilon^2/4. \end{aligned}$$

Therefore $\|(1-c)b\| \leq \varepsilon/2$ and, together with the estimate (2.5), this proves the lemma. □

LEMMA 2.4.3 ([**16**]). *There exists a sequence (x_i), $x_i \in \mathcal{M}$, such that the sequence of partial sums of the series $\sum_{i=1}^{k} \langle x_i, x_i \rangle$ is a countable approximate unit of the algebra A. If A is unital, then there exist a finite number k and elements $x_1, \dots, x_k \in \mathcal{M}$ such that $\sum_{i=1}^{k} \langle x_i, x_i \rangle = 1$.*

PROOF. Consider first the case of a unital C^*-algebra. Then, by Lemma 2.4.2, one can find an element $c \in S$ such that $\|1 - c\| < 1/2$. Therefore the element c is invertible and $c = \sum_{j=1}^{k} \langle y_j, y_j \rangle$ for some k and some $y_j \in \mathcal{M}$. Put $x_j = y_j \cdot c^{-1/2}$; then $\sum_{j=1}^{k} \langle x_j, x_j \rangle = 1$.

In the case of a nonunital C^*-algebra let $h \in A$ be a strictly positive element. By induction we shall construct a sequence (c_j) in S such that

$$\sum_{j=1}^{k} c_j \leq 1; \qquad \left\| \left(1 - \sum_{j=1}^{k} c_j\right) h \right\| < \frac{1}{2^k}.$$

By Lemma 2.4.2, we can find an element $c_1 \in S$ such that $\|(1 - c_1)h\| < \frac{1}{2}$. Under the assumption that the elements $c_1, \dots, c_k$ are already found, by Lemma 2.4.2 we can find an element $d \in S$ such that

$$\left\| (1-d)\Big(1-\sum_{j=1}^{k} c_j\Big)^{1/2} h \right\| < \frac{1}{2^{k+1}} \tag{2.6}$$

and put

$$c_{k+1} = \Big(1-\sum_{j=1}^{k} c_j\Big)^{1/2} d \Big(1-\sum_{j=1}^{k} c_j\Big)^{1/2}.$$

Since $\left\| \Big(1-\sum_{j=1}^{k} c_j\Big)^{1/2} \right\| \le 1$ and $d \in S$, we have $c_{k+1} \in S$. Since $\|d\| \le 1$ and $c_{k+1} \le 1-\sum_{j=1}^{k} c_j$, we have $\sum_{j=1}^{k+1} c_j \le 1$. Finally, inequality (2.6) implies that

$$\begin{aligned} \left\| \Big(1-\sum_{j=1}^{k+1} c_j\Big) h \right\| &= \left\| \Big(1-\sum_{j=1}^{k} c_j\Big)^{1/2} (1-d)\Big(1-\sum_{j=1}^{k} c_j\Big)^{1/2} h \right\| \\ &\le \left\| \Big(1-\sum_{j=1}^{k} c_j\Big)^{1/2} \right\| \left\| (1-d)\Big(1-\sum_{j=1}^{k} c_j\Big)^{1/2} h \right\| < \frac{1}{2^{k+1}}, \end{aligned}$$

thus completing the induction step. So we obtain that

$$\left\| \Big(1-\sum_{j=1}^{k} c_j\Big) h \right\| \longrightarrow 0$$

as $k \to \infty$. Since the strictly positive element h generates the whole C^*-algebra A [**104**], the lemma is proved. □

Let us continue the proof of the theorem. By Lemma 2.4.3, we can choose a sequence (x_i) in the module $\mathcal{M}$ such that the partial sums of $\sum \langle x_i, x_i \rangle$ form an approximate unit in A. Define the map $T : A \longrightarrow l_2(\mathcal{M})$ by the formula

$$T(a) = (x_1 a, \ldots, x_k a, \ldots), \qquad a \in A. \tag{2.7}$$

Since the series $\langle Ta, Ta \rangle = \sum_{i=1}^{\infty} a^* \langle x_i, x_i \rangle a = a^* a$ is convergent in A, one has $T(a) \in l_2(\mathcal{M})$. Moreover, the adjoint operator is well defined, $T^* : l_2(\mathcal{M}) \longrightarrow A$, $T^*(y_i) = \sum_{i=1}^{\infty} \langle x_i, y_i \rangle \in A$ for $(y_i) \in l_2(\mathcal{M})$. Since T^*T acts identically on A, the operator T is an isometry and

$$l_2(\mathcal{M}) = \operatorname{Im} T \oplus \operatorname{Ker} T^* \cong A \oplus \mathcal{N},$$

where $\mathcal{N} = \operatorname{Ker} T^*$. This completes the proof of statement (i) of the theorem for the case of a nonunital C^*-algebra. In the case of a unital C^*-algebra the previous reasoning can be applied literally if we replace the module $l_2(\mathcal{M})$ by $\mathcal{M}^k$ and replace the infinite sequence in (2.7) by a finite one.

We pass to the proof of statement (ii). Renumber the sequences in the module $l_2(\mathcal{M})$ with the help of a bijection $\mathbf{N} \longrightarrow \mathbf{N} \times \mathbf{N}$. Then elements of the module $l_2(\mathcal{M})$ can be viewed as sequences (m_{ij}), $m_{ij} \in \mathcal{M}$, and, for each $i \in \mathbf{N}$, the set of sequences $(m_{ij}), j \in \mathbf{N}$, is isomorphic to the module $l_2(\mathcal{M})$. Such a renumbering defines an isomorphism $l_2(\mathcal{M}) \cong l_2(l_2(\mathcal{M}))$. Taking into account the isomorphism $l_2(\mathcal{M}) \cong A \oplus \mathcal{N}$, we conclude that

$$l_2(\mathcal{M}) \cong l_2(l_2(\mathcal{M})) \cong l_2(A \oplus \mathcal{N}) = l_2(A) \oplus l_2(\mathcal{N}).$$

Note that the Hilbert module $l_2(\mathcal{N})$ is countably generated, hence

$$l_2(\mathcal{M}) \cong l_2(A) \oplus l_2(\mathcal{N}) \cong l_2(A)$$

by the Kasparov stabilization theorem. □

2.5. Dual modules. Self-duality

For a Hilbert A-module $\mathcal{M}$ let us denote by $\mathcal{M}'$ the set of all bounded A-linear maps from $\mathcal{M}$ to A. The structure of a vector space over the field $\mathbf{C}$ is introduced by the formula $(\lambda \cdot f)(x) := \overline{\lambda} f(x)$, where $\lambda \in \mathbf{C}$, $f \in \mathcal{M}'$, $x \in \mathcal{M}$. This definition may seem artificial, however it is convenient since it allows us to define a linear inclusion of the module $\mathcal{M}$ into $\mathcal{M}'$ (there is also the alternate approach: to define $\mathcal{M}'$ as the set of all *anti*-linear maps from $\mathcal{M}$ into A). The formula

$$(f \cdot a)(x) = a^* f(x),$$

where $a \in A$, introduces the structure of a right A-module on $\mathcal{M}'$. This module is complete with respect to the norm $\|f\| = \sup\{\|f(x)\| : \|x\| \leq 1\}$. Such modules are called *dual* (*Banach*) modules. The elements of the module $\mathcal{M}'$ are called *functionals* on the Hilbert module $\mathcal{M}$. Note that there is an obvious isometric inclusion $\mathcal{M} \subset \mathcal{M}'$, which is defined by the formula $x \mapsto \langle x, \cdot \rangle = \widehat{x}$. Sometimes we shall write $\langle f, x \rangle$ instead of $f(x)$.

DEFINITION 2.5.1. A Hilbert module $\mathcal{M}$ is called *self-dual* if $\mathcal{M} = \mathcal{M}'$.

The condition of self-duality is very strong. Below we shall see that there are quite a few self-dual modules: any Hilbert module over a C^*-algebra A is self-dual iff A is finite dimensional. Self-dual Hilbert C^*-modules behave quite like Hilbert spaces. In the same way as in the case of Hilbert spaces, the following statements can be easily checked.

PROPOSITION 2.5.2 ([**100**]). *Let $\mathcal{M}$ be a self-dual Hilbert A-module, $\mathcal{N}$ an arbitrary Hilbert A-module and $T : \mathcal{M} \longrightarrow \mathcal{N}$ a bounded operator, $T \in \mathrm{Hom}_A(\mathcal{M}, \mathcal{N})$. Then there exists an operator $T^* : \mathcal{N} \longrightarrow \mathcal{M}$ such that the equality $\langle x, T^* y \rangle = \langle Tx, y \rangle$ holds for all $x \in \mathcal{M}$, $y \in \mathcal{N}$.*

COROLLARY 2.5.3. *Let $\mathcal{M}$ be a self-dual Hilbert A-module. Then* $\mathrm{End}_A(\mathcal{M}) = \mathrm{End}^*_A(\mathcal{M})$.

PROPOSITION 2.5.4. *Let $\mathcal{M}$ be a self-dual Hilbert A-module and let $\mathcal{M} \subset \mathcal{N}$. Then $\mathcal{N} = \mathcal{M} \oplus \mathcal{M}^\perp$.*

PROOF. Since $\mathcal{M}$ is self-dual, $i : \mathcal{M} \to \mathcal{N}$ is an isometric adjointable inclusion. Therefore $\mathcal{M} = i\mathcal{M}$ has an orthogonal complement by Corollary 2.3.5. □

If A is a unital C^*-algebra, then the Hilbert module $L_n(A)$ is obviously self-dual. For an arbitrary module this is not true; moreover, the Banach module $\mathcal{M}'$ may not admit the structure of a Hilbert C^*-module at all. A description of the dual module for the standard Hilbert module H_A is given by the following statement.

PROPOSITION 2.5.5. *Consider the set of sequences $f = (f_i)$, $f_i \in A$, $i \in \mathbf{N}$, such that the norms of partial sums $\left\| \sum_{i=1}^N f_i^* f_i \right\|$ are uniformly bounded. If A is a* *unital C^*-algebra, then this set coincides with H'_A, the action of f on elements of the module H_A is defined by the formula*

$$f(x) = \sum_{i=1}^{\infty} f_i^* x_i, \tag{2.8}$$

where $x = (x_i) \in H_A$, and the norm of f is given by the equality

$$\|f\|^2 = \sup_N \left\| \sum_{i=1}^{N} f_i^* f_i \right\|. \tag{2.9}$$

PROOF. Let $f \in H'_A$ and let e_i be the standard basis in H_A. Put $f_i = (f(e_i))^*$. We shall show that the sequence (f_i) determines a functional f in a unique way. Let us assume that there exists a functional $g \neq f$, $g(e_i) = f(e_i)$. Choose $x \in H_A$ such that $\|f(x) - g(x)\| = C \neq 0$. Denote by $x^{(N)}$ the image of x under the projection onto the submodule $L_N(A) \subset H_A$, $x^{(N)} = \sum_{i=1}^N e_i x_i = (x_1, \ldots, x_N, 0, \ldots)$. Let us find a number N such that

$$\left\| x - x^{(N)} \right\| = \left\| \sum_{i=N+1}^{\infty} x^* x \right\|^{1/2} < \frac{C}{2(\|f\| + \|g\|)}.$$

Since $f(x^{(N)}) = g(x^{(N)})$, we have $\|f(x - x^{(N)}) - g(x - x^{(N)})\| = C$. But, on the other hand, we have

$$\begin{aligned} \left\| f(x - x^{(N)}) - g(x - x^{(N)}) \right\| &\le (\|f\| + \|g\|) \left\| x - x^{(N)} \right\| \\ &< (\|f\| + \|g\|) \frac{C}{2(\|f\| + \|g\|)} \le C/2. \end{aligned}$$

The obtained contradiction shows that $f = g$. The Cauchy–Bunyakovskii inequality

$$\left\| \sum_{i=1}^{N} f_i^* x_i \right\|^2 \le \left\| \sum_{i=1}^{N} f_i^* f_i \right\| \left\| \sum_{i=1}^{N} x_i^* x_i \right\| \tag{2.10}$$

shows that

$$\|f\|^2 \le \sup_N \left\| \sum_{i=1}^{N} f_i^* f_i \right\|. \tag{2.11}$$

Note that if we take $x_i = f_i / \left\| \sum_{i=1}^N f_i^* f_i \right\|^{1/2}$, then equality is obtained in (2.10). Put $f^{(N)} = (f_1, \ldots, f_N, 0, \ldots)$, $f^{(N)} \in L_N(A)' \cong L_N(A)$. It is clear that

$$\|f\| \ge \left\| f^{(N)} \right\|. \tag{2.12}$$

But $\|f^{(N)}\|^2 = \left\| \sum_{i=1}^N f_i^* f_i \right\|$, hence (2.11) and (2.12) imply (2.9). Since for any $\varepsilon > 0$ one can find a number N such that the estimate

$$\left\| \sum_{i=N}^{N+n} f_i^* x_i \right\| \le \left\| \sum_{i=N}^{N+n} f_i^* f_i \right\| \cdot \left\| \sum_{i=N}^{N+n} x_i^* x_i \right\| \le \|f\|^2 \left\| \sum_{i=N}^{N+n} x_i^* x_i \right\| < \|f\|^2 \varepsilon$$

holds for all $n > 0$, the convergence of the series (2.8) follows. □

Note that for the functional $f = (\varphi_i)$ from Example 2.1.2, the partial sums $\sum_{i=1}^N \varphi_i^* \varphi_i$ are uniformly bounded, however the corresponding series is not convergent.

Let us describe an interesting example of a dual module.

EXAMPLE 2.5.6 ([**44**]). Let $A = \mathcal{B}(H)$ be the algebra of all bounded operators on a separable Hilbert space H. Consider pairwise orthogonal projections $p_i \in A$, $i \in \mathbf{N}$, such that the series $\sum_i p_i$ converges w^*-weakly to $1 \in A$, and each projection p_i is equivalent to 1. We can consider $H = \bigoplus_i H_i$ as an orthogonal sum of Hilbert spaces isomorphic to H, $u_i : H \to H_i$ being isometries, so that

$$p_i = u_i u_i^*, \qquad 1 = u_i^* u_i.$$

As was shown above (see Proposition 2.5.5),

$$l_2(A)' = \left\{ \{a_i\} \middle| a_i \in A, i \in \mathbf{N}, \left\{ \sup_{N \in \mathbf{N}} \left\| \sum_{i=1}^N a_i a_i^* \right\| \right\} < \infty \right\}$$

is a C^*-Hilbert module over A with respect to the inner product

$$\langle \{a_i\}, \{b_i\} \rangle := w^* \text{-lim} \sum_i a_i b_i^*.$$

The maps

$$S : A \to l_2(A)', \qquad S : a \mapsto a \cdot \{u_i\},$$
$$S^{-1} : l_2(A)' \to A, \qquad S^{-1} : \{a_i\} \mapsto w^* \text{-lim} \sum_i a_i u_i^*$$

define an isometric isomorphism of A and $l_2(A)'$.

Let φ be a positive linear functional on A. If $\mathcal{M}$ is a Hilbert A-module and if $N_\varphi = \{x \in \mathcal{M} : \varphi(\langle x, x \rangle) = 0\}$ is a linear subspace of $\mathcal{M}$, then $\mathcal{M}/N_\varphi$ is a pre-Hilbert space with the inner product $(\cdot, \cdot)_\varphi$ given by the formula

$$(x + N_\varphi, y + N_\varphi)_\varphi = \varphi(\langle x, y \rangle), \quad x, y \in \mathcal{M}.$$

We denote the norm defined by this scalar-valued inner product by $\|\cdot\|_\varphi$ and the Hilbert space obtained by completion of $\mathcal{M}/N_\varphi$ with respect to this norm by H_φ. Let $f \in \mathcal{M}'$. By Theorem 2.1.4 and Corollary 2.1.6, we have

$$f(x)^* f(x) \le \|f\|^2 \langle x, x \rangle$$

for all $x \in \mathcal{M}$. Therefore, if $x \in N_\varphi$, then

$$\varphi(f(x)^* f(x)) = 0 = \varphi(f(x)).$$

Hence, the map

$$x + N_\varphi \longmapsto \varphi(f(x)) \tag{2.13}$$

defines a linear functional on $\mathcal{M}/N_\varphi$. Since

$$\begin{aligned} |\varphi(f(x))| &\le \|\varphi\|^{1/2} \varphi(f(x)^* f(x))^{1/2} \le \|\varphi\|^{1/2} \|f\| \varphi(\langle x, x \rangle)^{1/2} \\ &= \|\varphi\|^{1/2} \|f\| \|x + N_\varphi\|_\varphi , \end{aligned}$$

the functional (2.13) is bounded. Then there exists a unique vector $f_\varphi \in H_\varphi$ such that $\|f_\varphi\|_\varphi \le \|f\| \|\varphi\|^{1/2}$ and $(f_\varphi, x + N_\varphi)_\varphi = \varphi(f(x))$ for all $x \in \mathcal{M}$. For $x \in \mathcal{M}$, we shall denote by $\widehat{x}$ the functional $\langle x, \cdot \rangle \in \mathcal{M}'$. Note that $\widehat{y}_\varphi = y + N_\varphi$ for all $y \in \mathcal{M}$.

Let ψ be another positive linear functional on A such that $\psi \leq \varphi$. Then $N_\varphi \subset N_\psi$ and the natural map $x + N_\varphi \longmapsto x + N_\psi$ can be extended to the map

$$V_{\varphi,\psi} : H_\varphi \longrightarrow H_\psi, \qquad \|V_{\varphi,\psi}\| \leq 1.$$

It is easy to see that $V_{\varphi,\psi}(\widehat{x}_\varphi) = \widehat{x}_\psi$. It turns out that the same holds for all elements of $\mathcal{M}'$.

PROPOSITION 2.5.7. *Let $\mathcal{M}$ be a Hilbert A-module and let φ and ψ be positive linear functionals on A with $\psi \leq \varphi$. Then $V_{\varphi,\psi}(f_\varphi) = f_\psi$ for any functional $f \in \mathcal{M}'$.*

PROOF. Let $f \in \mathcal{M}'$. Since the quotient space $\mathcal{M}/N_\varphi$ is dense in H_φ, one can choose a sequence $\{y_n + N_\varphi\}$ of elements in $\mathcal{M}/N_\varphi$ such that $\|y_n + N_\varphi - f_\varphi\|_\varphi \to 0$. Then

$$V_{\varphi,\psi}(f_\varphi) = \lim_n V_{\varphi,\psi}(y_n + N_\varphi) = \lim_n (y_n + N_\psi).$$

To prove the statement, it is sufficient to show that $\psi(\langle y_n, x\rangle) \to \psi(f(x))$ for all $x \in \mathcal{M}$. But

$$\begin{aligned} |\psi(\langle y_n,x\rangle - f(x))|^2 & \\ &\leq \|\psi\| \cdot \psi(\langle y_n, x\rangle\langle x, y_n\rangle - \langle y_n, x\rangle f(x)^* - f(x)\langle x, y_n\rangle + f(x)f(x)^*) \\ &\leq \|\varphi\| \cdot \varphi(\langle y_n, x\rangle\langle x, y_n\rangle - \langle y_n, x\rangle f(x)^* - f(x)\langle x, y_n\rangle + f(x)f(x)^*) \end{aligned}$$

for each $n \in \mathbf{N}$. Since

$$\varphi(\langle y_n, x\rangle f(x)^*) = \varphi(\langle y_n, x \cdot (f(x))^*\rangle) \to \varphi(f(x \cdot (f(x))^*)) = \varphi(f(x)f(x)^*),$$

it is sufficient to show that $\varphi(\langle y_n, x\rangle\langle x, y_n\rangle - f(x)\langle x, y_n\rangle) \to 0$. Note that

$$\begin{aligned} \varphi(\langle y_n, x\rangle\langle x, y_n\rangle - f(x)\langle x, y_n\rangle) &= \varphi(\langle y_n, x \cdot \langle x, y_n\rangle\rangle - f(x \cdot \langle x, y_n\rangle)) \\ &= (y_n + N_\varphi - f_\varphi, x \cdot \langle x, y_n\rangle + N_\varphi)_\varphi, \end{aligned}$$

and the sequence $\{x \cdot \langle x, y_n\rangle + N_\varphi\}$ is bounded with respect to the norm $\|\cdot\|_\varphi$. Indeed,

$$\begin{aligned} \|x \cdot \langle x, y_n\rangle + N_\varphi\|_\varphi^2 &= \varphi(\langle x \cdot \langle x, y_n\rangle\rangle, \langle x \cdot \langle x, y_n\rangle\rangle) \\ &= \varphi(\langle y_n, x\rangle\langle x, x\rangle\langle x, y_n\rangle) \\ &\leq \|x\|^2 \cdot \varphi(\langle y_n, x\rangle\langle x, y_n\rangle) \\ &\leq \|x\|^2 \cdot \varphi(\|x\|^2 \cdot \langle y_n, y_n\rangle) \\ &= \|x\|^4 \cdot \|y_n + N_\varphi\|^2, \end{aligned}$$

and the sequence $\{y_n + N_\varphi\}$ is bounded. Since $\|y_n + N_\varphi - f_\varphi\|_\varphi \to 0$, the statement is proved. □

It turns out that the property of a Hilbert C^*-module over a unital C^*-algebra being self-dual is very close to the property of being finitely generated and projective. We complete this section with a complete description [**131**] of that interplay in the commutative case.

DEFINITION 2.5.8. A C^*-algebra A is called *module-infinite* (MI) if each countably generated Hilbert A-module is projective and finitely generated if and only if it is self-dual.

Recall that a projective finitely generated module over a unital C^*-algebra is always self-dual.

DEFINITION 2.5.9. A commutative unital C^* algebra $A = C(Y)$ is said to be DI (*divisibly infinite*) if for any infinite sequence u_i of elements of norm $1 \geq \|u_i\| \geq C > 0$ in A there exist a subsequence $i(k)$ and elements $0 \leq b_k \in A$ of norm 1 such that

(i) the supports of b_k in Y are pairwise disjoint;

(ii) for each k there exist points y_k, y'_k such that $b_k(y_k) = 1$, $y'_k \notin \bigcup_j \operatorname{supp} b_j$, $|u_{i(k)}(y'_k)| \geq \delta$, $|u_{i(k)}(y_k)| \geq \delta$, and the sequences $\{y_k\}$ and $\{y'_k\}$ have a common accumulation point. In particular, $\sum_k b_k^s$ is a discontinuous function for any integer $s \geq 1$.

THEOREM 2.5.10. *If a commutative unital C^*-algebra A is* DI, *then it is* MI.

PROOF. We have to prove that any countably generated self-dual Hilbert A-module $\mathcal{M}$ is finitely generated and projective. By the Kasparov stabilization theorem 1.4.2, one has $\mathcal{M} \oplus l_2(A) \cong l_2(A)$, where $l_2(A)$ is the standard Hilbert C^*-module (see p. 6). Denote by $p : l_2(A) \to \mathcal{M} \subset l_2(A)$ the corresponding orthogonal projection and let $p_j : l_2(A) \to E_j \cong A^j$ be the orthogonal projection onto the first j standard summands of $l_2(A)$. Then either 1) $\|(1 - p_j)p\| \to 0$ as $j \to \infty$ or 2) this is not the case.

1) Let us show that in this case $\mathcal{M}$ is finitely generated and projective. One can argue as in [**92**]: for a sufficiently large j, the operator

$$J(x) = \begin{cases} p_j(x) & \text{if } x \in \mathcal{M}, \\ x & \text{if } x \in \mathcal{M}^\perp \cong l_2(A), \end{cases}$$

is close to the identity, hence it is an isomorphism. It maps $\mathcal{M}$ isomorphically onto a direct summand of E_j.

2) In this case (changing the standard decomposition, if necessary) we can find a sequence $j(k)$ such that for each k there exists an element $z_k \in \mathcal{M}$ of norm 1 such that its $j(k)$-th entry $z_k^{j(k)}$ (with respect to the standard decomposition) has norm greater than some fixed $C > 0$. Let us choose functions b_k as in Definition 2.5.9 (for simplicity we assume that we do not need to pass to a subsequence for the second time). Then the formula $\beta(x) = \sum_k b_k x^{j_k}$ defines a functional on $l_2(A)$. It is evident that this functional is not adjointable on $l_2(A)$ since $b_k \not\to 0$ as $k \to \infty$. Let us show that there is no adjointable functional α on $l_2(A)$ with $\alpha|_{\mathcal{M}} = \beta|_{\mathcal{M}}$, so that $\beta|_{\mathcal{M}}$ is a nonadjointable functional on $\mathcal{M}$ and $\mathcal{M}$ is not a self-dual module. Indeed, suppose that there exists an element $a = (a_1, a_2, \dots) \in \mathcal{M} \subset l_2(A)$ such that $\alpha(x) := \sum_i a_i x^i = \beta(x)$ for any $x \in \mathcal{M}$. Then

$$\langle a, a \rangle \leq \sum_k b_k^2 \tag{2.14}$$

(the last element is a bounded measurable discontinuous function; see Definition 2.5.9 (ii)), since for any continuous positive function f_k equal to 1 on $\bigcup_{j \leq k} \operatorname{supp} b_j$ and equal to 0 on $\bigcup_{j > k} \operatorname{supp} b_j$ (the existence follows from normality of Y), one has $\beta f_k = \alpha f_k + (\beta - \alpha) f_k$. But βf_k is an element of $l_2(A)$ (or, more precisely, an adjointable functional $\beta f_k(x) = \sum_{j \leq k} b_k x^{j_k}$). Let $b f_k$ be the corresponding element. Then $b f_k = a f_k + (b - a) f_k$ is the decomposition corresponding to $l_2(A) = \mathcal{M} \oplus l_2(A)$. Indeed, $a \in \mathcal{M}$ and $\langle (b - a) f_k, x \rangle = 0$ for any $x \in M$, since

$\alpha(x)=\beta(x)$ for those x. The Pythagorean theorem for Hilbert C^*-modules shows that $\langle a,a\rangle f_k^2 \le \sum_{j\le k} b_j^2$. Taking all the f_k's, one obtains (2.14).

On the other hand,

$$\langle a,a\rangle(y_i) \ge \langle a,z_i\rangle\langle z_i,a\rangle(y_i) = \sum_{m,k} b_m z_i^{j(m)} \overline{z_i^{j(k)}} b_k(y_i) = b_i^2(y_i) z_i^{j(i)}(y_i) \ge \delta^2.$$

But (2.14) implies that $\langle a,a\rangle(y_i')=0$. Hence, $\langle a,a\rangle$ does not belong to A. This contradiction proves the theorem. □

Now we describe some consequences of this theorem.

THEOREM 2.5.11. *A commutative separable unital C^*-algebra A is* MI *if and only if its Gelfand spectrum Y has no isolated points.*

PROOF. If Y has an isolated point, then a separable Hilbert space arises as one of self-dual modules, hence, MI does not hold.

Now, suppose that Y has no isolated points, in particular it is infinite. Since Y is a compact Hausdorff separable space, the topology is generated by some metric ρ. For any given sequence u_i of elements with norm $\ge C$ we can find a sequence of different points $\widetilde{y}_i$ such that $|u_i(\widetilde{y}_i)| > 2C/3$.

Since Y is compact, one can pass to a convergent subsequence $y_k = \widetilde{y}_{i(k)}$. For a convergent sequence, by induction, we can choose ε_k such that the corresponding ε_k-neighborhoods U_k of y_k are pairwise disjoint. Then we can choose $y_k' \ne y_k$ inside these neighborhoods so close to y_k that $|u_{i(k)}| \ge C/2$. Finally one can choose functions b_k such that $b_k(y_k)=1$ and $\operatorname{supp} b_k \subset U_k \setminus y_k'$. It remains to take $\delta = C/2$. □

COROLLARY 2.5.12. *Suppose that X is a compact connected separable Hausdorff G-space. Then $C_G(X)$ is* MI *if and only if X/G has at least two separated points.*

PROOF. The Gelfand spectrum Y of $C_G(X)$ is a quotient space of X/G with respect to the equivalence of nonseparated points. Since X is connected, Y is connected too. So, by Theorem 2.5.11, $C_G(X)$ is MI if and only if Y is not reduced to one point. □

2.6. Banach-compact operators

DEFINITION 2.6.1. Let $\mathcal{M},\mathcal{N}$ be Hilbert A-modules and let $\mathcal{M}'$ be the dual module for $\mathcal{M}$. Consider the closure $\mathcal{B}K(\mathcal{M},\mathcal{N})$ of the linear span of all operators

$$\theta_{y,f}(x) = y\cdot f(x),$$

$x\in\mathcal{M}$, $y\in\mathcal{N}$, $f\in\mathcal{M}'$, in the Banach space $\operatorname{Hom}_A(\mathcal{M},\mathcal{N})$. We call the elements of the set $\mathcal{B}K(\mathcal{M},\mathcal{N})$ *Banach-compact operators.*

In the case $\mathcal{N}=\mathcal{M}$ the set $\mathcal{B}K(\mathcal{M},\mathcal{N})$ is equipped with the natural structure of a Banach algebra. If $T\in\operatorname{End}_A(\mathcal{M})$ is a (not necessarily adjointable) operator, then the equalities

$$\theta_{y,f}Tx = y\cdot f(Tx) = \theta_{y,f\circ T}(x), \qquad T\theta_{y,f}(x) = Ty\cdot f(x) = \theta_{Ty,f}(x)$$

show that $\mathcal{B}K(\mathcal{M})$ is a two-sided ideal in the Banach algebra $\operatorname{End}_A(\mathcal{M})$.

In the case of the standard Hilbert module over a unital C^*-algebra we give one more (geometric) description of compact and Banach-compact operators. Let

$S \subset H_A$ be a bounded set. We call it *A-pre-compact* if for each $\varepsilon > 0$ there exists a free finitely generated A-module $\mathcal{N} \cong L_n(A)$, $\mathcal{N} \subset H_A$, such that $\operatorname{dist}(S, \mathcal{N}) < \varepsilon$.

PROPOSITION 2.6.2. *Let $T \in \operatorname{End}_A(H_A)$ (resp. $T \in \operatorname{End}^*_A(H_A)$). Then the following conditions are equivalent:*

(i) *$T \in \mathcal{B}K(H_A)$ (resp., $T \in \mathcal{K}(H_A)$);*
(ii) *the image $T(B_1(H_A))$ of the unit ball $B_1(H_A)$ is A-pre-compact.*

PROOF. If statement (i) holds, it is sufficient to prove that one can find an approximating module $\mathcal{N} \cong L_n(A)$ for a finite set of elements in H_A and this can be easily done by the Dupré – Fillmore method, as in the proof of Theorem 1.4.5. So suppose that (ii) is fulfilled. Then for any $\varepsilon > 0$, one can find elements $b_1, \dots, b_k \in H_A$, generating the module $\mathcal{N} \subset H_A$, such that $\langle b_i, b_j \rangle = \delta_{ij}$ and $\operatorname{dist}(T(B_1(H_A)), \mathcal{N}) < \varepsilon$, where $B_1(H_A)$ is the unit ball of H_A. Denote by $P_\mathcal{N}$ the projection onto $\mathcal{N}$ and consider the operator $P_\mathcal{N} T$. It can be decomposed as

$$P_\mathcal{N} T x = b_1 \langle f_1, x \rangle + \dots + b_n \langle f_n, x \rangle, \tag{2.15}$$

where $f_i \in H'_A$. Since $x \in B_1(H_A)$, one can find an element $b \in \mathcal{N}$ such that $\|Tx - b\| < \varepsilon$. Therefore

$$\begin{aligned} \|Tx - P_\mathcal{N} Tx\| &= \|Tx - b + b - P_\mathcal{N} Tx\| \\ &= \|Tx - b\| + \|P_\mathcal{N}(b - Tx)\| \le \varepsilon + \|P_\mathcal{N}\|\, \varepsilon = 2\varepsilon, \end{aligned} \tag{2.16}$$

hence $\|T - P_\mathcal{N} T\| \le 2\varepsilon$ and T lies in the norm closure of the set of operators of the form (2.15). If T is adjointable, then $P_\mathcal{N} T$ is adjointable as well, hence $f_i \in H_A$ and T is compact. □

2.7. C^*-Fredholm operators and index. Mishchenko's approach

The material of this section is taken mainly from [**92**]. Recall first the definition of K-groups.

DEFINITION 2.7.1 ([**60**, § II.1]). Let M be an Abelian monoid. Consider the Cartesian product $M \times M$ and its quotient monoid with respect to the equivalence relation

$$(m, n) \sim (m', n') \Leftrightarrow \exists\, p, q : \ (m, n) + (p, p) = (m', n') + (q, q).$$

This quotient monoid is a group, which is denoted by $S(M)$ and is called the *symmetrization* of M. Consider now the additive category $\mathcal{P}(A)$ of projective modules over a unital C^*-algebra A and denote by $[\mathcal{M}]$ the isomorphism class of an object $\mathcal{M}$ from $\mathcal{P}(A)$. The set $\Phi(\mathcal{P}(A))$ of these classes has the structure of an Abelian monoid with respect to the operation $[\mathcal{M}] + [\mathcal{N}] = [\mathcal{M} \oplus \mathcal{N}]$. In this case the group $S(\Phi(\mathcal{P}(A)))$ is denoted by $K(A)$ or $K_0(A)$ and is called the *K-group* of A or the *Grothendieck group* of the category $\mathcal{P}(A)$. If A has no unit, then the natural map $A^+ \to \mathbf{C}$ induces a map of K-groups and $K_0(A)$ is defined by

$$K_0(A) := \operatorname{Ker}(K_0(A^+) \to K_0(\mathbf{C})).$$

The groups $K_n(A) := K_0(A \otimes C_0(\mathbf{R}^n))$ for integer $n > 0$ turn out to be 2-periodic in n and the definition can be extended to $n \in \mathbf{Z}$ by periodicity. For unital C^*-algebras one could use classes of stable homotopy of projections in A^n (i.e., allowing addition of direct summands) instead of classes of isomorphic projective modules.

More precisely, projections $p : A^n \to A^n$ and $q : A^m \to A^m$ are equivalent if there are m' and n' such that $n + n' = m + m' = s$ and the projections

$$A^s = A^n \oplus A^{n'} \xrightarrow{\begin{pmatrix} p & 0 \\ 0 & 0 \end{pmatrix}} A^n \oplus A^{n'} = A^s,$$

$$A^s = A^m \oplus A^{m'} \xrightarrow{\begin{pmatrix} q & 0 \\ 0 & 0 \end{pmatrix}} A^m \oplus A^{m'} = A^s$$

can be connected by a norm continuous homotopy in the set of all projections from $\operatorname{End}(A^s) = \operatorname{End}^*(A^s)$ (A is unital!). One can also consider the equivalence classes of projections in the algebraic sense. The details can be found in [**10, 60, 94, 134**].

Recall the following well-known statement.

LEMMA 2.7.2 (cf. [**115**, Theorem 4.15]). *The set of epimorphisms is open in the space of bounded linear maps of a Banach space E to a Banach space G.*

LEMMA 2.7.3 ([**92**, 1.4]). *Let $\mathcal{N}$ be a finitely generated Hilbert A-module over a unital C^*-algebra A and let $a_1, \dots, a_s$ be generators of $\mathcal{N}$. Then there exists a number $\varepsilon > 0$ such that if some elements $a'_1, \dots, a'_s \in \mathcal{N}$ satisfy*

$$\|a'_k - a_k\| < \varepsilon \qquad (k = 1, \dots, s),$$

then these elements also generate $\mathcal{N}$.

PROOF. The map

$$f : L_s(A) \to \mathcal{N}, \qquad (0, \dots, 0, \underset{i}{1}, 0, \dots, 0) \mapsto a_i,$$

is an epimorphism. Hence, by Lemma 2.7.2, there exists $\varepsilon > 0$ such that if $\|g - f\| < s\varepsilon$, then g is an epimorphism. Let

$$g : L_s(A) \to \mathcal{N}, \qquad (0, \dots, 0, \underset{i}{1}, 0, \dots, 0) \mapsto a'_i.$$

Then for any $\alpha = (\alpha_1, \dots, \alpha_s) \in L_s(A)$ with norm $\|\alpha\| \le 1$, one has

$$\|(g - f)\alpha\| = \left\| \sum_{i=1}^{s} (a'_i - a_i)\alpha_i \right\| \le s\varepsilon,$$

hence g is an epimorphism and the elements $a'_1, \dots, a'_s$ generate $\mathcal{N}$. □

In this section we always assume that the algebra A is unital. Recall the definition of a Fredholm operator suggested by Mishchenko in [**92**].

DEFINITION 2.7.4. A (bounded A-linear) operator $F : H_A \to H_A$ is called *A-Fredholm* if

(i) it is adjointable;
(ii) there exists a decomposition of the domain, $H_A = \mathcal{M}_1 \widetilde{\oplus} \mathcal{N}_1$, and the range, $H_A = \mathcal{M}_2 \widetilde{\oplus} \mathcal{N}_2$ (where $\mathcal{M}_1, \mathcal{M}_2, \mathcal{N}_1, \mathcal{N}_2$ are closed A-modules and $\mathcal{N}_1, \mathcal{N}_2$ have a finite number of generators), such that F has the matrix form $F = \begin{pmatrix} F_1 & 0 \\ 0 & F_2 \end{pmatrix}$ with respect to these decompositions and $F_1 : \mathcal{M}_1 \to \mathcal{M}_2$ is an isomorphism.

THEOREM 2.7.5 ([**92**]). *Let $H_A \cong \mathcal{M}\widetilde{\oplus}\mathcal{N}$, where $\mathcal{M}$ and $\mathcal{N}$ are closed A-modules and $\mathcal{N}$ has a finite number of generators $a_1, \dots, a_s$. Then $\mathcal{N}$ is a projective A-module of finite type.*

PROOF. By Lemma 2.7.3, there exists some $\varepsilon > 0$ such that the estimate

$$\|a'_k - a_k\| < \varepsilon, \quad a'_k \in \mathcal{N}, \quad k = 1, \dots, s,$$

implies that $\{a'_k\}$ generate $\mathcal{N}$. Let $P : H_A \to \mathcal{N}$ be the projection along the summand $\mathcal{M}$ in the module H_A. Then P is bounded and A-linear. Therefore there exists some $\delta > 0$ such that $\|x\| < \delta$ implies $\|Px\| < \varepsilon$. Choose a number n_0 such that

$$\|a_k - \overline{a}_k\| < \delta, \qquad k = 1, \dots, s,$$

where $\overline{a}_k$ is the projection of a_k onto L_{n_0} along $L_{n_0}^\perp$. Write $\overline{a}_k$ with respect to the decomposition $H_A = \mathcal{M}\widetilde{\oplus}\mathcal{N}$ as the sum

$$\overline{a}_k = a'_k + a''_k, \quad a'_k \in \mathcal{N}, \quad a''_k \in \mathcal{M}.$$

Then $a_k - a'_k = P(a_k - \overline{a}_k)$. Therefore $\|a_k - a'_k\| < \varepsilon$ and $\{a'_k\}$ generate $\mathcal{N}$. Let $\overline{\mathcal{N}}$ be the module generated by $\{\overline{a}_k\}$. Then H_A coincides with the sum $\mathcal{M} + \overline{\mathcal{N}}$. Indeed, if $x \in H_A$, then

$$x = x_{\mathcal{M}} + \sum \lambda^k a'_k = (x_{\mathcal{M}} - \sum \lambda^k a''_k) + \sum \lambda^k \overline{a}_k.$$

Consider the operator Q projecting onto L_{n_0} along $L_{n_0}^\perp$. Then

$$\begin{array}{ll} Q(a_k) = \overline{a}_k, & Q(\mathcal{N}) = \overline{\mathcal{N}}, \\ P(\overline{a}_k) = a'_k, & P(\overline{\mathcal{N}}) = \mathcal{N}. \end{array}$$

Since a_k are close to a'_k, the composition of operators

$$\mathcal{N} \xrightarrow{Q} \overline{\mathcal{N}} \xrightarrow{P} \mathcal{N}$$

is an isomorphism. Therefore $Q : \mathcal{N} \to \overline{\mathcal{N}}$ and $P : \overline{\mathcal{N}} \to \mathcal{N}$ are isomorphisms. In particular, if $\sum_{k=1}^{s} \lambda_k a'_k = 0$ holds, then $\sum_{k=1}^{s} \lambda_k \overline{a}_k = 0$. Therefore $\mathcal{M} \cap \mathcal{N} = 0$, i.e., $H_A = \mathcal{M}\widetilde{\oplus}\overline{\mathcal{N}}$. It is clear that $\overline{\mathcal{N}}$ is a closed A-submodule in H_A and

$$L_{n_0} = (\mathcal{M} \cap L_{n_0})\widetilde{\oplus}\overline{\mathcal{N}}.$$

Indeed, $(\mathcal{M} \cap L_{n_0}) \cap \overline{\mathcal{N}} = 0$; on the other hand, if $x \in L_{n_0}$, then $x = x' + x''$, $x' \in \overline{\mathcal{N}}$, $x'' \in \mathcal{M}$. Since $\overline{\mathcal{N}} \subset L_{n_0}$, one has $x'' \in L_{n_0}$, i.e., $x'' \in \mathcal{M} \cap L_{n_0}$. Thus $\mathcal{N}$ is isomorphic to a direct summand in the free finitely generated A-module L_{n_0}. $\square$

THEOREM 2.7.6 ([**129**]). *In the decomposition mentioned in the definition of a Fredholm operator (see 2.7.4), one always can assume that the modules $\mathcal{M}_0$ and $\mathcal{M}_1$ are orthogonally complementable. More precisely, there exist decompositions for F,*

$$\begin{pmatrix} F_3 & 0 \\ 0 & F_4 \end{pmatrix} : H_A = V_0\widetilde{\oplus}W_0 \to V_1\widetilde{\oplus}W_1 = H_A,$$

such that $V_0^\perp \oplus V_0 = H_A$, $V_1^\perp \oplus V_1 = H_A$ or (which is the same by Lemma 2.3.8) such that projections $V_0\widetilde{\oplus}W_0 \to V_0$ and $V_1\widetilde{\oplus}W_1 \to V_1$ are adjointable.

PROOF. Let $W_0 = \mathcal{N}_0$, $V_0 = W_0^\perp$. The orthogonal complement exists by Lemma 2.3.7. The restriction $F|_{W_0^\perp}$ is an isomorphism. Indeed, if $x_n \in W_0^\perp$, then let $x_n = x_1^n + x_2^n$, $x_1^n \in \mathcal{M}_0$, $x_2^n \in W_0$, $\|x_n\| = 1$. Let us assume that $\|Fx_n\| \to 0$; then $\|Fx_1^n + Fx_2^n\| \to 0$. Since $Fx_1^n \in \mathcal{M}_1$, $Fx_2^n \in \mathcal{N}_1$, $\mathcal{M}_1 \widetilde{\oplus} \mathcal{N}_1 = H_A$, this means that $\|Fx_1^n\| \to 0$ and $\|Fx_2^n\| \to 0$. The operator F_1 is an isomorphism, hence $\|x_1^n\| \to 0$. If $a_1, \dots, a_s$ are generators for $W_0 = \mathcal{N}_0$, then $0 = \langle x_n, a_j \rangle = \langle x_1^n, a_j \rangle + \langle x_2^n, a_j \rangle$,

$$\|\langle x_2^n, a_j \rangle\| = \|\langle x_1^n, a_j \rangle\| \le \|x_1^n\| \, \|a_j\| \to 0, \quad n \to \infty, \quad j = 1, \dots, s.$$

Since $x_2^n \in \mathcal{N}_0$, $x_2^n \to 0$ (as $n \to \infty$) and $x_n = x_1^n + x_2^n \to 0$. This contradiction with the condition $\|x_n\| = 1$ shows that $F|_{W_0^\perp}$ is an isomorphism. Let $V_1 = F(V_0)$. Since $W_0 = \mathcal{N}_0$, one can assume that $W_1 = \mathcal{N}_1$. Indeed, any $y \in H_A$ is of the form $y = m_1 + n_1 = F(m_0) + n_1$, where $m_1 \in \mathcal{M}_1$, $n_1 \in \mathcal{N}_1$, $m_0 \in \mathcal{M}_0$. Further, $m_0 = v_0 + n_0$, where $v_0 \in V_0$, $n_0 \in W_0 = \mathcal{N}_0$, and

$$y = F(v_0 + n_0) + n_1 = F(v_0) + (F(n_0) + n_1) \in V_1 + \mathcal{N}_1.$$

Thus $H_A = V_1 + W_1$.

Let $y \in V_1 \cap W_1 = V_1 \cap \mathcal{N}_1$, i.e., $n_1 = y = F(v_0)$, $n_1 \in \mathcal{N}_1$, $v_0 \in V_0$. Let us decompose v_0 into two summands, $v_0 = m_0 + n_0$, where $m_0 \in \mathcal{M}_0$, $n_0 \in \mathcal{N}_0$. Then

$$\begin{aligned} n_1 &= F(m_0) + F(n_0), \\ F(m_0) &= n_1 - F(n_0), \quad F(m_0) \in \mathcal{M}_1, \quad n_1 - F(n_0) \in \mathcal{N}_1, \end{aligned}$$

whence we obtain that $F(m_0) = 0$, $\quad n_1 - F(n_0) = 0$. Since $F : \mathcal{M}_0 \cong \mathcal{M}_1$, $m_0 = 0$. Further, $v_0 \in V_0 = \mathcal{N}_0^\perp$, hence

$$0 = \langle v_0, n_0 \rangle = \langle m_0 + n_0, n_0 \rangle = \langle n_0, n_0 \rangle, \quad n_0 = 0.$$

Thus $v_0 = m_0 + n_0 = 0$, $y = F(v_0) = 0$. Therefore $V_1 \cap W_1 = 0$ and $H_A = V_1 \widetilde{\oplus} W_1$.

By Corollary 2.3.5, the module V_1 has the orthogonal complement $V_1^\perp$, $V_1 \oplus V_1^\perp = H_A$ and the proof is complete. □

REMARK 2.7.7. If we do not require the operator F to be adjointable, then there still exists a decomposition $F : \mathcal{N}_0^\perp \oplus \mathcal{N}_0 \to \mathcal{M}_1 \widetilde{\oplus} L_n$, where $L_n = \operatorname{span}_A(e_1, \dots, e_n)$, but $\mathcal{M}_1$ does not necessarily have an orthogonal complement. This result was obtained in [**54**].

DEFINITION 2.7.8. Let the conditions of Definition 2.7.4 hold. By Theorem 2.7.5, $\mathcal{N}_1$ and $\mathcal{N}_2$ are projective A-modules and we can define the *index* of F by

$$\operatorname{index} F = [\mathcal{N}_1] - [\mathcal{N}_2] \in K(A).$$

In [**134**] this index is called Mindex. We suggest this notation be interpreted as emphasizing Mishchenko's role in introducing the index of C^*-Fredholm operators.

THEOREM 2.7.9. *The index is well defined.*

PROOF. It is necessary to check that the index does not depend on decompositions of the range and the domain in Definition 2.7.4. Let p_m be the projection onto L_m along $L_m^\perp$. Let F be an A-Fredholm operator and let $H_A = \mathcal{M}_1 \widetilde{\oplus} \mathcal{N}_1$ and

$H_A = \mathcal{M}_2\widetilde{\oplus}\mathcal{N}_2$ be decompositions of the domain and of the range, respectively. Then

$$F = \begin{pmatrix} F_1 & 0 \\ 0 & F_4 \end{pmatrix},$$

where $F_1 : \mathcal{M}_1 \to \mathcal{M}_2$ is an isomorphism. According to the proofs of Theorem 2.7.5 and Theorem 2.7.6, one can assume that

$$\mathcal{N}_1 \subset L_n, \qquad L_n = \mathcal{N}_1\widetilde{\oplus}\mathcal{P}_1, \qquad \mathcal{M}_1 = \mathcal{P}_1 \oplus L_n^{\perp},$$

where $\mathcal{P}_1$ is a projective finitely generated A-module. Let another decomposition of the domain and the range be given:

$$H_A = \mathcal{M}'_1\widetilde{\oplus}\mathcal{N}'_1, \qquad H_A = \mathcal{M}'_2\widetilde{\oplus}\mathcal{N}'_2.$$

Then there exists $m \geq n$ such that

$$L_m = \mathcal{P}'_1\widetilde{\oplus}p_m(\mathcal{N}'_1), \qquad p_m(\mathcal{N}'_1) \cong \mathcal{N}'_1,$$

where $\mathcal{P}'_1$ is a projective finitely generated A-module. This is exactly the result of the proof of Theorem 2.7.5.

Let us show that there exists $m \geq n$ such that if

$$L'_m = F(L_m) + \mathcal{N}_2, \qquad \text{and} \quad Q'_m : H_A \to H_A$$

is the projection on L'_m along $L''_m = F(L_m^{\perp})$,[1] then

$$L'_m = \mathcal{P}_2\widetilde{\oplus}Q'_m(\mathcal{N}_2), \qquad Q'_m(\mathcal{N}_2) \cong \mathcal{N}_2,$$

where $\mathcal{P}_2$ is a projective finitely generated A-module, and

$$L'_m = \mathcal{P}'_2\widetilde{\oplus}Q'_m(\mathcal{N}'_2), \qquad Q'_m(\mathcal{N}'_2) \cong \mathcal{N}'_2,$$

where $\mathcal{P}'_2$ is a projective finitely generated A-module. Indeed, $H_A = L'_m\widetilde{\oplus}L''_m$. If $a_1, \dots, a_k$ are generators of the module $\mathcal{N}_2$, then

$$a_j = a'_j + a''_j, \quad a'_j \in L'_m, \quad a''_j \in L''_m, \quad j = 1, \dots, k.$$

For $m \to \infty$ we have $\|a''_j\| \to 0$ since $a''_j = F(x_m^{\perp})$, where x is arbitrary, $x_m^{\perp}$ is a projection of x onto $L_m^{\perp}$ and $\|x_m^{\perp}\| \to 0$ as $m \to \infty$. Then for m sufficiently large, we have

$$L'_m = (L'_m \cap \mathcal{M}_2)\widetilde{\oplus}Q'_m(\mathcal{N}_2), \qquad Q'_m(\mathcal{N}_2) \cong \mathcal{N}_2$$

(the proof repeats the proof of Theorem 2.7.5). Similarly

$$L'_m = (L'_m \cap \mathcal{M}'_2)\widetilde{\oplus}Q'_m(\mathcal{N}'_2), \qquad Q'_m(\mathcal{N}'_2) \cong \mathcal{N}'_2.$$

Since $m \geq n$, we have $L_m \cong \mathcal{N}_1\widetilde{\oplus}\overline{\mathcal{P}}_1$, where $\overline{\mathcal{P}}_1$ is a finitely generated projective A-module. From the equalities

$$F(\overline{\mathcal{P}}_1) = F(L_m \cap \mathcal{M}_1) = \mathcal{P}_2, \qquad \overline{\mathcal{P}}_1 \subset \mathcal{M}_1,$$

we obtain that $F : \overline{\mathcal{P}}_1 \cong \mathcal{P}_2$ and it follows from relations $F(\mathcal{P}'_1) = \mathcal{P}'_2$, $\mathcal{P}'_1 \subset \mathcal{M}_1$ that $F : \mathcal{P}'_1 \cong \mathcal{P}'_2$. Therefore we have the following equalities in $K(A)$:

$$\begin{gathered} [\mathcal{N}_1] + [\overline{\mathcal{P}}_1] = [\mathcal{N}'_1] + [\mathcal{P}'_1] = [L_m], \quad [\overline{\mathcal{P}}_1] = [\mathcal{P}_2], \\ [\mathcal{N}_2] + [\mathcal{P}_2] = [\mathcal{N}'_2] + [\mathcal{P}'_2] = [L'_m], \quad [\mathcal{P}'_1] = [\mathcal{P}'_2]. \end{gathered}$$

Thus $[\mathcal{N}_1] - [\mathcal{N}_2] = [\mathcal{N}'_1] - [\mathcal{N}'_2]$ and we have proved that the index is well defined. □

[1]This projection is well defined since $L_m^{\perp} \subset \mathcal{M}_1$ for $m \geq n$ and hence $F|_{L_m^{\perp}}$ is an isomorphism, whence it follows that $L''_m \cong H_A$ is a closed A-module, $L'_m \cap L''_m = 0$, $L'_m + L''_m = H_A$. Therefore $H_A = L'_m\widetilde{\oplus}L''_m$ is a direct sum of closed A-modules and Q'_m is bounded and A-linear.

LEMMA 2.7.10. *Let an operator $F : H_A \to H_A$ be A-Fredholm. Then there exists a number $\varepsilon > 0$ such that any adjointable operator D satisfying the condition $\|F - D\| < \varepsilon$ is an A-Fredholm operator and* $\operatorname{index} D = \operatorname{index} F$.

PROOF. By the definition of Fredholmness

$$H_A = \mathcal{M}_1 \widetilde{\oplus} \mathcal{N}_1, \qquad H_A = \mathcal{M}_2 \widetilde{\oplus} \mathcal{N}_2, \qquad F = \begin{pmatrix} F_1 & 0 \\ 0 & F_4 \end{pmatrix},$$

$F_1 : \mathcal{M}_1 \cong \mathcal{M}_2$. Then $\|F_1\| \le \|F\|$; moreover, if $D : H_A \to H_A$ is an arbitrary operator, then

$$D = \begin{pmatrix} D_1 & D_2 \\ D_3 & D_4 \end{pmatrix},$$

and there exists a constant C such that $\|D_1\| \le C\|D\|$ (cf. [**92**]). Therefore, if D is an arbitrary operator satisfying the estimate $\|F - D\| < \varepsilon$, then $\|F_1 - D_1\| < C \cdot \varepsilon$. Since F_1 is an isomorphism, we can find $\delta > 0$ such that if $\|F_1 - D_1\| < \delta$ and D_1 is an operator, then D_1 is also an isomorphism. By taking $\varepsilon = \delta / C$ we obtain that for the operator

$$D = \begin{pmatrix} D_1 & D_2 \\ D_3 & D_4 \end{pmatrix},$$

the element D_1 is an isomorphism. Then

$$U_2 D U_1 = \begin{pmatrix} D_1 & 0 \\ 0 & D_4 - D_3 D_1^{-1} D_2 \end{pmatrix},$$

where

$$U_2 = \begin{pmatrix} 1 & 0 \\ -D_3 D_1^{-1} & 1 \end{pmatrix} : (\mathcal{M}_2 \widetilde{\oplus} \mathcal{N}_2) \to (\mathcal{M}_2 \widetilde{\oplus} \mathcal{N}_2),$$
$$U_1 = \begin{pmatrix} 1 & -D_1^{-1} D_2 \\ 0 & 1 \end{pmatrix} : (\mathcal{M}_1 \widetilde{\oplus} \mathcal{N}_1) \to (\mathcal{M}_1 \widetilde{\oplus} \mathcal{N}_1)$$

are isomorphisms. Using U_1 and U_2 we obtain a new decomposition of the domain and the range into direct sums:

$$H_A = \mathcal{M}_1' \widetilde{\oplus} \mathcal{N}_1', \quad \mathcal{M}_1' = U_1(\mathcal{M}_1), \quad \mathcal{N}_1' = U_1(\mathcal{N}_1),$$
$$H_A = \mathcal{M}_2' \widetilde{\oplus} \mathcal{N}_2', \quad \mathcal{M}_2' = U_2^{-1}(\mathcal{M}_2), \quad \mathcal{N}_2' = U_2^{-1}(\mathcal{N}_2).$$

With respect to the new decomposition the matrix of D is equal to $U_2 D U_1$. Thus D is A-Fredholm with the index

$$[U_1(\mathcal{N}_1)] - [U_2^{-1}(\mathcal{N}_2)] = [\mathcal{N}_1] - [\mathcal{N}_2] = \operatorname{index} F.$$

□

LEMMA 2.7.11. *Let F and D be A-Fredholm operators,*

$$F : H_A \to H_A, \qquad D : H_A \to H_A.$$

Then $DF : H_A \to H_A$ is an A-Fredholm operator and $\operatorname{index} DF = \operatorname{index} D + \operatorname{index} F$.

PROOF. For F and D, consider decompositions from the definition,

$$H_A = \mathcal{M}_1 \widetilde{\oplus} \mathcal{N}_1 \xrightarrow{F} \mathcal{M}_2 \widetilde{\oplus} \mathcal{N}_2 \cong H_A,$$

$$H_A = \mathcal{M}_1' \widetilde{\oplus} \mathcal{N}_1' \xrightarrow{D} \mathcal{M}_2' \widetilde{\oplus} \mathcal{N}_2' \cong H_A,$$

where

$$F = \begin{pmatrix} F_1 & 0 \\ 0 & F_4 \end{pmatrix}, \qquad D = \begin{pmatrix} D_1 & 0 \\ 0 & D_4 \end{pmatrix},$$

F_1 and D_1 are isomorphisms, and $\mathcal{N}_1, \mathcal{N}_2, \mathcal{N}_1', \mathcal{N}_2'$ are projective finitely generated A-modules. As earlier, without loss of generality we can assume that

$$\mathcal{N}_2 \subset L_n, \qquad L_n = \mathcal{N}_2 \widetilde{\oplus} \mathcal{P}, \qquad \mathcal{M}_2 = L_n^\perp \oplus \mathcal{P}.$$

Moreover, since $F_1 : \mathcal{M}_1 \to \mathcal{M}_2$ is an isomorphism, one can change the decomposition into direct sums by setting

$$\overline{\mathcal{M}}_1 = F_1^{-1}(L_n^\perp), \quad \overline{\mathcal{N}}_1 = F_1^{-1}(\mathcal{P}) \widetilde{\oplus} \mathcal{N}_1, \quad \overline{\mathcal{M}}_2 = L_n^\perp, \quad \overline{\mathcal{N}}_2 = L_n.$$

Thus a number n can be chosen as large as necessary. Choose n in such a way that

$$L_n = \mathcal{P}' \widetilde{\oplus} p_n(\mathcal{N}_1'), \qquad \mathcal{P}' = \mathcal{M}_1' \cap L_n, \qquad p_n(\mathcal{N}_1') \cong \mathcal{N}_1',$$

where, as before, $p_n : H_A \to H_A$ is the projection onto L_n along $L_n^\perp$. Then

$$H_A = L_n^\perp \widetilde{\oplus} \mathcal{P}' \widetilde{\oplus} p_n(\mathcal{N}_1').$$

Put $\overline{\overline{\mathcal{M}}}_2 = L_n^\perp$, $\overline{\overline{\mathcal{N}}}_2 = \mathcal{P}' \widetilde{\oplus} \mathcal{N}_1'$. With respect to the new decomposition, $H_A = \overline{\overline{\mathcal{M}}}_2 \widetilde{\oplus} \overline{\overline{\mathcal{N}}}_2$, the matrix of the operator F has the form

$$F = \begin{pmatrix} F_1 & F_2 \\ 0 & F_4 \end{pmatrix},$$

and F_1 is an isomorphism. Then

$$\begin{pmatrix} F_1 & F_2 \\ 0 & F_4 \end{pmatrix} \begin{pmatrix} 1 & -F_1^{-1} F_2 \\ 0 & 1 \end{pmatrix} = \begin{pmatrix} F_1 & 0 \\ 0 & F_4 \end{pmatrix}.$$

Denote by U the matrix $\begin{pmatrix} 1 & -F_1^{-1} F_2 \\ 0 & 1 \end{pmatrix}$ and put $\overline{\overline{\mathcal{M}}}_1 = U(\overline{\mathcal{M}}_1)$, $\overline{\overline{\mathcal{N}}}_1 = U(\overline{\mathcal{N}}_1)$. We obtained a new decomposition, $H_A = \overline{\overline{\mathcal{M}}}_1 \widetilde{\oplus} \overline{\overline{\mathcal{N}}}_1$, and the matrix F with respect to the decomposition

$$H_A = \overline{\overline{\mathcal{M}}}_1 \widetilde{\oplus} \overline{\overline{\mathcal{N}}}_1 \xrightarrow{F} \overline{\overline{\mathcal{M}}}_2 \widetilde{\oplus} \overline{\overline{\mathcal{N}}}_2 = H_A$$

has the previous diagonal form. Consider the projection

$$T : H_A = \overline{\overline{\mathcal{M}}}_2 \widetilde{\oplus} \mathcal{P}' \widetilde{\oplus} \mathcal{N}_1' \to \overline{\overline{\mathcal{M}}}_2 \widetilde{\oplus} \mathcal{P}'.$$

Since $H_A \cong \mathcal{M}_2' \widetilde{\oplus} \mathcal{N}_1'$, the restriction $T|_{\mathcal{M}_2'} : \mathcal{M}_2' \to \overline{\overline{\mathcal{M}}}_2 \widetilde{\oplus} \mathcal{P}'$ is an isomorphism. Consider the matrix D with respect to the decomposition

$$H_A = (\overline{\overline{\mathcal{M}}}_2 \widetilde{\oplus} \mathcal{P}') \widetilde{\oplus} \mathcal{N}_1' \xrightarrow{D} \mathcal{M}_2' \widetilde{\oplus} \mathcal{N}_2' = H_A.$$

This matrix has the form $D = \begin{pmatrix} D_1 & 0 \\ D_3 & D_4 \end{pmatrix}$, where D_1 is an isomorphism. Put

$$V D := \begin{pmatrix} 1 & 0 \\ -D_3 D_1^{-1} & 1 \end{pmatrix} \begin{pmatrix} D_1 & 0 \\ D_3 & D_4 \end{pmatrix} = \begin{pmatrix} D_1 & 0 \\ 0 & D_4 \end{pmatrix}.$$

Therefore one can change the decomposition of the range,

$$H_A = \overline{\mathcal{M}}_2' \widetilde{\oplus} \overline{\mathcal{N}}_2', \qquad \overline{\mathcal{M}}_2' = V(\mathcal{M}_2'), \qquad \overline{\mathcal{N}}_2' = V(\mathcal{N}_2'),$$

in such a way that the matrix of the operator D with respect to the new decomposition

$$H_A = \overline{\mathcal{M}}_2' \widetilde{\oplus} \overline{\mathcal{N}}_2', \qquad \overline{\mathcal{M}}_2' = V(\mathcal{M}_2'), \qquad \overline{\mathcal{N}}_2' = V(\mathcal{N}_2'),$$

has diagonal form. Let us change the decomposition of the range once again:

$$\overline{\overline{\mathcal{M}}}'_2 = D(\overline{\overline{\mathcal{M}}}_2), \qquad \overline{\overline{\mathcal{N}}}_2 = D(\mathcal{P}')\widetilde{\oplus}\overline{\overline{\mathcal{N}}}'_2.$$

The matrix D with the respect to the latter decomposition

$$H_A = \overline{\overline{\mathcal{M}}}_2\widetilde{\oplus}\overline{\overline{\mathcal{N}}}_2 \xrightarrow{D} \overline{\overline{\mathcal{M}}}'_2\widetilde{\oplus}\overline{\overline{\mathcal{N}}}'_2 = H_A$$

has diagonal form. Then the composition DF with the respect to the decomposition

$$H_A = \overline{\overline{\mathcal{M}}}_1\widetilde{\oplus}\overline{\overline{\mathcal{N}}}_1 \to \overline{\overline{\mathcal{M}}}'_2\widetilde{\oplus}\overline{\overline{\mathcal{N}}}'_2 = H_A$$

has the form $DF = \begin{pmatrix} (DF)_1 & 0 \\ 0 & (DF)_4 \end{pmatrix}$, and $(DF)_1$ is an isomorphism. Taking into account the fact that $\operatorname{End}^* H_A$ is a C^*-algebra, we conclude that DF is an A-Fredholm operator and

$$\operatorname{index} F = [\overline{\overline{\mathcal{N}}}_1] - [\overline{\overline{\mathcal{N}}}_2], \quad \operatorname{index} D = [\overline{\overline{\mathcal{N}}}_2] - [\overline{\overline{\mathcal{N}}}'_2],$$

$$\operatorname{index} DF = [\overline{\overline{\mathcal{N}}}_1] - [\overline{\overline{\mathcal{N}}}'_2].$$

It follows that $\operatorname{index} DF = \operatorname{index} D + \operatorname{index} F$. □

LEMMA 2.7.12. *Let $K : H_A \to H_A$ be a compact operator. Then $1 + K$ is an A-Fredholm operator and* $\operatorname{index}(1 + K) = 0$.

PROOF. It is clear that $1 + K$ is adjointable. Choose a number n such that the inequality $\|K|_{L_n^\perp}\| < 1$ is fulfilled. With respect to the decomposition $H_A = L_n^\perp \oplus L_n$, we have the following matrix presentation:

$$K = \begin{pmatrix} K_1 & K_2 \\ K_3 & K_4 \end{pmatrix}, \quad (1 + K) = \begin{pmatrix} 1 + K_1 & K_2 \\ K_3 & 1 + K_4 \end{pmatrix}.$$

By the estimate $\|K|_{L_n^\perp}\| < 1$, the operator $1 + K_1$ is invertible, hence, as before, there exist invertible operators U_1 and U_2 such that

$$U_2(1 + K)U_1 = \begin{pmatrix} 1 + K_1 & 0 \\ 0 & (1 + K_4) - K_3(1 + K_1)^{-1}K_2 \end{pmatrix}.$$

Then, with respect to the new decomposition, $H_A = \mathcal{M}_1\widetilde{\oplus}\mathcal{N}_1 \to \mathcal{M}_2\widetilde{\oplus}\mathcal{N}_2 = H_A$, where $\mathcal{M}_1 = U_1(L_n^\perp)$, $\mathcal{N}_1 = U_1(L_n)$, $\mathcal{M}_2 = U_2^{-1}(L_n^\perp)$, $\mathcal{N}_2 = U_2^{-1}(L_n)$, the operator $(1 + K)$ has the diagonal form and is thus an A-Fredholm operator with

$$\operatorname{index}(1 + K) = [U_1(L_n)] - [U_2^{-1}(L_n)] = 0.$$

□

LEMMA 2.7.13. *Consider an A-Fredholm operator $F : H_A \to H_A$ and let $K \in \mathcal{K}_A$. Then the operator $F + K$ is A-Fredholm and* $\operatorname{index}(F + K) = \operatorname{index} F$.

PROOF. Consider decompositions of the module H_A into direct sums such that the matrix F has the diagonal form:

$$H_A = \mathcal{M}_1\widetilde{\oplus}\mathcal{N}_1 \xrightarrow{F} \mathcal{M}_2\widetilde{\oplus}\mathcal{N}_2 = H_A.$$

Without loss of generality we can assume that

$$L_n = \mathcal{N}_1\widetilde{\oplus}\mathcal{P}_1, \quad \mathcal{M}_1 = L_n^\perp\widetilde{\oplus}\mathcal{P}_1,$$

where $\mathcal{P}_1$ is a finitely generated closed A-module. Choose a number n sufficiently large to provide that $\|K|_{L_n^\perp}\| < \|F_1^{-1}\|^{-1}$. Consider the new decomposition of the module H_A:

$$\overline{\mathcal{M}}_1 = L_n^\perp, \quad \overline{\mathcal{N}}_1 = L_n, \quad \overline{\mathcal{M}}_2 = FL_n^\perp, \quad \overline{\mathcal{N}}_2 = F(\mathcal{P}_1)\widetilde{\oplus}\mathcal{N}_2.$$

Let $F = \begin{pmatrix} F_1 & 0 \\ 0 & F_4 \end{pmatrix}$ and $K = \begin{pmatrix} K_1 & K_2 \\ K_3 & K_4 \end{pmatrix}$ be matrices of F and K with respect to the decomposition $H_A = \overline{\mathcal{M}}_1\widetilde{\oplus}\overline{\mathcal{N}}_1 \to \overline{\mathcal{M}}_2\widetilde{\oplus}\overline{\mathcal{N}}_2 = H_A$. Then

$$F + K = \begin{pmatrix} F_1 + K_1 & K_2 \\ K_3 & F_4 + K_4 \end{pmatrix},$$

and the operator $F_1 + K_1$ is invertible. By repeating the construction of Lemma 2.7.10 (about operators close to a Fredholm operator), we obtain

$$\begin{aligned} \operatorname{index}(F+K) &= [\overline{\mathcal{N}}_1] - [\overline{\mathcal{N}}_2] = [L_n] - [F(\mathcal{P}_1) + \mathcal{N}_2] \\ &= [\mathcal{N}_1] + [\mathcal{P}_1] - [\mathcal{P}_1] - [\mathcal{N}_2] = \operatorname{index} F. \end{aligned}$$

□

THEOREM 2.7.14. *Let*

$$F : H_A \to H_A, \quad D : H_A \to H_A, \quad D' : H_A \to H_A$$

be adjointable operators such that

$$FD = \mathrm{Id}_{H_A} + K_1, \quad D'F = \mathrm{Id}_{H_A} + K_2, \quad K_1,\, K_2 \in \mathcal{K}(H_A).$$

Then F is an A-Fredholm operator.

PROOF. Consider a decomposition of H_A, for which the operator $FD = 1_{H_A} + K_1$ has the diagonal form (Lemma 2.7.12)

$$H_A = \mathcal{M}_1\widetilde{\oplus}\mathcal{N}_1 \xrightarrow{1+K_1} \mathcal{M}_2\widetilde{\oplus}\mathcal{N}_2 = H_A,$$

and assume that the decomposition of H_A satisfies the conditions of Theorem 2.7.6. Consider the projection

$$P : H_A = \mathcal{M}_2\widetilde{\oplus}\mathcal{N}_2 \to \mathcal{N}_2.$$

Viewed as an operator on H_A, it is a compact operator. The image of the operator $(1-P)(1+K_1) = (1-P)FD$ is exactly equal to $\mathcal{M}_2$. It is easy to see that, up to an isomorphism,

$$(1-P)FD = (1-P)(1+K_1) = 1 + (-P(1+K_1)) + K_1 = 1 + \widetilde{K}_1,$$

$$D'(1-P)F = D'F - D'PF = 1 + \widetilde{K}_2,$$

where $\widetilde{K}_1 \in \mathcal{K}_A$, $\widetilde{K}_2 \in \mathcal{K}_A$. By Lemma 2.7.13, one can assume without loss of generality that $F : H_A \to \mathcal{M}_2$ is an epimorphism. Otherwise, we will pass to the operator $(1-P)F$. Consider now the decomposition for $1 + K_2$:

$$H_A = \overline{\mathcal{M}}_1\widetilde{\oplus}\overline{\mathcal{N}}_1 \xrightarrow{F} \mathcal{M}_2\widetilde{\oplus}\mathcal{N}_2 \xrightarrow{D'} \overline{\mathcal{M}}_2\widetilde{\oplus}\overline{\mathcal{N}}_2 = H_A.$$

The composition $D'F|_{\overline{\mathcal{M}}_1} : \overline{\mathcal{M}}_1 \to \overline{\mathcal{M}}_2$ is an isomorphism. Therefore, since $F : H_A \to \mathcal{M}_2$ is an epimorphism, $F : \overline{\mathcal{M}}_1\widetilde{\oplus}\overline{\mathcal{N}}_1 \to \mathcal{M}_2$ maps $\overline{\mathcal{M}}_1$ isomorphically into $\mathcal{M}_2$ and $\operatorname{Ker} F \subset \overline{\mathcal{N}}_1$, $\mathcal{M}_2 = F(\overline{\mathcal{M}}_1) + F(\overline{\mathcal{N}}_1)$. Let us show that $F(\overline{\mathcal{M}}_1) \cap F(\overline{\mathcal{N}}_1) = 0$. For this purpose, decompose F into a composition

$$\overline{\mathcal{M}}_1\widetilde{\oplus}\overline{\mathcal{N}}_1 \to (\overline{\mathcal{M}}_1\widetilde{\oplus}\overline{\mathcal{N}}_1)/\operatorname{Ker} F = \overline{\mathcal{M}}_1\widetilde{\oplus}(\overline{\mathcal{N}}_1/\operatorname{Ker} F) \xrightarrow{\widetilde{F}} \mathcal{M}_2,$$

where $\widetilde{F}$ is an isomorphism. Therefore

$$\mathcal{M}_2 = F(\overline{\mathcal{M}}_1)\widetilde{\oplus}\widetilde{F}(\overline{\mathcal{N}}_1/\operatorname{Ker} F) = F(\overline{\mathcal{M}}_1)\widetilde{\oplus}F(\overline{\mathcal{N}}_1).$$

Since the A-module $\overline{\mathcal{N}}_1$ is finitely generated, $F(\overline{\mathcal{N}}_1)$ is finitely generated too. We have obtained a decomposition

$$H_A = \overline{\mathcal{M}}_1\widetilde{\oplus}\overline{\mathcal{N}}_1 \to F(\overline{\mathcal{M}}_1)\widetilde{\oplus}[F(\overline{\mathcal{N}}_1)\widetilde{\oplus}\mathcal{N}_2] = H_A,$$

where $F|_{\overline{\mathcal{M}}_1} : \overline{\mathcal{M}}_1 \to F(\overline{\mathcal{M}}_1)$ is an isomorphism. □

LEMMA 2.7.15. *If adjointable operators D, D' and F are such that FD and $D'F$ are A-Fredholm operators, then F is an A-Fredholm operator.*

PROOF. By the definition of Fredholmness of FD and $D'F$, we can find adjointable operators T and T' such that

$$(FD)T = 1 + K,$$
$$T'(D'F) = 1 + K'.$$

By Theorem 2.7.14, the operator F is A-Fredholm. For T, for example, one can take an operator with the matrix $\begin{pmatrix} (F_1)^{-1} & 0 \\ 0 & 0 \end{pmatrix}$, where FD has the form $\begin{pmatrix} F_1 & 0 \\ 0 & F_2 \end{pmatrix}$ in the sense of Definition 2.7.4. □

REMARK 2.7.16. In the case where A is a W^*-algebra, the properties of A-Fredholm operators are more similar to the properties of usual Fredholm operators. This problem will be addressed in Proposition 3.6.8.

REMARK 2.7.17. For applications to elliptic operators it is important to develop the theory for modules of the form $l_2(\mathcal{P})$. This can be done similarly (see [**121**]).

2.8. Representations of groups on Hilbert modules

In this section we assume that G always denotes a compact group. First of all, we prove an equivariant version of the Kasparov stabilization theorem. We follow here the original proof [**63**]. For closely related problems see also [**87**].

DEFINITION 2.8.1. For a C^*-algebra B, put

$$\mathcal{H}_B := \sum_{i=1}^{\infty}(V_i \otimes_{\mathbf{C}} B),$$

where $\{V_i\}$ is a countable set of finite-dimensional spaces, in which all irreducible unitary representations G are realized (up to isomorphism) and each representation is repeated an infinite number of times; the B-Hilbert completion of the infinite algebraic sum is implemented with respect to the norm given by the following B-inner product on summands:

$$(x_1 \otimes b_1, x_2 \otimes b_2) := \langle x_1, x_2\rangle_{V_i} \cdot b_1^* b_2, \qquad x_1,\ x_2 \in V_i.$$

DEFINITION 2.8.2. A C^*-algebra B with a continuous action of a group G is called a *G-C^*-algebra.*

DEFINITION 2.8.3. A Hilbert C^*-module $\mathcal{E}$ over a G-C^*-algebra B equipped with a unitary action of G is called a *G-B-module* if this group action agrees with the Hilbert module structure, i.e.,

$$g(xb) = g(x)g(b), \quad \langle g(x), g(y)\rangle = g(\langle x, y\rangle), \quad x,\ y \in \mathcal{E},\ b \in B,\ g \in G.$$

THEOREM 2.8.4 ([**63**]). *Let B be a G-C^*-algebra and let $\mathcal{E}$ be a countably generated Hilbert G-B-module. Then there exists an equivariant B-linear isomorphism*

$$\mathcal{E} \oplus \mathcal{H}_B \cong \mathcal{H}_B,$$

which respects the inner product.

PROOF. Denote by $\mathcal{E}^+$ the module $\mathcal{E}$ viewed as a B^+-module and, for a B^+-module $\mathcal{N}$, denote by $(\mathcal{N})_B$ the same module $\mathcal{N}$ viewed as a B-module. Let the action of G on B^+ be extended from that on B by the formula $g(1) = 1$. Suppose that we have proved the theorem for unital C^*-algebras, so that there is an equivariant B-linear isomorphism

$$\mathcal{E}^+ \oplus \mathcal{H}_{B^+} \cong \mathcal{H}_{B^+}.$$

Then

$$\mathcal{E} \oplus \mathcal{H}_B \cong ((\mathcal{E} \oplus \mathcal{H}_B)^+)B \cong (\mathcal{E}^+ \oplus \mathcal{H}_{B^+})B \cong (\mathcal{H}_{B^+})B = \mathcal{H}_B,$$

so we can restrict ourselves to the unital case.

Let $\{x_k\}$ be a countable system of generators for $\mathcal{E}$ and let $\{e_k\}$ be an orthonormal basis of $\mathcal{H}_B$ such that each e_k is of the form $e_k = v_k \otimes 1_B$, where $v_k \in V_{s(k)}$. In other words, if $\{w_k\}$ is a union of some orthonormal bases of all V_j, then $e_k = w_k \otimes 1_B$. Let $\{y_i\}$ be a system of elements in $\mathcal{E} \oplus \mathcal{H}_B$, in which each element of the form $x_k \oplus 0$ or $0 \oplus e_k$ is repeated an infinite number of times. We can assume that $y_1 = 0 \oplus e_1$ and put $W_1 = 0 \oplus V_1 \otimes B$. Suppose that, by induction, we have already constructed subspaces $W_1, \dots, W_n$ satisfying the following conditions:

(i) W_i is a $\mathbf{C}$-finite-dimensional G-invariant subspace in $\mathcal{E} \oplus \mathcal{H}_B$,

(ii) each W_i has a basis $(z_i^1, \dots, z_i^{K(i)}) = (f_1, \dots, f_p)$ such that

$$\langle z_i^j, z_i^j \rangle = 1_B, \qquad \langle z_i^j, z_r^s \rangle = 0 \text{ for } i \neq r \text{ or } j \neq s,$$

(iii) there exists $m = m(n)$ such that

$$W_1 + \dots + W_n \subset \mathcal{E}_m := \mathcal{E} \oplus \left(\bigoplus_{i=1}^{m} V_i \otimes B \right),$$

and consequently $(W_1 + \dots + W_n)B \subset \mathcal{E}_m$,

(iv) the distance between y_n and $(W_1 + \dots + W_n)B$ does not exceed $1/n$.

Note that (i) and (ii) imply that the modules $W_i B$ are pairwise orthogonal and G-invariant and $(W_1 + \dots + W_n)B$ is also G-invariant. The last module is free, so, by Lemma 2.3.7, it has an orthogonal complement, which is G-invariant due to unitarity of the action.

Let us pass to the construction of W_{n+1}. Put

$$y'_{n+1} := \sum_{j=1}^{p} f_j \langle f_j, y_{n+1} \rangle, \qquad y''_{n+1} = y_{n+1} - y'_{n+1}.$$

Then

$$\begin{aligned} \langle w, y''_{n+1} \rangle &= \langle \sum_{j=1}^{p} f_j b_j, y_{n+1} - \sum_{j=1}^{p} f_j \langle f_j, y_{n+1} \rangle \rangle \\ &= \sum_{j=1}^{p} b_j^* [\langle f_j, y_{n+1} \rangle - \langle f_j, y_{n+1} \rangle] = 0 \end{aligned}$$

for any $w \in (W_1 + \cdots + W_n)B$. Since by the definition of the sequence y_j, the element y_{n+1} lies either in $\mathcal{E}$ or in some $V_i \otimes B$, we have $y''_{n+1} \in \mathcal{E}_{m'}$ for some $m' > m$. Consider the orthogonal complement $S_{n,m'}$ to $(W_1 + \cdots + W_n)B$ in $\mathcal{E}_{m'}$. It is an invariant module and $y''_{n+1} \in S_{n,m'}$. By the Mostow theorem on periodic vectors [**93**], the elements with $\mathbf{C}$-finite-dimensional orbits are dense in $S_{n,m'}$. Hence one can find a vector $z \in S_{n,m'}$ such that $\|z - y''_{n+1}\| < \frac{1}{2n+2}$ and the linear span R of the orbit Gz of z is an invariant finite-dimensional subspace of $S_{n,m'}$. Since z is a totalizing vector, R is an irreducible G-module. Therefore there is $m'' > m'$ such that there exists an equivariant isomorphism $\Gamma : R \to V_{m''}$. Let $\{h_1, \ldots, h_k\}$ be an orthonormal basis of $V_{m''}$ and $r_i := \Gamma^{-1}(h_i)$, $i = 1, \ldots, k$. Then for the corresponding irreducible matrix representation $T : G \to U(k)$, we have

$$g(h_i) = \sum_{j=1}^{k} T_i^j(g) h_j, \qquad g(r_i) = \sum_{j=1}^{k} T_i^j(g) r_j, \qquad g \in G.$$

Since $R \subset \mathcal{E}_{m'}$ and $m'' > m'$, R is orthogonal to $V_{m''}$. More precisely, each element of R is orthogonal to $V_{m''} \otimes B$ in $\mathcal{E} \otimes \mathcal{H}_B$. Hence $\langle r_i, h_j \rangle = 0$ for any i and j. Let

$$z := \sum_{i=1}^{k} r_i \alpha_i, \quad \alpha_i \in \mathbf{C}, \qquad \alpha := \left(\sum_{i=1}^{k} |\alpha_i|^2 \right)^{1/2} + 1,$$

$$r'_i := r_i + (h_i \otimes 1_B) \cdot \frac{1}{(2n+2)\alpha}.$$

Then $\langle r'_i, r'_j \rangle = \langle r_i, r_j \rangle + \{(2n+2)\alpha\}^{-2} \delta_{ij}$ and the matrix $L := \|\langle r'_i, r'_j \rangle\|_{i,j=1}^{k}$ is positive and invertible in $M_k(\mathbf{C}) \subset M_k(B)$. Let $D = \|d_{ji}\| := L^{-1/2} \in M_k(B)$ and $r''_i := \sum_{j=1}^{k} r'_j d_{ji}$. Set W_{n+1} to be the $\mathbf{C}$-linear span of vectors r''_i, $i = 1, \ldots, k$, or, equivalently, to the span of r'_i, since D has complex coefficients. Then $W_{m+1} \subset \mathcal{E}_{m''}$ and

$$\langle r''_i, r''_j \rangle = \sum_{p,q=1}^{k} \langle r'_p d_{pi}, r'_q d_{pj} \rangle = (D^* L D)_{ij} = \delta_{ij}.$$

Since all h_i and r_i are orthogonal to $W_1 + \cdots + W_n$, W_{n+1} is orthogonal to it as well. Further, put $F := L^{1/2}$, so that $r'_i := \sum_{j=1}^{k} r''_j F_{ji}$. Then

$$\begin{aligned}
g(r''_i) &= g\left(\sum_{j=1}^{k} r'_j d_{ji} \right) = \sum_{j=1}^{k} (g r'_j) d_{ji} = \sum_{j=1}^{k} \left(g r_j + (g h_j \otimes 1_B) \cdot \frac{1}{(2n+2)\alpha} \right) d_{ji} \\
&= \sum_{j=1}^{k} \left(\sum_{s=1}^{k} T_j^s(g) r_s + \left(\sum_{s=1}^{k} T_j^s(g) h_s \otimes 1_B \right) \cdot \frac{1}{(2n+2)\alpha} \right) d_{ji} \\
&= \sum_{j=1}^{k} \sum_{s=1}^{k} T_j^s(g) \left(r_s + (h_s \otimes 1_B) \cdot \frac{1}{(2n+2)\alpha} \right) d_{ji} = \sum_{j=1}^{k} \sum_{s=1}^{k} T_j^s(g) r'_s d_{ji} \\
&= \sum_{j=1}^{k} \sum_{s=1}^{k} T_j^s(g) \left(\sum_{t=1}^{k} r''_t F_{ts} \right) d_{ji} = \sum_{t=1}^{k} r''_t \left(\sum_{j=1}^{k} \sum_{s=1}^{k} T_j^s(g) F_{ts} d_{ji} \right) \in W_{n+1}.
\end{aligned}$$

Thus W_{n+1} is G-invariant. Let us estimate the distance by taking $z' = \sum_{i=1}^{k} r_i'\alpha_i$, so that

$$\rho(z, W_{n+1}) \le \rho(z, z') = \left\| \sum_{i=1}^{k} (r_i - r_i')\alpha_i \right\| = \left\| \sum_{i=1}^{k} (h_i \otimes 1_B) \cdot \frac{1}{(2n+2)\alpha}\alpha_i \right\|$$
$$= \frac{1}{(2n+2)\alpha} \cdot \left\| \sum_{i=1}^{k} h_i\alpha_i \right\| = \frac{1}{(2n+2)\alpha} \cdot \left(\sum_{i=1}^{k} |\alpha_i|^2 \right)^{1/2}$$
$$= \frac{1}{(2n+2)} \cdot \frac{\left(\sum_{i=1}^{k} |\alpha_i|^2 \right)^{1/2}}{\left(\sum_{i=1}^{k} |\alpha_i|^2 \right)^{1/2} + 1} < \frac{1}{(2n+2)}.$$

Therefore

$$\rho(y_{n+1}, (W_1 + \cdots + W_n)B) \le \rho(y''_{n+1}, z) + \rho(z, W_{n+1}B)$$
$$\le \frac{1}{2n+2} + \rho(z, z') \le \frac{1}{n+1} < \frac{1}{n}.$$

Thus, by induction, the **C**-subspaces W_i with properties (i)–(iv) are constructed for all i. From the explicit expression for r_i' we obtain that W_n is isomorphic to some V_i, i.e., is irreducible. Further, the B-Hilbert completion $\mathcal{M}$ (i.e., the closure in $\mathcal{E} \oplus \mathcal{H}_B$) of the algebraic orthogonal sum of modules W_nB gives the whole $\mathcal{E} \oplus \mathcal{H}_B$. Indeed, by property (iv), the algebraic sum is dense in $\mathcal{E} \oplus \mathcal{H}_B$. So, $\mathcal{M} \cong \mathcal{E} \oplus \mathcal{H}_B$. It remains to prove that $\mathcal{M}$ is isomorphic to $\mathcal{H}_B$, i.e., that each irreducible representation is repeated infinitely many times among the modules W_nB. Suppose the opposite, then $\mathcal{M} \oplus \mathcal{H}_B \cong \mathcal{H}_B$, or

$$\mathcal{E} \oplus \mathcal{H}_B = \mathcal{E} \oplus \mathcal{H}_B \oplus \mathcal{H}_B \cong \mathcal{M} \oplus \mathcal{H}_B \cong \mathcal{H}_B.$$

□

Let us now prove the theorem on decomposition of representations [**91**]. Let $\mathcal{M}$ be a Hilbert B-module with a strongly continuous unitary representation G,

$$T : G \to U(\mathcal{M}) \subset \mathrm{End}^*_B(\mathcal{M}), \qquad g \mapsto T_g,$$

and suppose that the group acts trivially on B. Now let $\{V_s\}$ be the complete collection of pairwise nonequivalent unitary representations of G of dimensions d_s and let D^s_{pq} be their matrix elements, which are continuous functions on G. For an invariant normalized Haar measure dg on G, we define the operator

$$P^s_{pq} : \mathcal{M} \to \mathcal{M}, \qquad P^s_{pq}(x) := d_s \int_G \overline{D^s_{pq}(g)} T_g(x)\, dg. \tag{2.17}$$

Since a product of continuous complex-valued functions by a continuous module-valued function is integrable for a fixed $x \in \mathcal{M}$ and since the group is compact, the integral is well defined and equals some element of $\mathcal{M}$. We obtain a bounded operator. Indeed,

$$\|P^s_{pq}x\| \le d_s \int_G \left|\overline{D^s_{pq}(g)}\right| \|T_g(x)\|\, dg \le d_s \sup_{g\in G} \left|\overline{D^s_{pq}(g)}\right| \|x\|.$$

Therefore

$$\|P^s_{pq}\| \le d_s \sup_{g\in G} \left|\overline{D^s_{pq}(g)}\right|.$$

It is well known [**7**, I, §7.1, Theorem 5] that

$$\int_G D^s_{ij}(g)\overline{D^{s'}_{mn}(g)} = \begin{cases} 0, & s \ne s', \\ \frac{1}{d_s}\delta_{im}\delta_{jn}, & s = s'. \end{cases} \tag{2.18}$$

We need the following Peter–Weil theorem (see, e.g., [**7**, I, §7.2, Theorem 1]).

THEOREM 2.8.5. *The functions $\sqrt{d_s}D^s_{jk}(g)$ form a complete orthonormal system in $L^2(G)$.*

LEMMA 2.8.6. *The operators P^s_{pq} have the following properties:*

(i) P^s_{pq} *is adjointable and*

$$(P^s_{pq})^* = P^s_{qp}; \tag{2.19}$$

(ii) *the following equality holds:*

$$P^s_{pq}P^{s'}_{p'q'} = \delta^{ss'}\delta_{qp'}P^s_{pq'}. \tag{2.20}$$

(iii) *the following equalities hold:*

$$T_gP^s_{jm} = \sum_{i=1}^{d_s} D^s_{ij}(g)\, P^s_{im}, \tag{2.21}$$

$$P^s_{jm}T_g = \sum_{i=1}^{d_s} D^s_{mi}(g)\, P^s_{ji}. \tag{2.22}$$

PROOF. First of all, note that for unitary operators in $\mathcal{M}$, the mapping $F \mapsto F^*$ is continuous in the strong operator topology. In other words, for unitary operators, the strong continuity implies the $*$-strong one. Indeed,

$$\|(F'^* - F^*)x\| = \|(F'^{-1} - F^{-1})x\| = \|F'(F'^{-1} - F^{-1})Fz\| = \|Fz - F'z\| \to 0.$$

Therefore one can take T_g^* instead of T_g in (2.17), and then take it outside the integral. More precisely, the first equality in the chain

$$\begin{aligned}(P^s_{pq})^* &= d_s\int_G D^s_{pq}(g)T^*_g(x)\,dg = d_s\int_G D^s_{pq}(g^{-1})T_g(x)\,d(g^{-1}) \\ &= d_s\int_G \overline{D^s_{qp}(g)}T_g(x)\,dg = P^s_{qp}\end{aligned}$$

has to be verified at first at the level of integral sums and then one has to take the limit, which is possible due to the $*$-strong continuity. Remaining equalities in the chain above follow from the invariance of Haar measure and the relations $T^*_g = T^{-1}_g = T_{g^{-1}}$. Statement (i) is proved.

It follows from (2.17) that

$$P^s_{pq}P^{s'}_{p'q'} = d_sd_{s'}\int_G\int_G \overline{D^s_{pq}(g)}\;\overline{D^{s'}_{p'q'}(g')}T_gT_{g'}\,dg\,dg'.$$

Since $T_gT_{g'} = T_{gg'}$, we obtain, by taking $\widetilde{g} := gg'$, from

$$D^s_{pq}(g) = D^s_{pq}(\widetilde{g}\,g'^{-1}) = D^s_{pr}(\widetilde{g})D^s_{rq}(g'^{-1}) = D^s_{pr}(\widetilde{g})\overline{D^s_{qr}(g')}$$

and from relations (2.18) that

$$\begin{aligned} P^s_{pq} P^{s'}_{p'q'} &= d_s d_{s'} \int_G D^s_{qr}(g') \, \overline{D^{s'}_{p'q'}(g')} \, dg' \cdot \int_G \overline{D^s_{pr}(\widetilde{g})} T_{\widetilde{g}} \, d\widetilde{g} \\ &= d_{s'} \delta^{ss'} \frac{1}{d_{s'}} \delta_{qp'} \delta_{rq'} P^s_{pr} = \delta^{ss'} \delta_{qp'} P^s_{pq'}. \end{aligned}$$

To prove statement (iii) let us note that

$$\begin{aligned} T_g P^s_{jm}(x) &= d_s \int_G \overline{D^s_{jm}(h)} T_{gh}(x) \, dh = d_s \int_G \overline{D^s_{jm}(g^{-1}h)} T_h(x) \, dh \\ &= d_s \int_G \sum_{i=1}^{d_s} \overline{D^s_{ji}(g^{-1})} \, \overline{D^s_{im}(h)} T_h(x) \, dh \\ &= \sum_{i=1}^{d_s} \overline{D^s_{ji}(g^{-1})} \, d_s \int_G \sum_{i=1}^{d_s} \overline{D^s_{im}(h)} T_h(x) \, dh \\ &= \sum_{i=1}^{d_s} \overline{D^s_{ji}(g^{-1})} \, P^s_{im}(x) = \sum_{i=1}^{d_s} D^s_{ij}(g) \, P^s_{im}(x). \end{aligned}$$

The second equality of this statement can be proved similarly. □

LEMMA 2.8.7. *The operators* $P^s_p := P^s_{pp}$ *are selfadjoint pairwise orthogonal projections.*

PROOF. If we rewrite the statement of the lemma as

$$(P^s_p)^* = P^s_p, \qquad P^s_p P^{s'}_{p'} = \delta^{ss'} \delta_{pp'} P^s_p, \tag{2.23}$$

then the proof follows immediately from (2.19) and (2.20). □

LEMMA 2.8.8. *Put*

$$P^s := \sum_{p=1}^{d_s} P^s_p = \sum_{p=1}^{d_s} P^s_{pp}.$$

The operators P^s *have the following properties:*

$$(P^s)^* = P^s, \tag{2.24}$$

$$P^s P^{s'} = \delta_{ss'} P^s, \tag{2.25}$$

$$T_g P^s = P^s T_g. \tag{2.26}$$

In other words, P^s *are selfadjoint invariant pairwise orthogonal projections in* $\mathcal{M}$.

PROOF. By the definition of P^s, formulas (2.24) and (2.25) immediately follow from (2.23). To verify the third relation consider the character of the representation V_s,

$$\chi^s(g) := \sum_{p=1}^{d_s} D^s_{pp}(g),$$

which, being a trace, satisfies the relation $\chi^s(g) = \chi^s(hgh^{-1})$. One also has

$$\begin{aligned} P^s &= d_s \int_G \chi^s(g) T_g \, dg, \\ T_g P^s &= d_s T_g \int_G \chi^s(g') T_{g'} \, dg' = d_s \int_G \chi^s(g') T_{gg'g^{-1}} T_g \, dg' \\ &= d_s \int_G \chi^s(gg'g^{-1}) T_{gg'g^{-1}} \, dg' T_g = P^s T_g. \end{aligned}$$

□

LEMMA 2.8.9. *Define*

$$\mathcal{M}^s := P^s(\mathcal{M}), \qquad \mathcal{M}^\bullet := \bigoplus_{s=1}^{\infty} \mathcal{M}^s, \tag{2.27}$$

where the sum is supposed to be completed either as a Hilbert sum or as a closure in $\mathcal{M}$ of the algebraic sum (which is the same). Then

$$\mathcal{M}^\bullet = \mathcal{M}. \tag{2.28}$$

PROOF. Assume that a **C**-linear functional f on $\mathcal{M}$ vanishes on $\mathcal{M}^\bullet$ and that $x \in \mathcal{M}$ is an arbitrary vector. Then, for any set of indices, we have $P^s_{ij}(x) \in \mathcal{M}^\bullet$, so that

$$0 = f(P^s_{ij}(x)) = d_s \int_G \overline{D^s_{ij}(g)} f(T_g(x)) \, dg.$$

Therefore, by the Peter–Weil Theorem 2.8.5, $f(T_g(x)) = 0$ holds almost everywhere and, by continuity, it vanishes everywhere. In particular, $f(T_e(x)) = f(x) = 0$. Hence, by the Hahn–Banach theorem, $\mathcal{M}^\bullet = \mathcal{M}$. □

THEOREM 2.8.10 ([**91**]). *Let $\mathcal{M}$ be a Hilbert B-module equipped with a strongly continuous unitary representation of G and let the group act trivially on B. Now let $\{V_s\}$ be the complete collection of pairwise nonequivalent unitary representations of G and let*

$$\mathcal{M}_s := \mathrm{Hom}_{G,\mathbf{C}}(V_s, \mathcal{M}) \subset \mathrm{Hom}_{\mathbf{C}}(V_s, \mathcal{M}) \cong V_s^* \otimes \mathcal{M}$$

be a Hilbert B-module with the inner B-valued product defined by the formula

$$\langle \varphi, \psi \rangle := \sum_{i,j=1}^{\dim V_s} \langle \varphi(h_i^s), \psi(h_j^s) \rangle_{\mathcal{M}}, \qquad h_1^s, \dots, h^s_{\dim V_s} \text{ orthonormal basis for } V_s.$$

Then, for the Hilbert sum, we have an equivariant B-linear isomorphism

$$\Gamma = \bigoplus_{s=1}^{\infty} \Gamma_s : \bigoplus_{s=1}^{\infty} V_s \otimes \mathcal{M}_s \cong \mathcal{M}, \qquad \Gamma_s : v \otimes \varphi \mapsto \varphi(v), \quad v \in V_s, \ \varphi \in \mathcal{M}_s,$$

and

$$\Gamma(V_s \otimes \mathcal{M}_s) = \mathcal{M}^s,$$

where $\mathcal{M}^s$ is defined by (2.27).

PROOF. To begin with, let us note that the Γ_s are algebraically injective. Indeed, let

$$0 = \Gamma_s \left(\sum_{j=1}^{d_s} h_j^s \alpha_j \otimes \varphi \right) = \varphi \left(\sum_{j=1}^{d_s} h_j^s \alpha_j \right).$$

Since, by the Schur lemma, φ is either an isomorphism or 0, the above equality can be true only if either $\sum_{j=1}^{d_s} h_j^s \alpha_j = 0$ or $\varphi = 0$. But then $\sum_{j=1}^{d_s} h_j^s \alpha_j \otimes \varphi = 0$.

By Lemma 2.8.9, it is sufficient to prove only that Γ_s maps $V_s \otimes \mathcal{M}_s$ bijectively onto $\mathcal{M}^s$.

Note that, by setting $\mathcal{M}_i^s := P_i^s(\mathcal{M}) = P_{ii}^s(\mathcal{M})$, we obtain, by relation (2.20), that the operators P_{ij}^s induce isomorphisms

$$P_{ij}^s : \mathcal{M}_j^s \to \mathcal{M}_i^s.$$

Thus $\mathcal{M}^s = \bigoplus_{j=1}^{d_s} \mathcal{M}_j^s$ is a sum of isomorphic modules.

Let $\{h_1^s, \dots, h_{d_s}^s\}$ be the orthonormal basis of V_s, with respect to which the matrix elements D_{ij}^s were defined. Define the homomorphism

$$\Phi^s : V_s \otimes [\mathcal{M}_1^s] \to \mathcal{M}^s, \qquad \Phi^s(h_j^s \otimes x) = P_{j1}^s(x), \tag{2.29}$$

where we have put $\mathcal{M}_1^s$ in square brackets to emphasize that there is no action of G on it. By the properties of the operators P_{j1}^s, the map Φ^s is an isomorphism. Since, by (2.21),

$$T_g \Phi^s(h_j^s \otimes x) = T_g P_{j1}^s(x) = \sum_{i=1}^{d_s} D_{ij}^s(g)\, P_{i1}^s(x)$$

and

$$\Phi^s(g(h_j^s) \otimes x) = \Phi^s\left(\sum_{i=1}^{d_s} D_{ij}^s(g) h_i^s \otimes x\right) = \sum_{i=1}^{d_s} D_{ij}^s(g) P_{i1}^s(x),$$

the map Φ^s is equivariant. Further, there is a map

$$\Psi^s : \mathcal{M}_1^s \to \mathcal{M}_s, \qquad \Psi^s(x)(v) := \Phi^s(v \otimes x).$$

Then

$$\Gamma_s \circ (\mathrm{Id}_{V_s} \otimes \Psi^s)(v \otimes x) = \Phi^s(v \otimes x).$$

Since we have an isomorphism on the right-hand side and since Γ_s is algebraically injective, Γ_s is an isomorphism (see Lemma 2.8.12), whence Ψ^s is an isomorphism. In particular, the image of Γ_s coincides with $\mathcal{M}^s$ and these images are orthogonal to each other. Hence Γ is topologically injective and its image coincides with $\mathcal{M}$. □

REMARK 2.8.11. Let the G-A-module $\mathcal{M}$ belong to the class $\mathcal{P}(A)$ of projective finitely generated modules. Then obviously $\mathcal{M}_s = \mathrm{Hom}_G(V_s, \mathcal{M}) \in \mathcal{P}(A)$. Let us show that only a finite number of summands do not vanish in the sum $\bigoplus_s$. Denote generators of $\mathcal{M}$ by $a_1, \dots, a_s$. Using the Mostow lemma [**93**], choose **C**-periodic vectors $b_1, \dots, b_s$ so close to $a_1, \dots, a_s$ that they generate $\mathcal{M}$ as an A-module (see Lemma 2.7.3). By decomposing the linear span of the orbit Gb_j, which is a finite-dimensional G-**C**-module, into irreducible modules, let us find a new system of generators $c_1, \dots, c_N$, now belonging to irreducible G-**C**-modules. Then it is evident that the number of nonzero summands does not exceed N.

LEMMA 2.8.12. *Let $F : L \to M$, $T : N \to L$ be continuous maps of Banach spaces, $S = FT$ an isomorphism and* $\operatorname{Ker} F = 0$. *Then F is an isomorphism.*

PROOF. Since S is an isomorphism and F is bounded, T is topologically injective and its image $T(N)$ is closed in L. Suppose it does not coincide with L. Choose a vector $0 \neq x \in L \backslash T(N)$. Then $0 \neq F(x) \notin FT(N)$. Indeed, let $F(x) = FT(y)$ for some $y \in N$. Since $z = Ty \in T(N)$, $z - x \neq 0$, while $F(z - x) = FT(y) - F(z) = 0$.

We get a contradiction to the condition $\operatorname{Ker} F = 0$. Hence, T is a topologically injective epimorphism, i.e., isomorphism, as well as $F = ST^{-1}$. $\square$

Let us recall some facts about integrating operator-valued functions [**62**]. Let X be a compact space, A a C^*-algebra and $\varphi : C(X) \to A$ an involutive homomorphism of unital algebras. Let $F : X \to A$ be a continuous map and let, for each $x \in X$, the element $F(x)$ commute with the image of φ. In this case the integral

$$\int_X F(x)\, d\varphi \in A$$

can be defined as follows. Let $X = \bigcup_{i=1}^n U_i$ be an open covering and let $\sum_{i=1}^n \alpha_i(x) = 1$ be a subordinate partition of unity. Choose points $\xi_i \in U_i$ and form the integral sum

$$\sum(F, \{U_i\}, \{\alpha_i\}, \{\xi_i\}) = \sum_{i=1}^n F(\xi_i)\varphi(\alpha_i).$$

If the limit of such an integral sums exists, then it is called an integral.

If X is a Lie group G, it is natural here to use a Haar measure $\varphi : C(X) \to \mathbf{C}$, $\varphi(\alpha) = \int_G \alpha(g)\, dg$ and to define

$$\int_G Q(g)\, dg := \lim \sum_i Q(\xi_i) \int_G \alpha_i(g)\, dg$$

for a norm continuous map $Q : G \to \mathcal{B}(H)$, where the C^*-algebra A is viewed as a subalgebra in the algebra $\mathcal{B}(H)$ of bounded operators on a Hilbert space H. If $Q : G \to P^+(A) \subset \mathcal{B}(H)$, then, since

$$\int_G \alpha_i(g)\, dg \geq 0,$$

we obtain that

$$\sum_i Q(\xi_i) \cdot \int_G \alpha_i(g)\, dg \in P^+(A) \quad \text{and} \quad \int_G Q(g)\, dg \in P^+(A)$$

(the positive cone $P^+(A)$ is convex and closed). Hence we have proved the following lemma.

LEMMA 2.8.13. *Let $Q : G \to P^+(A)$ be a continuous function. Then, for the integral in the sense of* [**62**], *the following inequality holds:*

$$\int_G Q(g)\, dg \geq 0.$$

THEOREM 2.8.14 ([**128**]). *Let* $\mathrm{GL} = \mathrm{GL}(A)$ *be the full general linear group, i.e., the group of invertible operators in* $\operatorname{End} l_2(A)$, *and suppose that for the group G, a representation $g \mapsto T_g$ ($g \in G, T_g \in \mathrm{GL}$) is given, and that the map*

$$G \times l_2(A) \to l_2(A), \quad (g, u) \mapsto T_g u$$

is continuous. Then there exists an A-valued inner product on $l_2(A)$ equivalent to the initial one (i.e., generating an equivalent norm) and such that the representation $g \mapsto T_g$ is unitary with respect to this new product.

PROOF. Let $\langle\, ,\, \rangle'$ be the initial inner product. For any $u, v \in l_2(A)$ there exists a continuous map $G \to A, \quad x \mapsto \langle T_x u, T_x v\rangle'$. Define the new product by the formula

$$\langle u, v\rangle = \int_G \langle T_x u, T_x v\rangle' \, dx,$$

where the integral can be considered in the sense of any of the two definitions in [**62**], since the map is norm continuous. It is easy to see that this new product gives a sesquilinear map $l_2(A) \times l_2(A) \to A$ satisfying properties (i)–(iv) of Definition 1.2.1 and that, by Lemma 2.8.13, $\langle u, u\rangle \geq 0$. Let us show that this map is continuous. Fix an arbitrary $u \in l_2(A)$. Then $x \mapsto T_x(u), \quad G \to l_2(A)$ is a continuous map defined on a compact space, thus the set $\{T_x(u) \mid x \in G\}$ is bounded. Therefore, by the uniform boundedness principle [**7**, v. 2],

$$\lim_{v\to 0} T_x(v) = 0 \tag{2.30}$$

uniformly in $x \in G$. If u is fixed, then

$$\|T_x(u)\| \leq M_u = \text{const}$$

and, by equality (2.30), one has

$$\begin{aligned}\|\langle u, v\rangle\| &= \|\int_G \langle T_x(u), T_x(v)\rangle' \, dx\| \\ &\leq M_u \cdot \operatorname{vol} G \cdot \sup_{x\in G} \|T_x(v)\| \to 0 \quad (v \to 0).\end{aligned}$$

We have obtained the continuity at the point 0, hence on the whole space $l_2(A) \times l_2(A)$. For $T_x u = (a_1(x), a_2(x), \dots) \in l_2(A)$, the equality $\langle u, u\rangle = 0$ takes the form

$$\int_G \sum_{i=1}^{\infty} a_i(x) a_i^*(x) \, dx = 0.$$

Let A be viewed as a subalgebra of the algebra of bounded operators on a Hilbert space L with an inner product $(\, ,\,)_L$. For each $p \in L$ we have

$$\begin{aligned} 0 &= \left(\left(\int_G \sum_{i=1}^{\infty} a_i(x) a_i^*(x) \, dx\right) p, p\right)_L \\ &= \int_G \left(\sum_{i=1}^{\infty} a_i(x) a_i^*(x) p, p\right)_L dx = \int_G \left(\sum_{i=1}^{\infty} (a_i^*(x) p, a_i^*(x) p)_L\right) dx \end{aligned}$$

(cf. [**62**]). Therefore $a_i(x) = 0$ almost everywhere, hence, by continuity, $a_i(x) = 0$ for all x and $T_x u = 0$. In particular, $u = 0$.

Since each operator T_y is an automorphism, we obtain

$$\langle T_y u, T_y v\rangle = \int_G \langle T_{xy} u, T_{xy} v\rangle' \, dx = \int_G \langle T_z u, T_z v\rangle' \, dz = \langle u, v\rangle.$$

Now we show the equivalence of the two norms, which, in particular, implies the continuity of the representation. There is a number $N > 0$ such that $\|T_x\|' \leq N$

for any $x \in G$. Hence, by [**62**], we have

$$\|u\|^2 = \|\langle u, u\rangle\|_A = \| \int_G \langle T_x u, T_x u\rangle' \, dx \|_A$$
$$\le \left(\sup_{x \in G} \|T_x u\|' \right)^2 \le N^2 (\|u\|')^2.$$

On the other hand, applying Theorem 2.1.4 and Lemma 2.8.13, we obtain that

$$\langle u, u\rangle' = \int_G \langle T_{g^{-1}} T_g u, T_{g^{-1}} T_g u\rangle' \, dg \le \int_G \|T_{g^{-1}}\|^2 \langle T_g u, T_g u\rangle' \, dg$$
$$\le \int_G N^2 \langle T_g u, T_g u\rangle' \, dg = N^2 \int_G \langle T_g u, T_g u\rangle' \, dg = N^2 \langle u, u\rangle.$$

Then $(\|u\|')^2 = \|\langle u, u\rangle'\|_A \le N^2 \|\langle u, u\rangle\|_A = N^2 \|u\|^2$. □

REMARK 2.8.15. Since $l_2(P)$ is a direct summand in $l_2(A)$, the previous theorem remains valid for $l_2(P)$ and for any other countably generated module $\mathcal{M}$.

DEFINITION 2.8.16. Denote by $\mathrm{GL}^* \subset \mathrm{GL}$ the *general linear group* formed by adjointable invertible operators.

REMARK 2.8.17. Before averaging we have had operators, which, in general, are nonadjointable, but after averaging we have obtained unitary operators out of them. In relation to this remark the following question arises: is it true that if a given operator represents an element of compact group, then it is adjointable? A negative answer to this question is contained in Example 2.3.2, since a decomposition into a direct (topological) sum defines a representation of the group $\mathbf{Z}/2\mathbf{Z}$.

COROLLARY 2.8.18 ([**129**]). *Let $\mathcal{M} = \mathcal{M}_1 \widetilde{\oplus} \mathcal{M}_2$ be a topological decomposition into a direct sum of closed Hilbert C^*-modules (not necessarily an orthogonal sum). Then there exists a new inner product on the module $\mathcal{M}$ equivalent to the initial one, with respect to which the given decomposition is orthogonal.*

PROOF. Define the operator $J : \mathcal{M} \longrightarrow \mathcal{M}$ by the formula

$$Jx = \begin{cases} x, & \text{if } x \in \mathcal{M}_1, \\ -x, & \text{if } x \in \mathcal{M}_2. \end{cases}$$

One can consider the operator J as a representation of the group $\mathbf{Z}/2\mathbf{Z}$ on the module $\mathcal{M}$ and, by Theorem 2.8.14, the inner product $\langle x, y\rangle_\beta = \langle x, y\rangle + \langle Jx, Jy\rangle$ is equivalent to the initial one. Orthogonality of $\mathcal{M}_1$ and $\mathcal{M}_2$ with respect to this inner product is evident. □

In Theorem 5.7.13 we shall show how the averaging Theorem 2.8.14 can be generalized from the case of a compact group to the case of an amenable group, but only for Hilbert W^*-modules.

2.9. Equivariant Fredholm operators

In this section we transfer the main definitions and statements related to A-Fredholm operators to the G-equivariant case, where G is a compact group. We will present only a list of results of [**128**]. The proofs (very similar to the nonequivariant case) and details can be found, e.g., in [**121**].

By the averaging Theorem 2.8.14 we can restrict ourselves to a unitary action of G on $l_2(P)$. Suppose the action of G on A is trivial and apply the Kasparov stabilization Theorem 2.8.4 to $l_2(P)$:

$$l_2(P) \oplus \mathcal{H}_A \cong \mathcal{H}_A, \tag{2.31}$$

where $\mathcal{H}_A = \sum_{i=1}^{\infty}(A \otimes_{\mathbf{C}} V_i)$ and $\{V_i\}$ is a countable collection of finite-dimensional vector spaces of all irreducible representations of G (up to their equivalence), and each representation is repeated countably many times. The isomorphism (2.31) is a G-isomorphism of Hilbert A-modules and the sum on the left-hand side of (2.31) is orthogonal. Put

$$R_m = \sum_{i=m+1}^{\infty} (A \otimes_{\mathbf{C}} V_i), \qquad R_m^{\perp} = \sum_{i=1}^{m} (A \otimes_{\mathbf{C}} V_i).$$

For any bounded G-A-operator $F : l_2(P_1) \to l_2(P_2)$ denote by $S(F) : \mathcal{H}_A \to \mathcal{H}_A$ the following (stabilized) operator:

$$\begin{pmatrix} 1 & 0 \\ 0 & F \end{pmatrix} : \mathcal{H}_A \cong \mathcal{H}_A \oplus l_2(P_1) \to \mathcal{H}_A \oplus l_2(P_2) \cong \mathcal{H}_A.$$

Note that this stabilization is defined up to a G-A-isomorphism.

THEOREM 2.9.1 ([**128**]). *Let $\mathcal{H}_A \cong \mathcal{M}\widetilde{\oplus}\mathcal{N}$, where $\mathcal{M}$ and $\mathcal{N}$ are closed G-A-modules, and let $\mathcal{N}$ have a finite system of generators $a_1, \dots, a_s$. Then $\mathcal{N}$ is a projective finitely generated G-A-module.*

DEFINITION 2.9.2. A bounded G-A-operator

$$F : l_2(\mathcal{P}_1) \to l_2(\mathcal{P}_2)$$

is called *G-A-Fredholm* if

1) the operator F is adjointable;

2) there exist decompositions of the domain $\mathcal{H}_A = \mathcal{M}_1\widetilde{\oplus}\mathcal{N}_1$ and the range $\mathcal{H}_A = \mathcal{M}_2\widetilde{\oplus}\mathcal{N}_2$ of the operator $S(F)$ (where $\mathcal{M}_1, \mathcal{M}_2, \mathcal{N}_1, \mathcal{N}_2$ are closed G-A-modules and $\mathcal{N}_1, \mathcal{N}_2$ are finitely generated), such that the matrix of $S(F)$ with respect to these decompositions has the form $S(F) = \begin{pmatrix} F_1 & 0 \\ 0 & F_2 \end{pmatrix}$, where $F_1 : M_1 \to M_2$ is an isomorphism of G-A-modules. By the previous theorem, $\mathcal{N}_1$ and $\mathcal{N}_2$ are projective G-A-modules and we can define the index of F by

$$\operatorname{index} F = [\mathcal{N}_1] - [\mathcal{N}_2] \in K^G(A).$$

REMARK 2.9.3. Since the action of G is unitary, an analog of 2.7.6 evidently holds.

THEOREM 2.9.4 ([**128**]). *The index is well defined.*

LEMMA 2.9.5 ([**128**]). *Let an operator $F : l_2(A) \to l_2(A)$ be G-A-Fredholm. Then there exists a number $\varepsilon > 0$ such that any bounded adjointable G-A-operator D satisfying the inequality $\|F - D\| < \varepsilon$ is a G-A-Fredholm operator and* $\operatorname{index} D = \operatorname{index} F$.

LEMMA 2.9.6 ([**128**]). *Let F and D be G-A-Fredholm operators:*

$$F : l_2(\mathcal{P}_1) \to l_2(\mathcal{P}_2), \qquad D : l_2(\mathcal{P}_2) \to l_2(\mathcal{P}_3).$$

Then $DF : l_2(\mathcal{P}_1) \to l_2(\mathcal{P}_3)$ is a G-A-Fredholm operator and

$$\operatorname{index} DF = \operatorname{index} D + \operatorname{index} F.$$

LEMMA 2.9.7 ([**128**]). *Let* $K : l_2(\mathcal{P}_1) \to l_2(\mathcal{P}_2)$ *be a compact* G*-operator. Then* $1 + K$ *is a* G-A*-Fredholm operator and* $\operatorname{index}(1 + K) = 0$.

LEMMA 2.9.8 ([**128**]). *Suppose* $F : l_2(\mathcal{P}_1) \to l_2(\mathcal{P}_2)$ *is a* G-A*-Fredholm operator and an operator* $K \in \mathcal{K}(l_2(\mathcal{P}_1), l_2(\mathcal{P}_2))$ *is* G*-equivariant. Then the operator* $F + K$ *is* G-A*-Fredholm and* $\operatorname{index}(F + K) = \operatorname{index} F$.

LEMMA 2.9.9 ([**128**]). *Suppose* $F : l_2(\mathcal{P}_1) \to l_2(\mathcal{P}_2)$ *is a bounded adjointable* G-A*-operator,* $D \in \operatorname{End}^* \mathcal{H}_A$ *and* $K \in \mathcal{K}(\mathcal{H}_A)$ *are* G-A*-operators, and for* D *there exists a decomposition of* $\mathcal{H}_A$*, as in the definition of a* G-A*-Fredholm operator, and* $D = S(F) + K$. *Then* F *is a* G-A*-Fredholm operator.*

THEOREM 2.9.10 ([**128**]). *Suppose*

$$F : l_2(\mathcal{P}_1) \to l_2(\mathcal{P}_2), \qquad D : l_2(\mathcal{P}_2) \to l_2(\mathcal{P}_1), \qquad D' : l_2(\mathcal{P}_2) \to l_2(\mathcal{P}_1)$$

are bounded adjointable G-A*-operators and*

$$S(FD) = 1_{\mathcal{H}_A} + K_1, \quad S(D'F) = 1_{\mathcal{H}_A} + K_2, \quad K_1, K_2 \in \mathcal{K}(\mathcal{H}_A).$$

Then F *is a* G-A*-Fredholm operator.*

LEMMA 2.9.11 ([**128**]). *If, for bounded adjointable* G-A*-operators* D, D' *and* F*, the operators* FD *and* $D'F$ *are* G-A*-Fredholm operators, then* F *is a* G-A*-Fredholm operator too.*

CHAPTER 3

Hilbert Modules over W^*-Algebras

3.1. W^*-algebras

Detailed information about W^*-algebras can be found in the books [**123, 13, 32, 116, 58**]. We also recommend the original papers of Murray and von Neumann [**96**], which are still current. We list here only the basic definitions and some necessary facts.

Topologies on $\mathcal{B}(H)$. Besides the norm topology on the algebra $\mathcal{B}(H)$ of bounded linear operators on a Hilbert space H, a number of other locally convex topologies is useful to consider. These topologies are defined by various sets of seminorms on $\mathcal{B}(H)$. Let $a \in \mathcal{B}(H)$, $\xi, \xi_i, \eta, \eta_i \in H$.

1. The *σ-weak topology* is defined by the seminorms

$$p(a) = \left|\sum_{i=1}^{\infty}(a\xi_i, \eta_i)\right|, \qquad \sum_{i=1}^{\infty}\|\xi_i\|^2 < \infty, \quad \sum_{i=1}^{\infty}\|\eta_i\|^2 < \infty.$$

2. The *σ-strong topology* is defined by the seminorms

$$p(a) = \sum_{i=1}^{\infty}\|a\xi_i\|, \qquad \sum_{i=1}^{\infty}\|\xi_i\|^2 < \infty.$$

3. The *σ-strong* topology* is defined by the seminorms

$$p(a) = \sum_{i=1}^{\infty}(\|a\xi_i\| + \|a^*\xi_i\|)^{1/2}, \qquad \sum_{i=1}^{\infty}\|\xi_i\|^2 < \infty.$$

4. The *weak topology* is defined by the seminorms $p(a) = |(a\xi, \eta)|$.

5. The *strong topology* is defined by the seminorms $p(a) = \|a\xi\|$.

6. The *strong* topology* is defined by the seminorms $p(a) = (\|a\xi\| + \|a^*\xi\|)^{1/2}$.

When restricted to bounded subsets in $\mathcal{B}(H)$, the σ-weak topology coincides with the weak topology, the σ-strong topology coincides with the strong topology and the σ-strong* topology coincides with the strong* topology. We will use the notation w-lim and s-lim for weak and strong convergence, respectively.

A *commutant* of a subset $R \subset \mathcal{B}(H)$ is the set $R^! := \{a \in \mathcal{B}(H) : ar = ra \text{ for any } r \in R\}$. A *bicommutant* of a set R is the set $R^{!!} = (R^!)^!$.

THEOREM 3.1.1 (von Neumann bicommutant theorem). *Let $\mathcal{A} \subset \mathcal{B}(H)$ be an involutive subalgebra. Then the following conditions are equivalent:*

(i) *the algebra $\mathcal{A}$ contains the identity operator and is closed with respect to the σ-weak topology;*

(ii) *the algebra $\mathcal{A}$ contains the identity operator and is closed with respect to the σ-strong topology;*

(iii) *the algebra $\mathcal{A}$ coincides with its bicommutant, $\mathcal{A}^{!!} = \mathcal{A}$.*

In particular, it follows from here that if $A \subset B \subset \mathcal{B}(H)$ are two subsets, then $A^{!!} \subset B^{!!}$.

DEFINITION 3.1.2. An involutive subalgebra in $\mathcal{B}(H)$ is called a *von Neumann algebra* if it satisfies the conditions of Theorem 3.1.1.

Any von Neumann algebra is a C^*-algebra.

Unlike general C^*-algebras, there exist polar decompositions in von Neumann algebras: any element $a \in \mathcal{A}$ can be written in a unique way as $a = uh$, where u is a partial isometry, h is a positive element of the algebra $\mathcal{A}$ and $\operatorname{Ker} u = \operatorname{Ker} h$.

Universal enveloping von Neumann algebra. Let ω be a positive linear functional on a C^*-algebra A and let (π_ω, H_ω) be a cyclic representation of A on a Hilbert space H_ω obtained from the GNS-construction. Put

$$(\pi, H) = \bigoplus_{\omega \in A^*_+} (\pi_\omega, H_\omega),$$

where A^*_+ denotes the set of all positive linear functionals on the C^*-algebra A. The representation (π, H) is called *universal*. The universal representation of a C^*-algebra contains any representation of this algebra as a subrepresentation.

Let us recall that the $\sigma(E^*, E)$*-topology* on the dual space (of functionals) E^* for a Banach space E is defined by the system of seminorms $p_x(f) = |f(x)|$, $f \in E^*$, $x \in E$. The *weak** ($\sigma(E, E^*)$-)*topology* on E is defined by the system of seminorms $q_f(x) = |f(x)|$.

THEOREM 3.1.3. *The second dual space A^{**} of a C^*-algebra A equipped with the $\sigma(A^{**}, A^*)$-topology is homeomorphic to the bicommutant $A^{!!}$ of the algebra $A \subset \mathcal{B}(H)$ with respect to the universal representation equipped with the σ-weak topology.*

The von Neumann algebra $A^{!!}_u$, where the bicommutant is taken with respect to the universal representation, is called a *universal enveloping von Neumann algebra* for the C^*-algebra A and is denoted by A^{**}. Any homomorphism of C^*-algebras $A \longrightarrow B$ canonically extends to a homomorphism of the second dual spaces $A^{**} \longrightarrow B^{**}$. If $A \subset \mathcal{B}(H_0)$ and $A \longrightarrow \mathcal{B}(H)$ is the universal representation, then $H_0 \subset H$ and $A^{!!} \subset A^{!!}_u \cong A^{**}$.

W^*-algebras. The notion of W^*-algebra allows us to talk about von Neumann algebras without any relation to the concrete Hilbert space where they act.

DEFINITION 3.1.4. A C^*-algebra $\mathcal{A}$, which, as a Banach space, is dual to some Banach space F, $\mathcal{A} = F^*$, is called a *W^*-algebra*.

A Banach space F is called *pre-dual* for $\mathcal{A}$.

DEFINITION 3.1.5. A linear functional φ on a von Neumann algebra $\mathcal{A}$ is called *normal* if for any increasing net $a_\lambda \in \mathcal{A}$, $\lambda \in \Lambda$, with the least upper bound $a \in \mathcal{A}$, the value $\varphi(a)$ is the least upper bound of the set $\varphi(a_\lambda)$.

THEOREM 3.1.6. *Let $\mathcal{A}$ be a W^*-algebra. Then there exists a unique (up to an isomorphism) pre-dual space for $\mathcal{A}$, which coincides with the space of all normal linear functionals on $\mathcal{A}$.*

Denote by $\mathcal{A}_*$ the pre-dual space of a W^*-algebra $\mathcal{A}$. Let P denote the set of normal positive functionals on $\mathcal{A}$, $P \subset \mathcal{A}_*$. The pre-dual space $\mathcal{A}_*$ is the linear span of the set P.

DEFINITION 3.1.7. A W^*-algebra is called *σ-finite* if it contains no more than a countable set of mutually orthogonal projections.

The condition of σ-finiteness is equivalent to the existence of a faithful normal positive linear functional on $\mathcal{A}$.

Traces and classification of W^*-algebras. Let $\mathcal{A}$ be a W^*-algebra and $P^+(\mathcal{A})$ the cone of positive elements in $\mathcal{A}$.

A *trace* on $\mathcal{A}$ is a function $\tau : P^+(\mathcal{A}) \to [0, +\infty]$ such that

(i) if $s, t \in P^+(\mathcal{A})$, then $\tau(s+t) = \tau(s) + \tau(t)$;
(ii) if $s \in P^+(\mathcal{A})$, $\lambda \in \mathbf{C}$, $\lambda \geq 0$, then $\tau(\lambda s) = \lambda\tau(s)$;
(iii) if $s \in P^+(\mathcal{A})$ and if $u \in \mathcal{A}$ is unitary, then $\tau(u^* s u) = \tau(s)$.

Being defined on the set $P^+(\mathcal{A})$, a trace can be easily extended to a linear functional on the whole algebra $\mathcal{A}$.

A trace τ is called *finite* if $\tau(s) < +\infty$ for any $s \in P^+(\mathcal{A})$.

A trace τ is called *semifinite* if for any $s \in P^+(\mathcal{A})$, the number $\tau(s)$ is the exact upper bound of the values $\tau(t)$ for all $t \in P^+(\mathcal{A})$ satisfying $t \leq s$ and $\tau(t) < +\infty$.

A trace τ is called *faithful* if the conditions $s \in P^+(\mathcal{A})$ and $\tau(s) = 0$ imply that $s = 0$.

A trace τ is called *normal* if the functional on $\mathcal{A}$ obtained by extending τ is normal.

A W^*-algebra $\mathcal{A}$ is called *finite* (resp. *semifinite*) if for any nonzero element $t \in P^+(\mathcal{A})$ there exists a finite (resp. semifinite) trace τ on $\mathcal{A}$ such that $\tau(t) \neq 0$. A W^*-algebra $\mathcal{A}$ is called *properly infinite* (resp. purely infinite) if there is no nonzero finite (resp. semifinite) trace on $\mathcal{A}$.

THEOREM 3.1.8. *Any (semi)finite W^*-algebra possesses a faithful normal (semi) finite trace.*

A W^*-algebra $\mathcal{A}$ is called an algebra of *type* I if it is isomorphic to a von Neumann algebra $\mathcal{B}$ with commutative commutant $\mathcal{B}^!$.

Let p be a projection in a W^*-algebra $\mathcal{A}$. Denote by $\mathcal{A}p \subset \mathcal{A}$ the set of elements of the form pap, $a \in \mathcal{A}$. It is easy to check that $\mathcal{A}p$ is also a W^*-algebra.

A W^*-algebra $\mathcal{A}$ is called *continuous* if there is no projection in the center $\mathcal{Z}(\mathcal{A})$ of $\mathcal{A}$, for which the algebra $\mathcal{A}p$ is of type I. A semifinite continuous W^*-algebra $\mathcal{A}$ is called an algebra of *type* II. A finite type II W^*-algebra is called an algebra of *type* II_1, and a properly infinite type II W^*-algebra is called an algebra of *type* II_∞.

A purely infinite W^*-algebra is called an algebra of *type* III.

THEOREM 3.1.9. *For a W^*-algebra $\mathcal{A}$ there exist mutually orthogonal central projections $p_{\mathrm{I}}, p_{\mathrm{II}_1}, p_{\mathrm{II}_\infty}, p_{\mathrm{III}} \in \mathcal{Z}(\mathcal{A})$ such that their sum equals* 1 *and the algebras $\mathcal{A}p_{\mathrm{I}}$, $\mathcal{A}p_{\mathrm{II}_1}$, $\mathcal{A}p_{\mathrm{II}_\infty}$ and $\mathcal{A}p_{\mathrm{III}}$ are algebras of type* I, II_1, II_∞ *and* III, *respectively.*

THEOREM 3.1.10. *For a W^*-algebra $\mathcal{A}$ the following conditions are equivalent:*

(i) *$\mathcal{A}$ is finite;*
(ii) *there exists a linear map T from $\mathcal{A}$ into its center $\mathcal{Z}(\mathcal{A})$ such that:*
  (a) *$T(x^*x) = T(xx^*) \geq 0$ for any $x \in \mathcal{A}$,*
  (b) *$T(ax) = aT(x)$ for any $a \in \mathcal{Z}(\mathcal{A})$, $x \in \mathcal{A}$,*
  (c) *$T(1) = 1$,*
  (d) *$T(x^*x) \neq 0$ for any nonzero element $x \in \mathcal{A}$.*

In this case this map (a center-valued trace) T is unique and σ-weakly continuous.

A W^*-algebra, whose center is isomorphic to $\mathbf{C}$, is called a *factor*. Like the direct integral of Hilbert spaces, one can define the direct integral of von Neumann algebras (see, e.g., [**123**]).

THEOREM 3.1.11 (central decomposition). *Let $\mathcal{A}$ be a W^*-algebra that admits a faithful representation in a separable Hilbert space. Then there exists a Borel space Γ with a measure $d\gamma$ and factors $\mathcal{A}_\gamma$, $\gamma \in \Gamma$, such that*

$$\mathcal{A} = \int_\Gamma^\oplus \mathcal{A}_\gamma \, d\gamma.$$

Let $\mathcal{A}$ be a von Neumann algebra. Two projections $e, f \in \mathcal{A}$ are called *equivalent*, $p \sim q$, if there exists a partial isometry $v \in \mathcal{A}$ such that $e = v^*v$, $f = vv^*$. If there is a partial isometry $v \in \mathcal{A}$ such that $e = v^*v$ and $vv^* \leq f$, we write $e \preceq f$. It can be easily checked that the relation $\preceq$ is an order relation on the set of all projections in a von Neumann algebra. In the case where $e \preceq f$, $e \not\sim f$, we write $e \prec f$.

A projection $e \in \mathcal{A}$ is called *finite* if the conditions $e \sim f$ and $f \leq e$ imply $e = f$. Otherwise a projection e is called *infinite*. If there are no finite projections $f \in \mathcal{A}$ with $f \leq e$, then the projection e is called *purely infinite*. If the projection $ze \neq 0$ is infinite for any central projection z, then the projection e is called *properly infinite*. If a von Neumann algebra $e\mathcal{A}e$ is commutative, then the projection e is called *Abelian*. For a projection $e \in \mathcal{A}$ the least projection $z \in \mathcal{Z}(\mathcal{A})$ from the center $\mathcal{Z}(\mathcal{A})$ such that $e \preceq z$ is called the *central support* for a projection e and is denoted by $z(e)$.

3.2. Inner product on dual modules

Hilbert C^*-modules over W^*-algebras will be called Hilbert W^*-modules. Some aspects of the theory of Hilbert C^*-modules become simpler in the W^*-case.

THEOREM 3.2.1 ([**100**]). *Let $\mathcal{M}$ be a Hilbert $\mathcal{A}$-module. An $\mathcal{A}$-valued inner product $\langle\cdot,\cdot\rangle$ admits an extension to the Banach module $\mathcal{M}'$, making it a self-dual Hilbert $\mathcal{A}$-module. In particular, the extended inner product satisfies the equality $\langle f, \widehat{x}\rangle = f(x)$ for all $x \in \mathcal{M}$, $f \in \mathcal{M}'$.*

PROOF. Let $f, g \in \mathcal{M}'$. Our aim is to define the inner product of these functionals $\langle f, g\rangle$. For this purpose define the map $\Gamma : P \longrightarrow \mathbf{C}$ by $\Gamma(\varphi) = (f_\varphi, g_\varphi)_\varphi$, where $\varphi \in P$ is a normal positive functional on $\mathcal{A}$. Let us show that the map Γ admits an extension to the set $\mathcal{A}_*$ of all normal functionals on $\mathcal{A}$. To do this we require the following two technical lemmas.

LEMMA 3.2.2. *If $\lambda_1, \dots, \lambda_n \in \mathbf{C}$ and $\varphi_1, \dots, \varphi_n \in P$ satisfy $\sum_{i=1}^n \lambda_i \varphi_i = 0$, then $\sum_{i=1}^n \lambda_i \Gamma(\varphi_i) = 0$.*

PROOF. Consider a normal positive functional $\varphi = \sum_{i=1}^n \varphi_i$. Then $\varphi \geq \varphi_i$, $i = 1, \dots, n$. If $x, y \in \mathcal{M}$, then, by assumption,

$$\sum_{i=1}^n \lambda_i (V^*_{\varphi,\varphi_i} V_{\varphi,\varphi_i}(x + N_\varphi), y + N_\varphi)_\varphi = \sum_{i=1}^n \lambda_i (x + N_{\varphi_i}, y + N_{\varphi_i})_{\varphi_i}$$
$$= \sum_{i=1}^n \lambda_i \varphi_i(\langle x, y\rangle) = 0;$$

therefore $\sum_{i=1}^n \lambda_i V^*_{\varphi,\varphi_i} V_{\varphi,\varphi_i} = 0$. Note that, by Proposition 2.5.7,

$$\sum_{i=1}^n \lambda_i \Gamma(\varphi_i) = \sum_{i=1}^n \lambda_i (f_{\varphi_i}, g_{\varphi_i})_{\varphi_i} = \sum_{i=1}^n \lambda_i (V_{\varphi,\varphi_i} f_\varphi, V_{\varphi,\varphi_i} g_\varphi)_{\varphi_i}$$
$$= \sum_{i=1}^n \lambda_i (V^*_{\varphi,\varphi_i} V_{\varphi,\varphi_i} f_\varphi, g_\varphi)_\varphi = 0.$$

□

This means that the map Γ can be extended to $\mathcal{A}_*$. To avoid extra notation we denote this extension still by Γ.

LEMMA 3.2.3. *The map Γ is bounded.*

PROOF. One can write an arbitrary normal functional $\psi \in \mathcal{A}_*$ as $\psi = \varphi_1 - \varphi_2 + i(\varphi_3 - \varphi_4)$, where $\varphi_i \in P$, $i = 1, \dots, 4$, and $\sum_{i=1}^4 \|\varphi_i\| \leq 2\|\psi\|$. Then

$$|\Gamma(\psi)| \leq \sum_{i=1}^4 |(f_{\varphi_i}, g_{\varphi_i})_{\varphi_i}| \leq \sum_{i=1}^4 \|f_{\varphi_i}\|_{\varphi_i} \|g_{\varphi_i}\|_{\varphi_i} \leq \sum_{i=1}^4 \|\varphi_i\| \|f\| \|g\|$$
$$\leq 2\|f\| \|g\| \|\psi\|,$$

as required. □

Let us continue the proof of the theorem. Since $\mathcal{A}$ is isomorphic to the space of linear functionals on the pre-dual space $\mathcal{A}_*$ [**123**], there exists a unique element $\langle f, g\rangle \in \mathcal{A}$ such that $\Gamma(\psi) = \psi(\langle f, g\rangle)$ for any $\psi \in \mathcal{A}_*$, in particular, $(f_\varphi, g_\varphi)_\varphi = \varphi(\langle f, g\rangle)$ for all $\varphi \in P$. The sesquilinearity of the defined map $\langle\cdot,\cdot\rangle : \mathcal{M}' \times \mathcal{M}' \longrightarrow \mathcal{A}$ follows from the linearity of the map $f \longmapsto f_\varphi$ from $\mathcal{M}'$ to H_φ for $\varphi \in P$. Let us show that $\langle\cdot,\cdot\rangle$ satisfies properties (i)–(iv) of Definition 1.2.1.

(i) Since $\varphi(\langle f, f\rangle) = (f_\varphi, f_\varphi)_\varphi \geq 0$ for any $\varphi \in P$, the inequality $\langle f, f\rangle \geq 0$ follows.

(ii) Let $\langle f, f\rangle = 0$ for $f \in \mathcal{M}'$. Then $f_\varphi = 0$ for all $\varphi \in P$, hence $\varphi(f(x)) = 0$ for all $x \in \mathcal{M}$. Therefore $f = 0$.

(iii) Since

$$\varphi(\langle f, g\rangle) = (f_\varphi, g_\varphi)_\varphi = \overline{(g_\varphi, f_\varphi)_\varphi} = \overline{\varphi(\langle g, f\rangle)} = \varphi(\langle g, f\rangle^*)$$

for any $\varphi \in P$, we conclude that $\langle f, g\rangle = \langle g, f\rangle^*$.

(iv) Let $a \in \mathcal{A}$, $\varphi \in P$. Let us define the functional φ_a on the algebra $\mathcal{A}$ by $\varphi_a(b) = \varphi(a^* b)$, $b \in \mathcal{A}$. Then $\varphi_a \in \mathcal{A}_*$ and $\varphi_a = \sum_{i=1}^4 \lambda_i \varphi_i$, where $\lambda_i \in \mathbf{C}$, $\varphi_i \in P$. Let us put $\psi = \varphi + \sum_{i=1}^4 \varphi_i$; then ψ is a positive functional and $\psi \geq \varphi, \varphi_1, \dots, \varphi_4$.

It follows from Proposition 2.5.7 that

$$\varphi(a^*\langle f,g\rangle)=\sum_{i=1}^4\lambda_i\varphi_i(\langle f,g\rangle)=\sum_{i=1}^4\lambda_i(f_{\varphi_i},g_{\varphi_i})_{\varphi_i}=\sum_{i=1}^4\lambda_i(f_{\varphi_i},V_{\psi,\varphi_i}g_\psi)_{\varphi_i}.$$

But

$$\begin{aligned}\sum_{i=1}^4\lambda_i(f_{\varphi_i},V_{\psi,\varphi_i}(x+N_\psi))_{\varphi_i}&=\sum_{i=1}^4\lambda_i(f_{\varphi_i},x+N_{\varphi_i})_{\varphi_i}=\sum_{i=1}^4\lambda_i\varphi_i(f(x))\\&=\varphi_a(f(x))=\varphi(a^*f(x))=\varphi((f\cdot a)(x))\\&=((f\cdot a)_\varphi,x+N_\varphi)_\varphi=((f\cdot a)_\varphi,V_{\psi,\varphi}(x+N_\psi))_\varphi\end{aligned}$$

for any $x\in\mathcal{M}$. Since the subspace $\mathcal{M}/N_\psi$ is dense in H_ψ, the equality

$$\begin{aligned}\varphi(a^*\langle f,g\rangle)&=\sum_{i=1}^4\lambda_i(f_{\varphi_i},V_{\psi,\varphi_i}g_\psi)_{\varphi_i}=((f\cdot a)_\varphi,V_{\psi,\varphi}g_\psi)_\varphi\\&=((f\cdot a)_\varphi,g_\varphi)_\varphi=\varphi(\langle f\cdot a,g\rangle)\end{aligned}$$

holds for any $\varphi\in P$, hence $a^*\langle f,g\rangle=\langle f\cdot a,g\rangle$. Passing to adjoints, we also obtain $\langle f,g\rangle a=\langle f,g\cdot a\rangle$.

The obtained inner product on $\mathcal{M}'$ is an extension of the inner product from $\mathcal{M}$. Indeed, if $x,y\in\mathcal{M}$, $\varphi\in P$, then

$$\varphi(\langle\widehat{x},\widehat{y}\rangle)=(\widehat{x}_\varphi,\widehat{y}_\varphi)_\varphi=(x+N_\varphi,y+N_\varphi)_\varphi=\varphi(\langle x,y\rangle).$$

Hence $\langle\widehat{x},\widehat{y}\rangle=\langle x,y\rangle$. Further, if $f\in\mathcal{M}'$, then

$$\varphi(\langle f,\widehat{x}\rangle)=(f_\varphi,\widehat{x}_\varphi)_\varphi=\varphi(f(x));$$

therefore $\langle f,\widehat{x}\rangle=f(x)$. Let us show that $\mathcal{M}'$ is complete with respect to the norm $\|\cdot\|_{\mathcal{M}'}$ defined by the inner product constructed above. There also exists the norm $\|\cdot\|$ on $\mathcal{M}'$ defined as the norm of linear maps from $\mathcal{M}$ to $\mathcal{A}$, with respect to which the space $\mathcal{M}'$ is complete. Let us prove that $\|\cdot\|_{\mathcal{M}'}=\|\cdot\|$. Since

$$f(x)^*f(x)=\langle\widehat{x},f\rangle\langle f,\widehat{x}\rangle\le\|f\|^2_{\mathcal{M}'}\cdot\langle x,x\rangle,$$

we obtain that $\|f\|\le\|f\|_{\mathcal{M}'}$. But since $\|f_\varphi\|\le\|f\|\,\|\varphi\|^{1/2}$ for each $\varphi\in P$,

$$\|f\|^2_{\mathcal{M}'}=\|\langle f,f\rangle\|=\sup\{\|f_\varphi\|^2_\varphi:\varphi\in P,\|\varphi\|=1\}\le\|f\|^2,$$

and $\|\cdot\|_{\mathcal{M}'}=\|\cdot\|$.

So, we have proved that $\mathcal{M}'$ is a Hilbert $\mathcal{A}$-module. It remains to verify that it is self-dual. Let $F\in(\mathcal{M}')'$. The restriction of F to $\mathcal{M}\subset\mathcal{M}'$ is an element of the module $\mathcal{M}'$, hence one can find a functional $f\in\mathcal{M}'$ such that $F(\widehat{x})=f(x)$ for all $x\in\mathcal{M}$. Let us define the functional $F_0\in(\mathcal{M}')'$ by

$$F_0(g)=F(g)-\langle f,g\rangle,\qquad g\in\mathcal{M}'.$$

It is clear that $F_0(\widehat{x})=0$ for all $x\in\mathcal{M}$. We should verify that $F_0(g)=0$ for all $g\in\mathcal{M}'$. Let $\varphi\in P$. Choose a sequence $\{y_n+N_\varphi\}$ in $\mathcal{M}/N_\varphi$ converging to g_φ. Since F_0 is bounded, we can find a number K such that $F_0(h)^*F_0(h)\le K\langle h,h\rangle$ for all $h\in\mathcal{M}'$. For all $n=1,2,\dots$ one has

$$\varphi(F_0(g)^*F_0(g))=\varphi(F_0(g-\widehat{y}_n)^*F_0(g-\widehat{y}_n))\le K\varphi(\langle g-\widehat{y}_n,g-\widehat{y}_n\rangle).$$

But since

$$\begin{aligned}\varphi(\langle g-\widehat{y}_n, g-\widehat{y}_n\rangle) &= (g_\varphi, g_\varphi)_\varphi - (y_n+N_\varphi, g_\varphi)_\varphi \\ &\quad - (g_\varphi, y_n+N_\varphi)_\varphi + (y_n+N_\varphi, y_n+N_\varphi)_\varphi \\ &= \|g_\varphi - (y_n+N_\varphi)\|^2_\varphi ,\end{aligned}$$

we have $\varphi(\langle g-\widehat{y}_n, g-\widehat{y}_n\rangle) \to 0$. Therefore

$$\varphi(F_0(g)^* F_0(g)) = 0. \tag{3.1}$$

Since the equality (3.1) is valid for any normal functional $\varphi \in P$, we obtain that $F_0(g) = 0$. □

3.3. Hilbert W^*-modules and dual Banach spaces

PROPOSITION 3.3.1. *Let $\mathcal{M}, \mathcal{N}$ be Hilbert C^*-modules over a W^*-algebra $\mathcal{A}$ and let $T : \mathcal{M} \longrightarrow \mathcal{N}$ be a bounded operator, $T \in \mathrm{Hom}_{\mathcal{A}}(\mathcal{M}, \mathcal{N})$. Then there exists a unique extension of the operator T to an operator $\widetilde{T} : \mathcal{M}' \longrightarrow \mathcal{N}'$.*

PROOF. Define the operator $T^\# : \mathcal{N} \longrightarrow \mathcal{M}'$ by $(T^\# y)(x) := \langle y, Tx\rangle$, $x \in \mathcal{M}$, $y \in \mathcal{N}$. Since $\|(T^\# y)(x)\| \le \|T\|\,\|x\|\,\|y\|$, the operator $T^\#$ is bounded, $\|T^\# y\| \le \|T\|\,\|y\|$. For any $a \in \mathcal{A}$ we have

$$(T^\#(y\cdot a))(x) = \langle y\cdot a, Tx\rangle = a^*\langle y, Tx\rangle = ((T^\# y)\cdot a)(x);$$

therefore the map $T^\#$ is $\mathcal{A}$-linear. Define the map $\widetilde{T} : \mathcal{M}' \longrightarrow \mathcal{N}'$ by $(\widetilde{T} f)(y) = \langle f, T^\# y\rangle$, where $y \in \mathcal{N}$, $f \in \mathcal{M}'$. Since $\widetilde{T} = (T^\#)^\#$, the map $\widetilde{T}$ is also a bounded $\mathcal{A}$-linear map. The equality $(\widetilde{T}\widehat{x})(y) = (Tx)\widehat{\ }(y)$ shows that the operator $\widetilde{T}$ is an extension of the operator T.

Let us show uniqueness of this extension. Let $S : \mathcal{M}' \longrightarrow \mathcal{N}'$ be a bounded $\mathcal{A}$-linear map that coincides with $\widetilde{T}$ on the submodule $\widehat{\mathcal{M}} = \{\widehat{x} : x \in \mathcal{M}\} \subset \mathcal{M}'$. Then their difference $V = \widetilde{T} - S$ vanishes on $\widehat{\mathcal{M}}$. Since the module $\mathcal{M}'$ is self-dual, the operator V has an adjoint operator $V^* : \mathcal{N}' \longrightarrow \mathcal{M}'$. If $g \in \mathcal{N}'$, $x \in \mathcal{M}$, then

$$(V^* g)(x) = \langle V^* g, \widehat{x}\rangle = \langle g, V\widehat{x}\rangle = 0,$$

i.e., $V^* = 0$. Therefore $V = 0$, hence $S = \widetilde{T}$. □

COROLLARY 3.3.2. *Let $\mathcal{M}$ be a Hilbert $\mathcal{A}$-module. Then the map $T \longmapsto \widetilde{T}$ defines a monomorphism $\mathrm{End}^*_{\mathcal{A}}(\mathcal{M}) \subseteq \mathrm{End}^*_{\mathcal{A}}(\mathcal{M}') = \mathrm{End}_{\mathcal{A}}(\mathcal{M}')$.*

Let us show that self-dual Hilbert W^*-modules are dual Banach spaces, as well as the C^*-algebras of operators acting on them.

PROPOSITION 3.3.3 ([**100**]). *Let $\mathcal{M}$ be a self-dual Hilbert W^*-module. Then $\mathcal{M}$ is a dual Banach space.*

PROOF. Let us introduce the notation $\mathcal{M}^o$ for the Hilbert module $\mathcal{M}$ viewed as a Banach space with multiplication by scalars given by the formula $\lambda \cdot x := \overline{\lambda} x$, $x \in \mathcal{M}^o$. Consider an algebraic tensor product $\mathcal{A}_* \otimes \mathcal{M}^o$ over $\mathbf{C}$, where $\mathcal{A}_*$ is a pre-dual space of normal functionals on $\mathcal{A}$. Equip the space $\mathcal{A}_* \otimes \mathcal{M}^o$ with the maximal cross-norm and define for $x \in \mathcal{M}$ the linear functional $\check{x}$ on $\mathcal{A}_* \otimes \mathcal{M}^o$ by the formula

$$\check{x}\left(\sum_{i=1}^n \varphi_i \otimes y_i\right) = \sum_{i=1}^n \varphi_i(\langle y_i, x\rangle),$$

where $\varphi_1, \dots, \varphi_n \in \mathcal{A}_*$, $y_1, \dots, y_n \in \mathcal{M}^o$. It is easy to see that this functional is well defined. Since

$$\left| \breve{x} \left(\sum_{i=1}^{n} \varphi_i \otimes y_i \right) \right| \leq \|x\| \sum_{i=1}^{n} \|\varphi_i\| \, \|y_i\| ,$$

it follows from the definition of the maximal cross-norm [**66**] that $\|\breve{x}\| \leq \|x\|$. Let us show that in fact the equality $\|\breve{x}\| = \|x\|$ holds. Let $\{\psi_n\}$ be a sequence of functionals of norm 1 in $\mathcal{A}_*$ such that $|\psi_n(\langle x, x\rangle)| \to \|x\|^2$. For each element of the form $\psi_n \otimes x \in \mathcal{A}_* \otimes \mathcal{M}^o$ we have $\|\psi_n\| \, \|x\| = \|x\|$ and $|\breve{x}(\psi_n \otimes x)| \to \|x\|^2$, hence $\|x\| \leq \|\breve{x}\|$. Therefore the map $x \longmapsto \breve{x}$ defines an isometric inclusion $\mathcal{M} \subset (\mathcal{A}_* \otimes \mathcal{M}^o)^*$. To prove the statement it is sufficient to check that the set $\breve{\mathcal{M}} = \{\breve{x} : x \in \mathcal{M}\}$ is closed in $(\mathcal{A}_* \otimes \mathcal{M}^o)^*$ with respect to the weak* topology, since it would mean that $\mathcal{M}$ is isometric to the dual space of some quotient space of $\mathcal{A}_* \otimes \mathcal{M}^o$. Let $\{\breve{x}_\alpha\}$ be a net in $\mathcal{M}$ that converges to some element $F \in (\mathcal{A}_* \otimes \mathcal{M}^o)^*$ with respect to the weak* topology. For $y \in \mathcal{M}$, let us define the linear functional on $\mathcal{A}_*$ by $\Phi_y(\psi) = F(\psi \otimes y)$, where $\psi \in \mathcal{A}_*$. The functional Φ is bounded, $\|\Phi\| \leq \|F\| \, \|y\|$, hence there exists a unique element $f(y) \in \mathcal{A}$ satisfying the properties $\|f(y)\| \leq \|F\| \, \|y\|$ and $F(\psi \otimes y) = \psi(f(y)^*)$ for all $\psi \in \mathcal{A}_*$. The map f is obviously linear. Let us show that it is $\mathcal{A}$-linear as well. Let $y \in \mathcal{M}$, $a, b \in \mathcal{A}$, $\varphi \in \mathcal{A}_*$. Define the normal functional $\psi \in \mathcal{A}_*$ by $\psi(b) = \varphi(a^* b)$. Then it follows from the equalities

$$\begin{aligned} \varphi(f(y \cdot a)^*) &= F(\varphi \otimes (y \cdot a)) = \lim_\alpha \breve{x}_\alpha(\varphi \otimes (y \cdot a)) = \lim_\alpha \varphi(\langle y \cdot a, x_\alpha\rangle) \\ &= \lim_\alpha \varphi(a^* \langle y, x_\alpha\rangle) = \lim_\alpha \psi(\langle y, x_\alpha\rangle) = F(\psi \otimes y) \\ &= \psi(f(y)^*) = \varphi(a^* f(y)^*) = \varphi((f(y)a)^*), \end{aligned}$$

which hold for any $\varphi \in \mathcal{A}$, that $f(y \cdot a) = f(y)a$. Since the module $\mathcal{M}$ is self-dual, we can find an element $x_0 \in \mathcal{M}$ such that $f(y) = \langle x_0, y\rangle$. Thus $F = \breve{x}_0$, hence $\breve{\mathcal{M}}$ is closed in $(\mathcal{A}_* \otimes \mathcal{M}^o)^*$. □

Consider the weak* topology on the dual Banach space $\mathcal{M}$. A net $\{x_\alpha\}$ in $\mathcal{M}$ obviously converges to an element $x \in \mathcal{M}$ with respect to this topology iff $\varphi(\langle y, x_\alpha\rangle) \longrightarrow \varphi(\langle y, x\rangle)$ for every $\varphi \in \mathcal{A}_*$ and for every $y \in \mathcal{M}$.

Some modification of the previous argument allows us to also obtain the following statement.

PROPOSITION 3.3.4 ([**100**]). *Let $\mathcal{M}$ be a self-dual Hilbert W^*-module. Then the C^*-algebra* $\operatorname{End}^*_{\mathcal{A}}(\mathcal{M})$ *is a W^*-algebra.*

3.4. Properties of Hilbert W^*-modules

The elements of self-dual Hilbert W^*-modules admit the following convenient representation (an analog of the polar decomposition).

PROPOSITION 3.4.1 ([**100**]). *Let $\mathcal{M}$ be a self-dual Hilbert W^*-module. Any element $x \in \mathcal{M}$ can be written as $x = z \cdot \langle x, x\rangle^{1/2}$, where $z \in \mathcal{M}$ is such that $\langle z, z\rangle$ is the projection onto the image of $\langle x, x\rangle^{1/2}$. Such a decomposition is unique: if $x = z' \cdot a$, where $a \geq 0$, and if $\langle z', z'\rangle$ is the projection onto the image of a, then $z' = z$ and $a = \langle x, x\rangle^{1/2}$.*

PROOF. For $x \in \mathcal{M}$ and $n \in \mathbf{N}$ put

$$a_n = (\langle x, x\rangle + 1/n)^{1/2}, \qquad x_n = x \cdot a_n^{-1}.$$

Since $\langle x_n, x_n\rangle = \langle x, x\rangle(\langle x, x\rangle + 1/n)^{-1}$, $\|x_n\| \le 1$. Let $y \in \mathcal{M}$ be an accumulation point for the sequence $\{x_n\}$ in the weak* topology (which exists due to the compactness of the unit ball). Since $\|a_n - \langle x, x\rangle^{1/2}\| \to 0$ and $x_n \cdot a_n = x$, one has $x = y \cdot \langle x, x\rangle^{1/2}$. Let p be the projection onto the image of $\langle x, x\rangle^{1/2}$. Then

$$p\langle x, x\rangle^{1/2} = \langle x, x\rangle^{1/2} p = \langle x, x\rangle^{1/2},$$

hence $x = y \cdot p\langle x, x\rangle^{1/2}$ and

$$\langle x, x\rangle = \langle x, x\rangle^{1/2} p\langle y, y\rangle p\langle x, x\rangle^{1/2}.$$

Therefore

$$\langle x, x\rangle^{1/2}(p - p\langle y, y\rangle p)\langle x, x\rangle^{1/2} = 0.$$

Since $\|y\| \le 1$, we have $p - p\langle y, y\rangle p \ge 0$. Therefore

$$\langle x, x\rangle^{1/2}(p - p\langle y, y\rangle p)^{1/2} = 0,$$

whence it follows that $p(p - p\langle y, y\rangle p)^{1/2} = 0$, hence $p = p\langle y, y\rangle p$. Put $z = y \cdot p$. Then $z \cdot \langle x, x\rangle^{1/2} = y \cdot p\langle x, x\rangle^{1/2} = x$, $\langle z, z\rangle = p\langle y, y\rangle p = p$ and $z \cdot p = z$.

In order to prove the uniqueness of the decomposition suppose that $x = z' \cdot a$, where $a \ge 0$, and that $\langle z', z'\rangle$ is the projection onto the image of a. Then $\langle x, x\rangle = a\langle z', z'\rangle a = a^2$, hence $a = \langle x, x\rangle^{1/2}$ and $\langle z', z'\rangle = p$. Since $\langle z' - z' \cdot p, z' - z' \cdot p\rangle = 0$, we obtain $z' = z'p$. Also one has

$$\langle z, x\rangle = \langle x, x\rangle^{1/2} = \langle z, z'\rangle\langle x, x\rangle^{1/2},$$

i.e., $(p - \langle z, z'\rangle)\langle x, x\rangle^{1/2} = 0$, whence we obtain that $(p - \langle z, z'\rangle)p = p - \langle z, z' \cdot p\rangle = p - \langle z, z'\rangle = 0$. Now it can be easily seen that $\langle z - z', z - z'\rangle = 0$, hence $z' = z$. $\square$

Let $\{p_\alpha\}$ be a set of projections in a W^*-algebra $\mathcal{A}$. For each of them the set $\mathcal{M}_\alpha = p_\alpha \mathcal{A} \subset \mathcal{A}$ has the natural structure of a singly generated projective Hilbert $\mathcal{A}$-module. Similarly to the definition of the standard Hilbert module we can define the module $\bigoplus_\alpha \mathcal{M}_\alpha$ as the set of sequences (m_α), $m_\alpha \in \mathcal{M}_\alpha \subset \mathcal{A}$, such that the series $\sum_\alpha m_\alpha^* m_\alpha$ is norm-convergent in $\mathcal{A}$. The dual Hilbert module $(\bigoplus_\alpha \mathcal{M}_\alpha)'$ is called the *ultra weak direct sum* of the modules $\mathcal{M}_\alpha$. For self-dual Hilbert W^*-modules we have the following structural result.

THEOREM 3.4.2 ([**100**]). *Let $\mathcal{M}$ be a self-dual Hilbert W^*-module over $\mathcal{A}$. Then there exists a set $\{p_\alpha\}$ of projections in $\mathcal{A}$ such that the module $\mathcal{M}$ is isomorphic to the ultra weak direct sum of the modules $p_\alpha \mathcal{A}$.*

PROPOSITION 3.4.3. *Let $\mathcal{N} \subset H_\mathcal{A}$ be a Hilbert submodule over a W^*-algebra $\mathcal{A}$. If $\mathcal{N}^\perp = 0$, then the dual module $\mathcal{N}'$ coincides with $H'_\mathcal{A}$.*

PROOF. Let $j : \mathcal{N} \longrightarrow H_\mathcal{A}$ be an inclusion of modules. The restriction of functionals $f \longmapsto f|_\mathcal{N}$, $f \in H'_\mathcal{A}$, defines the map $j' : H'_\mathcal{A} \longrightarrow \mathcal{N}'$ dual to j. If $f \in H'_\mathcal{A}$ is such that $f|_\mathcal{N} = 0$, then $f \perp \mathcal{N}$ and, by assumption, one has $f = 0$. Therefore the map j' is injective. Consider the composition of maps

$$i = j' \circ \widehat{\ } \circ j : \mathcal{N} \hookrightarrow H_\mathcal{A} \hookrightarrow H'_\mathcal{A} \longrightarrow \mathcal{N}'.$$

If $n \in \mathcal{N}$, then $i(n) = j'(\widehat{j(n)}) = \widehat{j(n)}|_{\mathcal{N}} = \widehat{n}$. Therefore the map i coincides with the inclusion map $\widehat{\ }: \mathcal{N} \hookrightarrow \mathcal{N}'$. The dual map (after identification of the first and second dual modules)

$$(i' \circ i)' = i' \circ i'' : \mathcal{N}' = \mathcal{N}'' \longrightarrow H'_{\mathcal{A}} \longrightarrow \mathcal{N}'$$

is an isomorphism, hence the map i' should be surjective; then the map j' is surjective as well. □

PROPOSITION 3.4.4. *Let $\mathcal{A}$ be a W^*-algebra and $\mathcal{R} \subset H_{\mathcal{A}}$ an $\mathcal{A}$-submodule without orthogonal complement, i.e., $\mathcal{R}^{\perp} = 0$ in $H_{\mathcal{A}}$. Then $\mathcal{R}' = H'_{\mathcal{A}}$.*

PROOF. It is sufficient to show that if the orthogonal complement to a submodule $\mathcal{R}$ in $H_{\mathcal{A}}$ is equal to zero, then the orthogonal complement to $\mathcal{R}$ in $H'_{\mathcal{A}}$ is equal to zero too. Let us assume the contrary. Suppose that there exists a functional $f \in H'_{\mathcal{A}}$ such that $f(r) = \langle f, r\rangle \neq 0$ for some $r \in \mathcal{R}$. Since the series $\sum_{i=1}^{\infty} f_i^* r_i$ is norm-convergent in $\mathcal{A}$, there is a number n such that $f^{(n)}(z) \neq 0$ for $f^{(n)} = (f_1, \dots, f_n, 0, \dots)$. But, as $f^{(n)} \in H_{\mathcal{A}}$, we get a contradiction. □

Even Hilbert C^*-modules over commutative W^*-algebras behave differently from Hilbert spaces. The following example shows what kind of pathology can occur.

Let $A = l_\infty$ be the W^*-algebra of bounded sequences and let $c_0 \subset l_\infty$ be its subalgebra of sequences vanishing at infinity. Consider the Hilbert C^*-module $\mathcal{M} = l_2(c_0) \oplus l_2(l_\infty)$ over A.

PROPOSITION 3.4.5 ([**85**]). *There is a submodule $\mathcal{N} \subset \mathcal{M}$ such that it coincides with its bi-orthogonal module, but is not complementable.*

PROOF. Put

$$\mathcal{N} = \{(f_1, f_2, f_3, \dots), (f_1, 2f_2, 3f_3, \dots)\},$$

where $f_k \in c_0$ for all $k \in \mathbf{N}$ and the series $\sum_{k=1}^{\infty} k^2 |f_k|^2$ is convergent in c_0. It is easy to see that the submodule $\mathcal{N} \subset \mathcal{M}$ is closed and coincides with its bi-orthogonal complement, $\mathcal{N}^{\perp\perp} = \mathcal{N}$. Let us show that $\mathcal{N}$ is noncomplementable. Assume the contrary: let $\mathcal{M} = \mathcal{N} \widetilde{\oplus} \mathcal{K}$ for some submodule $\mathcal{K} \subset \mathcal{M}$. Then there exists an A-linear (nonorthogonal) bounded projection $P : \mathcal{M} \to \mathcal{N}$. Since P commutes with the action of A, it can be written as a sequence $(P_k)_{k \in \mathbf{N}}$, where P_k is a projection acting on the k-th coordinate. More precisely, if $p_k \in A$ is the projection onto the k-th coordinate, then $P_k = p_k P$. Since P is bounded, the norms of the projections P_k should be uniformly bounded. But any projection P_k projects $l_2 \oplus l_2$ onto $l_2 \oplus N(l_2)$, where $N : l_2 \to l_2$ is an unbounded operator, $N(a_1, a_2, \dots) = (a_1, 2a_2, \dots)$. For each k there is a decomposition $\mathcal{M}_k = l_2 \oplus l_2 = \mathcal{N}_k \oplus \mathcal{K}_k$, where $\mathcal{K}_k = p_k \mathcal{K}$, $\mathcal{N}_k = p_k \mathcal{N}$. The uniform boundedness of the projections P_k implies that the complements $\mathcal{K}_k$ should consist of the pairs $\{(x, y) : x, y \in l_2\}$ such that $|x_n| > \varepsilon |y_n|$ for some $\varepsilon > 0$ independent of k.

Consider an arbitrary element $(f_1, f_2, \dots), (g_1, g_2, \dots) \in \mathcal{M}$, where $f_i \in c_0$, $g_i \in l_\infty$ for any i. If $\mathcal{M} = \mathcal{N} \widetilde{\oplus} \mathcal{K}$, then this element can be written as the sum

$$(h_1, h_2, \dots), (h_1, 2h_2, \dots) \widetilde{\oplus} (f_1 - h_1, f_2 - h_2, \dots), (g_1 - h_1, g_2 - 2h_2, \dots).$$

For a fixed n, we use a superscript to denote the number of an element in the sequence $f_n \in A = l_\infty$, $f_n = (f_n^k)_{k \in \mathbf{N}}$. For sufficiently large n the elements of the submodule $\mathcal{K}$ satisfy $|f_n^k - h_n^k| > \varepsilon |g_n^k - n h_n^k|$. But the left-hand side of this inequality vanishes as $k \to \infty$ (when n is fixed) and the right-hand side does not vanish, since $h_n \in c_0$ (it is sufficient to take the constant sequences $g_n^k = g_n \neq 0$). This contradiction shows that the submodule $\mathcal{N}$ is not complementable. □

3.5. Topological characterization of self-dual Hilbert W^*-modules

Let $\mathcal{M}$ be a Hilbert C^*-module over a W^*-algebra $\mathcal{A}$ and let $P \in \mathcal{A}_*$ be the set of normal states on $\mathcal{A}$. Let us define (see [**42**]) two topologies on $\mathcal{M}$ as follows. Denote by τ_1 the topology given by the system of seminorms $\varphi(\langle\cdot,\cdot\rangle)^{1/2}$, $\varphi \in P$, and denote by τ_2 the topology given by the system of seminorms $\varphi(\langle y,\cdot\rangle)$, $y \in \mathcal{M}$, $\varphi \in P$. In the case where $\mathcal{A} = \mathbf{C}$ and $\mathcal{M}$ is a Hilbert space, the topology τ_1 is the norm topology and the topology τ_2 coincides with the weak topology, so in general these two topologies do not coincide.

THEOREM 3.5.1 ([**42**]). *Let $\mathcal{M}$ be a Hilbert W^*-module. Then the following conditions are equivalent:*

(i) *the module $\mathcal{M}$ is self-dual;*
(ii) *the unit ball $B_1(\mathcal{M})$ is complete with respect to the topology τ_1;*
(iii) *the unit ball $B_1(\mathcal{M})$ is complete with respect to the topology τ_2.*

PROOF. Let us prove the implication (i) $\Rightarrow$ (ii). Assume, for this purpose, that the unit ball $B_1(\mathcal{M})$ is not complete with respect to the topology τ_1. Let us denote by L the linear span of the completion of $B_1(\mathcal{M})$ with respect to the topology τ_1. For the extensions of seminorms from $\mathcal{M}$ to L we use the same notation. By assumption, there exists an element $r \in L \setminus \mathcal{M}$ and a bounded net $\{y_\alpha\}$, $\alpha \in \Lambda$, such that for any $\varphi \in P$ and for any $\varepsilon > 0$ there exists some $\alpha \in \Lambda$, for which $\varphi(\langle r - y_\beta, r - y_\beta\rangle) < \varepsilon$ for all $\beta \in \Lambda$, $\beta \geq \alpha$. For arbitrary $x \in \mathcal{M}$ we have

$$\begin{aligned}|\varphi(\langle y_\beta, x\rangle) - \varphi(\langle y_\gamma, x\rangle)| &= |\varphi(\langle y_\beta - y_\gamma, x\rangle)| \\ &\leq \varphi(\langle x,x\rangle)^{1/2}\varphi(\langle y_\beta - y_\gamma, y_\beta - y_\gamma\rangle)^{1/2} \\ &\leq (2\varepsilon\, \varphi(\langle x,x\rangle))^{1/2}\end{aligned}$$

for all $\beta, \gamma \geq \alpha$. Therefore there exists the limit (with respect to the $\sigma(\mathcal{A},\mathcal{A}_*)$-topology)

$$R(x) = \lim_\alpha \langle y_\alpha, x\rangle \in \mathcal{A}$$

in the W^*-algebra $\mathcal{A}$ for each $x \in \mathcal{M}$. The inequality

$$|\varphi(\langle y_\beta, x\rangle)| \leq \|x\| \sup_\alpha\{\|y_\alpha\| : \alpha \in \Lambda\}$$

shows the continuity of the map $R : \mathcal{M} \longrightarrow \mathcal{A}$, $x \longmapsto R(x)$. It is clear that the map R is an $\mathcal{A}$-module map, so R is a functional on $\mathcal{M}$. By assumption, the module $\mathcal{M}$ is self-dual, so there exists an element $z \in \mathcal{M}$ such that $R(x) = \langle z, x\rangle$. Then $\lim_\alpha \langle y_\alpha, x\rangle = \langle z, x\rangle$ (where the limit is taken in the $\sigma(\mathcal{A},\mathcal{A}_*)$-topology). Therefore the net $\{y_\alpha\}$ converges to $z \in \mathcal{M}$ in the topology τ_1 and $r = z$, giving a contradiction to our assumption.

Let us now prove the implication (ii) $\Rightarrow$ (i). The extension of the inner product to the dual module $\mathcal{M}'$ we again denote by $\langle\cdot,\cdot\rangle$. It is easy to see that the ideal $\langle\mathcal{M},\mathcal{M}\rangle \in \mathcal{A}$ is norm dense in the ideal $\langle\mathcal{M}',\mathcal{M}'\rangle \in \mathcal{A}$, hence $\langle\mathcal{M},\mathcal{M}\rangle \in \mathcal{A} = \langle\mathcal{M}',\mathcal{M}'\rangle \in \mathcal{A}$. Let us first consider the case where a W^*-algebra $\mathcal{A}$ is σ-unital. Then there exists an exact normal state $\psi \in \mathcal{A}_*$ (see [**13**, Prop. 2.5.6]). Let $\{H, \pi, \xi\}$ be the cyclic representation associated with ψ. The vector $\xi \in H$ is simultaneously cyclic and separating. The linear space $\mathcal{M}$ with the inner product $(\cdot,\cdot) = \psi(\langle\cdot,\cdot\rangle)$ becomes a pre-Hilbert space and the map $\psi(f(\cdot)) : \mathcal{M} \longrightarrow \mathbf{C}$, where $f \in \mathcal{M}'$, becomes a linear functional on $\mathcal{M}$. Then one can find an element f_ψ in the completion of the space $\mathcal{M}$ with respect to the norm $\psi(\langle\cdot,\cdot\rangle)^{1/2}$ such that $(f_\psi, x) = \psi(f(x))$ for all $x \in \mathcal{M}$. This means that there exists a sequence (x_i), $x_i \in \mathcal{M}$, $i \in \mathbf{N}$, such that

$$\begin{aligned} 0 &= \lim_{i\to\infty} (x_i - f_\psi, x_i - f_\psi) = \lim_{i\to\infty} \psi(\langle \widehat{x}_i - f, \widehat{x}_i - f\rangle) \\ &= \lim_{i\to\infty} \left\| \left(\pi(\langle \widehat{x}_i - f, \widehat{x}_i - f\rangle)\right)^{1/2} \xi \right\|^2, \end{aligned}$$

where the image of the element x under the canonical inclusion $\mathcal{M} \subset \mathcal{M}'$ is denoted by $\widehat{x}$. Since the vector ξ is cyclic and separating, there exists the limit (in the $\sigma(\mathcal{A}, \mathcal{A}_*)$-topology on $\mathcal{A}$) $\lim_i \langle \widehat{x}_i - f, \widehat{x}_i - f\rangle = 0$ (see [**13**, Lemmas 2.5.38, 2.5.39]). Therefore $f \in \mathcal{M}$ and the module $\mathcal{M}$ is self-dual. Let us pass to the general case. If a W^*-algebra $\mathcal{A}$ is not σ-unital, then one can choose a directed set of projections $\{p_\alpha\}$, $\alpha \in \Lambda$, $p_\alpha \in \mathcal{A}$, such that for each $\alpha \in \Lambda$, the algebra $p_\alpha \mathcal{A} p_\alpha$ is a σ-unital W^*-algebra and $\lim_\alpha p_\alpha = 1$, where the limit is taken in the $\sigma(\mathcal{A}, \mathcal{A}_*)$-topology. As was proved earlier, the functional fp_α on the Hilbert $p_\alpha\mathcal{A}p_\alpha$-module $p_\alpha\mathcal{M}$ is an element of the module $p_\alpha\mathcal{M}$ for all $\alpha \in \Lambda$. But then there exists the limit (in the τ_1 topology) of the net $\{fp_\alpha\}$, which belongs to $\mathcal{M}$ and is equal to f, so self-duality of the module $\mathcal{M}$ is proved.

It remains to show the equivalence of conditions (ii) and (iii). If $B_1(\mathcal{M})$ is complete with respect to the topology τ_1, then $\mathcal{M}$ is self-dual. Therefore it is a dual Banach space with respect to the topology τ_2 (see Prop. 3.3.3), hence $B_1(\mathcal{M})$ is complete with respect to the topology τ_2. Let us assume now that $B_1(\mathcal{M})$ is complete with respect to the topology τ_2 and $\{x_\alpha\}$, $\alpha \in \Lambda$, is a bounded Cauchy τ_1-net. For all $y \in \mathcal{M}$, $\beta, \gamma \in \Lambda$, $\varphi \in P$ we have

$$|\varphi(\langle y, x_\beta\rangle) - \varphi(\langle y, x_\gamma\rangle)|^2 \le \varphi(\langle y, y\rangle)\varphi(\langle x_\beta - x_\gamma, x_\beta - x_\gamma\rangle). \tag{3.2}$$

As before, let L denote the linear span of the τ_1-completion of $B_1(\mathcal{M})$. There exists the limit $\lim_\alpha x_\alpha = t \in L$ in L (with respect to the topology τ_1). It follows from inequality (3.2) that $\{x_\alpha\}$ is a Cauchy net with respect to the topology τ_2 as well. But since the topology τ_2 is weaker than the topology τ_1, one has $L \subseteq \mathcal{M}$. Therefore the limits $\lim_i x_i$ with respect to the topologies τ_1 and τ_2 coincide and are equal to $t \in \mathcal{M}$, implying the completeness of $\mathcal{M}$ with respect to the topology τ_1. □

3.6. Fredholm operators over W^*-algebras

Throughout this section, let $\mathcal{A}$ denote an arbitrary W^*-algebra. We show here that the properties of Fredholm operators in the W^*-case are more similar to those of the usual Fredholm operators (over $\mathbf{C}$) than in the general C^*-case, which may be more pathological (see Example 3.6.3 below). We present here slightly modified results of [**48**].

LEMMA 3.6.1. *Let $\mathcal{M}$ be a self-dual Hilbert C^*-module over a W^*-algebra $\mathcal{A}$. For each closed submodule $\mathcal{N} \subseteq \mathcal{M}$ the biorthogonal set $\mathcal{N}^{\perp\perp} \subseteq \mathcal{M}$ is a Hilbert $\mathcal{A}$-submodule and is a direct summand of $\mathcal{M}$, as well as its orthogonal complement $\mathcal{N}^\perp$.*

PROOF. It is clear, by the definition of the orthogonal complement, that $\mathcal{N}^{\perp\perp} \subseteq \mathcal{M}$ is an $\mathcal{A}$-submodule. Consider the inclusion $i : \mathcal{N} \to \mathcal{M}$ and its adjoint map $i^* : \mathcal{M} = \mathcal{M}' \to \mathcal{N}'$. Since dual W^*-modules are self-dual, i^* admits an adjoint and, consequently, its kernel is an image of a selfadjoint projection, i.e., it has an orthogonal complement in $\mathcal{M}$. But

$$i^*(m) = 0 \Leftrightarrow i^*(m)(n) = 0 \,\forall n \in \mathcal{N} \Leftrightarrow \langle i(n), m\rangle = 0 \,\forall n \in \mathcal{N} \Leftrightarrow m \in \mathcal{N}^{\perp}.$$

□

LEMMA 3.6.2. *Let ϕ be a bounded $\mathcal{A}$-module morphism of a self-dual module $\mathcal{M}$. Then the kernel* $\mathrm{Ker}(\phi)$ *of the map ϕ is a direct summand in $\mathcal{M}$ and satisfies the equality* $\mathrm{Ker}(\phi) = \mathrm{Ker}(\phi)^{\perp\perp}$.

PROOF. By Proposition 3.3.4, the algebra $\mathrm{End}_{\mathcal{A}}(\mathcal{M}) = \mathrm{End}^*_{\mathcal{A}}(\mathcal{M})$ is a W^*-algebra. Therefore there exists a polar decomposition $\phi = US$, where S is positive and U is a partial isometry and $\mathrm{Ker}\,\phi = \mathrm{Ker}\,U$. There also exists a selfadjoint projection $p \in \mathrm{End}^*_{\mathcal{A}}(\mathcal{M})$ onto $\mathrm{Ker}\,\phi = \mathrm{Ker}\,U$. We are done. □

EXAMPLE 3.6.3. Note that the kernel of a bounded A-linear operator on a Hilbert A-module over an arbitrary C^*-algebra A need not be a direct summand. For example, consider the C^*-algebra $A = C([0,1])$ of all continuous functions on the segment [0,1] as a Hilbert A-module over itself equipped with the standard inner product $\langle a, b\rangle_A = a^*b$. Define the mapping ϕ_g by the formula $\phi_g(f) = g \cdot f$ for the fixed function

$$g(x) = \begin{cases} -2x+1 & \text{if } x \le 1/2, \\ 0 & \text{if } x \ge 1/2 \end{cases}$$

and for every $f \in A$. Then $\mathrm{Ker}(\phi_g)$ equals the Hilbert A-submodule and the (left) ideal $\{f \in A : f(x) = 0 \text{ for } x \in [0, 1/2]\}$ is not a direct summand of A, but coincides, nevertheless, with its bi-orthogonal complement in A.

COROLLARY 3.6.4. *Let $\phi : \mathcal{M} \to \mathcal{N}$ be a bounded A-linear mapping of self-dual modules. Then the kernel* $\mathrm{Ker}(\phi)$ *of ϕ is a direct summand of $\mathcal{M}$ and has the property* $\mathrm{Ker}(\phi) = \mathrm{Ker}(\phi)^{\perp\perp}$.

PROOF. Consider the self-dual Hilbert A-module $\mathcal{L} = \mathcal{M} \oplus \mathcal{N}$. The mapping ϕ can be identified with a bounded A-linear mapping ϕ' on $\mathcal{L}$ that acts on $\mathcal{M}$ as ϕ and on $\mathcal{N}$ as the zero operator. Since the kernel of ϕ' is a direct summand of $\mathcal{L}$, which contains $\mathcal{N}$ by Lemma 3.6.2, its orthogonal complement is a direct summand of $\mathcal{M}$. □

EXAMPLE 3.6.5. Let $\mathcal{A}$ be the set of all bounded operators $\mathcal{B}(H)$ on a separable Hilbert space H with the basis $\{e_i : i \in \mathbf{N}\}$. Denote by k the operator $k(e_i) = \lambda_i e_i$ for a sequence $\{\lambda_i : i \in \mathbf{N}\} \in c_o(\mathbf{R})$. Then the mapping $\phi_k : \mathcal{A} \to \mathcal{A}$, $\phi_k : a \to k \cdot a$ is a bounded $\mathcal{A}$-linear mapping on the right projective Hilbert $\mathcal{A}$-module $\mathcal{A}$. But its image is not a direct summand of this $\mathcal{A}$-module and is not even a Hilbert module, since all direct summands of $\mathcal{A}$ are of the form $p\mathcal{A}$ for some projection p of $\mathcal{A}$, and $k \cdot 1_{\mathcal{A}}$ should equal p. The image of ϕ_k is a subset of the set of all compact operators on H. Note that the mapping ϕ_k is not injective.

The following statement under some restrictions can be proved in the C^*-case as well [**71, 74**].

PROPOSITION 3.6.6. *Let $\mathcal{M}$ be a self-dual $\mathcal{A}$-Hilbert module and let $\{\mathcal{N}, \langle .,. \rangle\}$ be an arbitrary Hilbert $\mathcal{A}$-module. Suppose there exists an injective bounded module mapping $\alpha : \mathcal{M} \to \mathcal{N}$ such that $\alpha(\mathcal{M})^{\perp\perp} = \mathcal{N}$. Then the operator $\alpha(\alpha^*\alpha)^{-1/2}$ is a bounded module isomorphism between $\mathcal{M}$ and $\mathcal{N}$. In particular, these modules are isomorphic as Hilbert $\mathcal{A}$-modules.*

PROOF. The mapping α possesses an adjoint bounded $\mathcal{A}$-module mapping $\alpha^* : \mathcal{N} \to \mathcal{M}$ due to the self-duality of $\mathcal{M}$. Since $\alpha^*\alpha$ is a positive element of the C^*-algebra $\mathrm{End}_A(\mathcal{M})$ of all bounded (adjointable) module mappings on the Hilbert $\mathcal{A}$-module $\mathcal{M}$, its square root $(\alpha^*\alpha)^{1/2}$ is well defined by the series

$$(\alpha^*\alpha)^{1/2} = \lim_{n\to\infty} \|(\alpha^*\alpha)\|^{1/2} \left(\mathrm{id}_{\mathcal{M}} - \sum_{k=1}^{n} \lambda_k \left(\mathrm{id}_{\mathcal{M}} - \frac{(\alpha^*\alpha)}{\|(\alpha^*\alpha)\|} \right)^k \right),$$

where $\{\lambda_k\}$ are the coefficients of the Taylor series of the function $f(x) = \sqrt{1-x}$ at $x = 0$. Moreover, since

$$\langle (\alpha^*\alpha)^{1/2}(x), (\alpha^*\alpha)^{1/2}(x) \rangle = \langle \alpha(x), \alpha(x) \rangle,$$

due to the injectivity of α the mapping $(\alpha^*\alpha)^{1/2}$ has trivial kernel. Note that the range of $(\alpha^*\alpha)^{1/2}$ is τ_1-dense in $\mathcal{M}$. Indeed, for every $\mathcal{A}$-linear bounded functional $r(\cdot) = \langle ., y \rangle$ on the self-dual Hilbert $\mathcal{A}$-module $\mathcal{M}$ that maps the range of $(\alpha^*\alpha)^{1/2}$ into the origin one has

$$0 = \langle (\alpha^*\alpha)^{1/2}(x), y \rangle = \langle x, (\alpha^*\alpha)^{1/2}(y) \rangle$$

for every $x \in \mathcal{M}$. Hence $y = 0$, since $(\alpha^*\alpha)^{1/2}$ is injective and $x \in \mathcal{M}$ was chosen arbitrarily.

Now consider the mapping $\alpha(\alpha^*\alpha)^{-1/2}$ defined on $\mathcal{M}$. Since, by the assumptions on α, the range of $(\alpha^*\alpha)^{1/2}$ is τ_1-dense and the kernel is trivial, one concludes that its inverse unbounded module operator $(\alpha^*\alpha)^{-1/2}$ is τ_1-densely defined. One gets

$$\langle \alpha(\alpha^*\alpha)^{-1/2}(x), \alpha(\alpha^*\alpha)^{-1/2}(y) \rangle = \langle x, y \rangle$$

for every x, y from the (τ_1-dense) domain of $(\alpha^*\alpha)^{-1/2}$. Consequently the operator $\alpha(\alpha^*\alpha)^{-1/2}$ can be extended to a bounded isometric module operator on $\mathcal{M}$ by τ_1-continuity. Its range is τ_1-closed (i.e., is a self-dual direct summand of $\mathcal{N}$), hence it equals $\mathcal{N}$ by assumption. □

COROLLARY 3.6.7. *Let $\mathcal{M}$ be a self-dual Hilbert $\mathcal{A}$-module and let $\{\mathcal{N}, \langle .,. \rangle\}$ be an arbitrary Hilbert $\mathcal{A}$-module. Every injective module mapping from $\mathcal{M}$ into $\mathcal{N}$ is a Hilbert $\mathcal{A}$-module isomorphism between $\mathcal{M}$ and a direct summand of $\mathcal{N}$.*

PROPOSITION 3.6.8. *Let $\mathcal{M}$ and $\mathcal{N}$ be countably generated Hilbert $\mathcal{A}$-modules and let $F : \mathcal{M} \to \mathcal{N}$ be an $\mathcal{A}$-Fredholm operator. Then* $\mathrm{Ker}\, F$ *and* $(\mathrm{Im}\, F)^\perp$ *are projective finitely generated $\mathcal{A}$-submodules and* $\mathrm{index}\, F = [\mathrm{Ker}\, F] - [(\mathrm{Im}\, F)^\perp]$ *in $K_0(\mathcal{A})$.*

PROOF. Let $\mathcal{M} = \mathcal{M}_0 \hat{\oplus} \mathcal{M}_1$, $\mathcal{N} = \mathcal{N}_0 \oplus \mathcal{N}_1$ be the decompositions from the definition of $\mathcal{A}$-Fredholm operator,

$$F = \begin{pmatrix} F_0 & 0 \\ 0 & F_1 \end{pmatrix} : \mathcal{M}_0 \hat{\oplus} \mathcal{M}_1 \to \mathcal{N}_0 \oplus \mathcal{N}_1,$$

$F_0 : \mathcal{M}_0 \cong \mathcal{N}_0$, $F_1 : \mathcal{M}_1 \to \mathcal{N}_1$, where $\mathcal{M}_1$ and $\mathcal{N}_1$ are projective finitely generated modules. Let $x = x_0 + x_1$, $x_0 \in \mathcal{M}_0$, $x_1 \in \mathcal{M}_1$ and $F(x) = 0$, so $0 = F_0(x_0) +$

$F_1(x_1) \in \mathcal{N}_0 \oplus \mathcal{N}_1$. Thus $F_0(x_0) = 0$, $F_1(x_1) = 0$, so $x_0 = 0$ and $x \in \mathcal{M}_1$. Thus $\operatorname{Ker} F = \operatorname{Ker} F_1 \subset \mathcal{M}_1$. By Lemma 3.6.2, $\operatorname{Ker} F$ is a projective finitely generated $\mathcal{A}$-module and has an orthogonal complement. So, by Corollary 3.6.7,

$$F = \begin{pmatrix} F_0 & 0 & 0 \\ 0 & F_1' & 0 \\ 0 & 0 & 0 \end{pmatrix} : \mathcal{M}_0 \hat{\oplus} \mathcal{M}_1' \oplus \operatorname{Ker} F \to \left(\mathcal{N}_0 \oplus \overline{F(\mathcal{M}_1')}\right) \hat{\oplus} (\operatorname{Im} F)^\perp$$

and $\operatorname{index} F = [\operatorname{Ker} F] - [(\operatorname{Im} F)^\perp]$. □

The following example shows that the case of general Hilbert C^*-modules and injective mappings between them is more complicated.

EXAMPLE 3.6.9. Consider the C^*-algebra $A = C([0,1])$ of all continuous functions on the segment [0,1] as a self-dual Hilbert A-module over itself equipped with the standard A-valued inner product $\langle a, b\rangle_A = a^*b$. The mapping $\phi : f(x) \to x \cdot f(x)$ $(x \in [0,1])$ is an injective bounded module mapping. Its range has trivial orthogonal complement, but it is not norm closed and, consequently, not a direct summand of A. Nevertheless, the bi-orthogonal complement to the range of ϕ with respect to A equals A.

LEMMA 3.6.10. *Let $\mathcal{P}$ and $\mathcal{Q}$ be self-dual Hilbert $\mathcal{A}$-submodules of $\mathcal{M}$. Then $\mathcal{P} \cap \mathcal{Q}$ is a self-dual Hilbert $\mathcal{A}$-module and a direct summand of $\mathcal{M}$. Moreover, $\mathcal{P} + \mathcal{Q} \subseteq \mathcal{M}$ is a self-dual Hilbert $\mathcal{A}$-submodule. If $\mathcal{P}$ is projective and finitely generated, then the intersection $\mathcal{P} \cap \mathcal{Q}$ is projective and finitely generated too. If both $\mathcal{P}$ and $\mathcal{Q}$ are projective and finitely generated, then the sum $\mathcal{P} + \mathcal{Q}$ is projective and finitely generated too.*

PROOF. Let $p : \mathcal{M} = \mathcal{P} \oplus \mathcal{P}^\perp \to \mathcal{P}^\perp$ be the canonical orthogonal projection, which exists by Proposition 2.5.4. Let $p_Q = p : \mathcal{Q} \to \mathcal{P}^\perp$. Since $\mathcal{Q}$ is a self-dual Hilbert $\mathcal{A}$-module, p_Q admits an adjoint operator and $\operatorname{Ker} p_Q \subseteq \mathcal{Q}$ is a direct summand by Lemma 3.6.2. Consequently it is a self-dual Hilbert $\mathcal{A}$-submodule of $\mathcal{Q} \subseteq \mathcal{M}$. But $\operatorname{Ker} p_Q = \mathcal{P} \cap \mathcal{Q}$. To obtain the second assertion, one has to again apply the fact that every self-dual Hilbert $\mathcal{A}$-submodule is a direct summand by Proposition 2.5.4.

If $\mathcal{P}$ is projective and finitely generated, then any direct summand of $\mathcal{P}$ is projective and finitely generated, whence the remaining assertion follows. □

3.7. Dupré – Fillmore theorem for Hilbert modules over finite W^*-algebras

Let $\mathbf{A}$ be a finite W^*-algebra. Denote by τ the normal faithful finite trace on the algebra $\mathbf{A}$ normalized by $\tau(1) = 1$.

Let $\mathcal{M} \subset H_{\mathbf{A}}'$ be a Hilbert $\mathbf{A}$-submodule. By $\mathcal{M}^\perp$ we denote its orthogonal complement in the module $H_{\mathbf{A}}'$. It is known from Theorem 1.4.5 above [**36**] that if $\mathcal{M}$ is a finitely generated projective Hilbert $\mathbf{A}$-submodule in $H_{\mathbf{A}}'$, then it admits an orthogonal complement, $H_{\mathbf{A}}' = \mathcal{M} \oplus \mathcal{M}^\perp$. In the case where $\mathcal{M}$ is a submodule in the module $H_{\mathbf{A}}$ instead of $H_{\mathbf{A}}'$, the orthogonal complement to $\mathcal{M}$ in the module $H_{\mathbf{A}}$ is isomorphic to $H_{\mathbf{A}}$. However, in the general case, nothing is known about the relation between the modules $\mathcal{M}^\perp$ and $H_{\mathbf{A}}'$. The following theorem describes this relation in the case of finite W^*-algebras.

THEOREM 3.7.1 ([**81**]). *Let* $\mathbf{A}$ *be a* W^**-algebra that admits a decomposition into a direct integral of finite factors. If* $\mathcal{M}$ *is a finitely generated projective* $\mathbf{A}$*-submodule in* $H'_{\mathbf{A}}$*, then the modules* $\mathcal{M}^{\perp}$ *and* $H'_{\mathbf{A}}$ *are isomorphic.*

PROOF. The idea of the proof is contained in [**36**]. Let $g_1, \dots, g_n$ be generators of the module $\mathcal{M}$. Without loss of generality we can assume that the operators $\langle g_i, g_i\rangle \in \mathbf{A}$ are projections, $\langle g_i, g_i\rangle = p_i$. Let $\{e_m\}$ be the standard basis of the module $H_{\mathbf{A}} \subset H'_{\mathbf{A}}$. Fix $\varepsilon > 0$ and define the elements $e'_m \in \mathcal{M}^{\perp}$ by

$$e'_m = e_m - \sum_{i=1}^{n} g_i \langle g_i, e_m\rangle.$$

Then

$$\langle e'_m, e'_m\rangle = 1 - \sum_{i=1}^{n} \langle g_i, e_m\rangle^* \langle g_i, e_m\rangle.$$

The inequality

$$\tau\Big(\langle g_i, g_i\rangle\Big) \geq \tau\Big(\sum_m \langle g_i, e_m\rangle^* \langle g_i, e_m\rangle\Big)$$

shows that the series $\sum_m \tau\Big(\langle g_i, e_m\rangle^* \langle g_i, e_m\rangle\Big)$ is convergent, hence there exists a number m_0 such that for all $m > m_0$, one has

$$\tau\Big(\langle g_i, e_m\rangle^* \langle g_i, e_m\rangle\Big) < \frac{\varepsilon}{2n};$$

$$\tau\Big(\langle e'_m, e'_m\rangle\Big) > 1 - \frac{\varepsilon}{2}.$$

To proceed with the proof, the following lemma is required.

LEMMA 3.7.2. *If* $x \in H'_{\mathbf{A}}$, $\|x\| = 1$ *and if* $\tau(\langle x, x\rangle) > 1 - \frac{\varepsilon}{2}$*, then there exists a projection* $p \in \mathbf{A}$ *such that* $\tau(p) > 1 - \varepsilon$ *and the operator* $p\langle x, x\rangle p$ *is invertible in the* W^**-algebra* $p\mathbf{A}p$*.*

PROOF. Let us denote by $dP(\lambda)$ a projection-valued measure defined by the positive operator $a = \langle x, x\rangle \in \mathbf{A}$, $a = \int_0^1 \lambda dP(\lambda)$. Put

$$f(\lambda) = \begin{cases} 0, & \lambda \leq \lambda_0, \\ 1, & \lambda > \lambda_0, \end{cases}$$

where $\lambda_0 \in [0; 1]$. Then $f(a) = p$ is a projection. Denote $d\tau\left(P(\lambda)\right)$ by $d\mu(\lambda)$. It is a $\mathbf{C}$-valued measure on $[0; 1]$ and, by [**96**], one has

$$\tau(a) = \int_0^1 \lambda\, d\mu(\lambda).$$

Then

$$1 - \frac{\varepsilon}{2} < \tau(a) = \int_0^1 \lambda\, d\mu(\lambda) \leq \lambda_0\, \mu\left([0; \lambda_0)\right) + \mu\left([\lambda_0; 1]\right). \tag{3.3}$$

Since $P(1) = 1$, we obtain

$$\mu\left([0; \lambda_0)\right) + \mu\left([\lambda_0; 1]\right) = 1. \tag{3.4}$$

From (3.3) and (3.4) we get the inequality

$$\mu\left([\lambda_0; 1]\right) > 1 - \frac{\varepsilon}{2\left(1 - \lambda_0\right)}.$$

Choosing in an appropriate way a number $\lambda_0 \neq 0$, we get $\mu([\lambda_0; 1]) > 1 - \varepsilon$. It follows from the definition of the function $f(\lambda)$ that

$$\tau(p) = \tau(f(\lambda)) = \int_0^1 f(\lambda)\, d\mu(\lambda)$$
$$= \int_{\lambda_0}^1 d\mu(\lambda) = \mu([\lambda_0; 1]) > 1 - \varepsilon.$$

Now consider an operator $pap \in p\mathbf{A}p$. From the spectral theorem we obtain that

$$pap = \int_0^1 \lambda\, f(\lambda)\, dP(\lambda).$$

Therefore the spectrum of the element pap of the W^*-algebra $p\mathbf{A}p$ lies within the segment $[\lambda_0; 1]$, hence it is separated from zero, so this operator is invertible in $p\mathbf{A}p$. □

Now we can proceed with the proof of Theorem 3.7.1. Let

$$\left(p\langle e'_m, e'_m\rangle p\right)^{-1/2} = pbp = b \in p\mathbf{A}p$$

and $e''_m = e'_m \cdot b$. Then

$$\langle e''_m, e''_m\rangle = pbp\langle e'_m, e'_m\rangle pbp = P.$$

Let us now take an element $y \in \mathcal{M}^\perp$ such that $y \neq 0$, $\|y\| \leq 1$. Since the series $\sum_m \tau\Big(\langle e_m, y\rangle^* \langle e_m, y\rangle\Big)$ converges, for any $\varepsilon > 0$ we have

$$\tau\Big(\langle e''_m, y\rangle^* \langle e''_m, y\rangle\Big) = \tau\left(\langle e_m, y\rangle^* b^2 \langle e_m, y\rangle\right)$$
$$\leq \|b\|^2 \tau\Big(\langle e_m, y\rangle^* \langle e_m, y\rangle\Big) < \frac{\varepsilon^2}{2}$$

for sufficiently large numbers m.

Denote an operator $\langle e''_m, y\rangle^* \langle e''_m, y\rangle \in \mathbf{A}$ by c. Then $\tau(c) < \frac{\varepsilon^2}{2}$; $\|c\| \leq 1$ and $c \geq 0$. If $dQ(\lambda)$ is a projection-valued measure and if we denote a measure $d\tau(Q(\lambda))$ by $d\nu(\lambda)$, then $\int_0^1 \lambda\, d\nu(\lambda) < \frac{\varepsilon^2}{2}$. If $\lambda_1 \in [0; 1]$, then

$$\lambda_1 \cdot \int_{\lambda_1}^1 d\nu(\lambda) \leq \int_0^1 \lambda\, d\nu(\lambda) = \tau(c) < \frac{\varepsilon^2}{2},$$

hence $\int_{\lambda_1}^1 d\nu(\lambda) < \frac{\varepsilon^3}{2}$. Taking $\lambda_1 = \frac{\varepsilon^2}{2}$, we obtain that $\nu([\lambda_1; 1]) < \varepsilon$. Put

$$g(\lambda) = \begin{cases} 1, & \lambda < \lambda_1, \\ 0, & \lambda \geq \lambda_1. \end{cases}$$

Then the operator $q = g(c)$ is a projection such that $\tau(q) = \nu([0; \lambda_1)) > 1 - \varepsilon$ and

$$\|qcq\| \leq \lambda_1 = \frac{\varepsilon^2}{2}. \tag{3.5}$$

Denote by $p \vee q$ (resp. $p \wedge q$) the least upper (resp. greatest lower) bound of the projections p and q. Put $p' = p \wedge q$. Since (see [**123**])

$$\tau(p) + \tau(q) = \tau(p \vee q) + \tau(p \wedge q),$$

we have

$$\tau(p') = \tau(p) + \tau(q) - \tau(p \vee q) > (1 - \varepsilon) + (1 - \varepsilon) - 1 = 1 - 2\varepsilon,$$

due to the inequality $\tau(p \vee q) \leq \tau(1) = 1$. We get from the estimate (3.5) that

$$\|p'cp'\| \leq \frac{\varepsilon^2}{2}. \tag{3.6}$$

Put $e_m''' = e_m'' \cdot p'$. Then $\langle e_m''', e_m''' \rangle = p'$. Put also $y' = y + \varepsilon e_m''' \in \mathcal{M}^\perp$. The element y' can be decomposed into the sum of two mutually orthogonal summands: $y' = u + v$, where $u = y - e_m'''\langle e_m''', y\rangle$, $v = e_m'''\Big(\langle e_m''', y\rangle + \varepsilon \cdot 1\Big)$, $u, v \in \mathcal{M}^\perp$. Then

$$\langle y', y' \rangle = \langle u, u \rangle + \Big(\langle e_m''', y\rangle + \varepsilon p'\Big)^* \Big(\langle e_m''', y\rangle + \varepsilon p'\Big)$$

and

$$\begin{aligned} p'\langle y', y' \rangle p' &= p'\langle u, u \rangle p' + \Big(p'\langle e_m''', y\rangle p' + \varepsilon p'\Big)^* \Big(p'\langle e_m''', y\rangle p' + \varepsilon p'\Big) \\ &= p'\langle u, u \rangle p' + \Big(\langle e_m''', y\rangle p' + \varepsilon p'\Big)^* \Big(\langle e_m''', y\rangle p' + \varepsilon p'\Big). \end{aligned} \tag{3.7}$$

Since

$$\begin{aligned} \Big(\langle e_m''', y\rangle p'\Big)^* \langle e_m''', y\rangle p' &= p'\langle e_m'', y\rangle^* p' \langle e_m'', y\rangle p' \\ &\leq p'\langle e_m'', y\rangle^* \langle e_m'', y\rangle p' = p'cp', \end{aligned}$$

it follows from (3.6) that $\|\langle e_m'', y\rangle p'\| \leq \frac{\varepsilon}{\sqrt{2}} < \varepsilon$. Therefore the operator $\langle e_m'', y\rangle p' + \varepsilon p'$ is invertible in the W^*-algebra $p'\mathbf{A}p'$. Invertibility of the operator $p'\langle y', y'\rangle p'$ follows from (3.7). Let us consider a trace norm on the module $\mathcal{M}^\perp$ (and on $H'_{\mathbf{A}}$) defined by the formula

$$\|x\|_\tau = \tau\Big(\langle x, x\rangle\Big)^{1/2}.$$

The inequality

$$\tau\Big(\langle y' - y, y' - y\rangle\Big) = \tau\Big(\varepsilon^2 \langle e_m''', e_m''' \rangle\Big) = \tau(\varepsilon^2 p') < \varepsilon^2$$

implies the estimate $\|y' - y\|_\tau < \varepsilon$. Thus we have proved that the set of the elements of the module $\mathcal{M}^\perp$, for which there exists a projection p' such that $\tau(p') > \frac{1}{2}$ and the operator $p'\langle x, x\rangle p'$ is invertible in the algebra $p'\mathbf{A}p'$, is dense in the module $\mathcal{M}^\perp$ with respect to the trace norm. In particular the following statement holds.

LEMMA 3.7.3. *There exists an element $x \in \mathcal{M}^\perp$ such that*

$$\|x\|_\tau > \frac{1}{2} \quad \textit{and} \quad \|x\| \leq 1.$$

Let us come back to the proof of Theorem 3.7.1. Let $\{y_n\}$ be a sequence that contains each e_m an infinite number of times. Put $y = y_1 - \sum_{k=1}^n g_k\langle g_k, y_1\rangle$. Then for $\varepsilon_1 = 1$ one can find an element $y' \in \mathcal{M}^\perp$ such that $\|y'\| \leq 1$ and

$$\|y - y'\|_\tau < \varepsilon_1, \tag{3.8}$$

and a projection p_1 with $\tau(p_1) > \frac{1}{2}$ such that the operator $p_1\langle y', y'\rangle p_1$ is invertible in the algebra $p_1\mathbf{A}p_1$. Then, taking $h_1 = y'b'$, where $b' = \Big(p_1\langle y', y'\rangle p_1\Big)^{-1/2} \in p_1\mathbf{A}p_1$, we obtain from (3.8) the inequality

$$\operatorname{dist}_\tau(y, B_1(h_1\mathbf{A})) \leq \operatorname{dist}_\tau(y, h_1(b')^{-1}) = \operatorname{dist}_\tau(y, y') < \varepsilon_1,$$

where B_1 denotes the unit ball of the Hilbert module in the initial norm. Therefore

$$\operatorname{dist}_\tau(y_1, B_1(\operatorname{Span}_{\mathbf{A}}(\mathcal{M}, h_1))) < \varepsilon_1.$$

Further, taking $\varepsilon_2 = \frac{1}{2}$, we can find an element $h_2 \in (\mathrm{Span}_{\mathbf{A}}(\mathcal{M}, h_1))^\perp$ such that the operator $\langle h_2, h_2 \rangle = p_2$ is a projection, $\tau(p_2) > \frac{1}{2}$ and

$$\mathrm{dist}_\tau(y_2, B_1(\mathrm{Span}_{\mathbf{A}}(\mathcal{M}, h_1, h_2))) < \varepsilon_2.$$

Repeating this procedure and taking $\varepsilon_k = \frac{1}{k}$, we obtain a set of mutually orthogonal elements $h_i \in \mathcal{M}^\perp$ such that the operator $\langle h_i, h_i \rangle = p_i$ is a projection, $\tau(p_i) > \frac{1}{2}$ and for which one has

$$\mathrm{dist}_\tau(y_k, B_1(\mathrm{Span}_{\mathbf{A}}(\mathcal{M}, h_1, h_2, \dots, h_k))) < \varepsilon_k. \tag{3.9}$$

Such elements h_i generate a $\mathbf{A}$-module $\mathcal{N} \subset \mathcal{M}^\perp$ and it follows from (3.9) that

$$\mathrm{dist}_\tau(y_k, B_1(\mathcal{M} \oplus \mathcal{N})) < \frac{1}{k}.$$

Therefore the closure of $B_1(\mathcal{M} \oplus \mathcal{N})$ with respect to the trace norm contains the unit ball of the whole module $H'_{\mathbf{A}}$ and the closure of $B_1(\mathcal{N})$ with respect to the trace norm contains $B_1(\mathcal{M}^\perp)$.

Thus the constructed basis $\{h_i\}$ of the module $\mathcal{N}$ is not convenient to work with, since the elements $\langle h_i, h_i \rangle$ are not equal to one. Therefore we should pass to another basis. Denote by T the standard center-valued trace on $\mathbf{A}$ [**123**].

LEMMA 3.7.4 ([**48**]). *For any number C there exists n such that $T\left(\sum_{i=1}^n p_i\right) \geq C$.*

PROOF. Suppose that there is a normal state f on the center Z of the algebra $\mathbf{A}$ such that for some number C the inequality $(f \circ T)\left(\sum_{i=1}^\infty p_i\right) < C$ holds. Then one can find a central projection $z \in Z$ such that

$$T\left(\sum_{i=1}^\infty p_i z\right) < C. \tag{3.10}$$

Consider the W^*-algebra $z\mathbf{A}z$. Multiplication by z turns Hilbert modules over $\mathbf{A}$ into Hilbert modules over $z\mathbf{A}z$ and respects the orthogonality of submodules. Therefore $\mathcal{N}z \subset \mathcal{M}^\perp z$ and the set $B_1(\mathcal{N}z)$ is dense in $B_1(\mathcal{M}^\perp z)$ with respect to the trace norm $\|\cdot\|_{\tau_z}$ defined by the faithful trace τ_z on $z\mathbf{A}z$ induced by the trace τ. The inequality (3.10) means that for any $\varepsilon > 0$, one can find (substituting, if necessary, z by a smaller central projection) a number k such that the inequality $T\left(\sum_{i>k} p_i z\right) < \varepsilon$ holds. The module $\mathcal{N}z$ is decomposable, $\mathcal{N}z = \mathcal{L}_k \oplus \mathcal{R}_k$, where $\mathcal{L}_k$ is a $z\mathbf{A}z$-module generated by the elements $h_1 z, \dots, h_k z$ and the module $\mathcal{R}_k$ is the orthogonal complement to $\mathcal{L}_k$ in $\mathcal{N}z$. Since $B_1(\mathcal{N}z)$ is dense in $B_1(\mathcal{M}^\perp z)$, the set $B_1(\mathcal{R}_k)$ should be dense in $B_1((\mathcal{M}z \oplus \mathcal{L}_k)^\perp)$ with respect to the trace norm $\|\cdot\|_{\tau_z}$. Let $x = \sum_{i>k} h_i x_i \in B_1(\mathcal{R}_k)$. Estimate its trace norm:

$$\begin{aligned}
\|x\|_{\tau_z}^2 &= \tau_z\left(\sum_{i>k} x_i^* \langle h_i z, h_i z \rangle x_i\right) = \tau_z\left(\sum_{i>k} x_i^* p_i z x_i\right) \\
&= \tau_z\left(\sum_{i>k} p_i z x_i x_i^*\right) \leq \tau_z\left(\sum_{i>k} p_i z \cdot \|x_i\|^2\right) \\
&\leq \tau_z\left(\sum_{i>k} p_i z \cdot \|x\|^2\right) \leq \tau_z\left(\sum_{i>k} p_i z\right).
\end{aligned}$$

Since $T\left(\sum_{i>k} p_i z\right) < \varepsilon$, we obtain $\tau_z\left(\sum_{i>k} p_i z\right) < \varepsilon$. Therefore $\|x\|_{\tau_z}^2 < \varepsilon$ for all $x \in B_1(\mathcal{R}_k)$. But since the set $B_1(\mathcal{R}_k)$ is dense in $B_1((\mathcal{M}z \oplus \mathcal{L}_k)^\perp)$, one has the estimate $\|y\|^2_{\tau_z} < \varepsilon$ for all $y \in B_1((\mathcal{M}z \oplus \mathcal{L}_k)^\perp)$. On the other hand, if we apply Lemma 3.7.3 to the module $(\mathcal{M}z \oplus \mathcal{L}_k)^\perp$ instead of to the module $\mathcal{M}^\perp$, then we can find in $B_1((\mathcal{M}z \oplus \mathcal{L}_k)^\perp)$ an element y satisfying the inequality $\|y\|_{\tau_z} > \frac{1}{2}$. This contradiction completes the proof of the lemma. □

To complete the proof of Theorem 3.7.1, let us choose a projection q in the algebra $\mathbf{A}$ with properties

$$T(q) = \min(T(p_1 + p_2); 1) - T(p_1) \tag{3.11}$$

and $q \perp p_1$. It follows from (3.11) that $T(q) \leq T(p_2)$, hence there exists another projection q' equivalent to the projection q and such that $q' \leq p_2$. Equivalence of projections q and q' implies the existence of a unitary element $u \in \mathbf{A}$ such that $qu = uq'$. Put $r = h_2 q' u^* \in \mathcal{N}$. Then r is orthogonal to h_1 and

$$\langle r, r \rangle = uq' \langle h_2, h_2 \rangle q' u^* = uq' p_2 q' u^* = uq' u^* = q.$$

Put $h_1^{(1)} = h_1 + r$. Then

$$\langle h_1^{(1)}, h_1^{(1)} \rangle = \langle h_1, h_1 \rangle + \langle r, r \rangle = p_1 + q.$$

Note that $T(\langle h_1^{(1)}, h_1^{(1)} \rangle) = \min(T(p_1 + p_2); 1)$. Passing to the next element, h_3, we can obtain an element $h_1^{(2)}$ such that $\langle h_1^{(2)}, h_1^{(2)} \rangle$ is a projection and $T(\langle h_1^{(2)}, h_1^{(2)} \rangle) = \min(T(p_1 + p_2 + P_3); 1)$. By Lemma 3.7.4, repeating this procedure and increasing the value $T(\langle h_1^{(n)}, h_1^{(n)} \rangle)$, we construct an element h_1^∞ such that $\langle h_1^\infty, h_1^\infty \rangle = 1$. The orthogonal complement to h_1^∞ in $\mathcal{N}$ is generated by the elements $h_i q_i$, $i > 1$, where q_i are some projections. Applying the construction described above to these generators, we can construct, by induction, a set of elements h_i^∞ such that they satisfy the equality $\langle h_i^\infty, h_i^\infty \rangle = 1$ and generate the module $\mathcal{N}$. Therefore $\{h_i^\infty\}$ is a basis in $\mathcal{N}$ and the module $\mathcal{N}$ is isomorphic to $H_{\mathbf{A}}$.

It remains to prove that $\mathcal{N}' = \mathcal{M}^\perp$. Since the module $\mathcal{N} \subset \mathcal{M}^\perp$ is closed in the initial norm, the restriction $f|_\mathcal{N}$ of any $f \in (\mathcal{M}^\perp)'$ belongs to $\mathcal{N}'$. Note that the module $\mathcal{M}^\perp$ is self-dual, $(\mathcal{M}^\perp)' = \mathcal{M}^\perp$, because of the self-duality of the modules $H'_{\mathbf{A}}$ and $\mathcal{M}$. Suppose that $f|_\mathcal{N} = 0$. Since the submodule $\mathcal{N}$ is dense in $\mathcal{M}^\perp$ with respect to the trace norm, one has $f = 0$ on $\mathcal{M}^\perp$ due to the continuity of the map $f : \mathcal{M}^\perp \longrightarrow \mathbf{A}$ with respect to this norm because of the inequality

$$\tau\Big((f(y))^* f(y)\Big) \leq \|f\|^2 \cdot \tau\Big(\langle y, y \rangle\Big), \tag{3.12}$$

where $y \in \mathcal{M}^\perp$. This proves injectivity of the map $\mathcal{M}^\perp \longrightarrow \mathcal{N}'$. Now let $\phi \in \mathcal{N}'$. This functional can be extended up to a map from $\mathcal{M}^\perp$ to $\mathbf{A}$. If $\{y_n\} \subset \mathcal{N}$ is a sequence converging to the element $y \in \mathcal{M}^\perp$ with respect to the trace norm, then put $\phi(y) = \lim \phi(y_n)$. It is well defined due to the estimate (3.12) with ϕ instead of f. Thus, the $\mathbf{A}$-modules $\mathcal{M}^\perp$ and $\mathcal{N}'$ coincide and the theorem is proved since the modules $\mathcal{N}'$ and $H'_{\mathbf{A}}$ are isomorphic. □

CHAPTER 4

Reflexive Hilbert C^*-Modules

4.1. Inner product on bidual modules

For a Hilbert C^*-module $\mathcal{M}$ over a C^*-algebra A we define the *bidual* Banach right A-module $\mathcal{M}''$ as the set of all bounded A-homomorphisms from the dual module $\mathcal{M}'$ into A. It turns out that the A-valued inner product on $\mathcal{M}$ can always be extended to the bidual module, as opposed to the dual module, which admits an extension of the inner product only in the case of W^*-algebras.

Let $x \in \mathcal{M}$, $f \in \mathcal{M}'$. Put

$$\dot{x}(f) := f(x)^*.$$

The map $x \longmapsto \dot{x}$ is an isometric map from the A-module $\mathcal{M}$ into the A-module $\mathcal{M}''$:

$$\|\dot{x}\| = \sup\{\|f(x)\| : f \in \mathcal{M}',\ \|f\| \leq 1\} \leq \|f\|\,\|x\| \leq \|x\|\,;$$

$$\|\dot{x}\| \geq \frac{1}{\|\widehat{x}\|}\,\|\widehat{x}(x)\| = \frac{1}{\|\widehat{x}\|}\,\|\langle x, x\rangle\| = \|x\|\,.$$

For a functional $F \in \mathcal{M}''$ we define the functional $\widetilde{F} \in \mathcal{M}'$ by the formula

$$\widetilde{F}(x) := F(\widehat{x}).$$

Identifying $\mathcal{M}$ and $\widehat{\mathcal{M}} = \{\widehat{x} : x \in \mathcal{M}\} \subset \mathcal{M}'$, we obtain that $\widetilde{F}$ is the restriction of F onto $\mathcal{M} \subset \mathcal{M}'$. Note that $(\dot{x})^{\sim} = \widehat{x}$ for all $x \in \mathcal{M}$. It is clear that the map $F \longmapsto \widetilde{F}$ is an A-module map from $\mathcal{M}''$ to $\mathcal{M}'$ and $\|\widetilde{F}\| \leq \|F\|$. We shall check soon that this map is an isometry.

Let us define the inner product $\langle \cdot, \cdot \rangle : \mathcal{M}'' \times \mathcal{M}'' \longrightarrow A$ by the equality

$$\langle F, G\rangle := F(\widetilde{G}), \qquad F, G \in \mathcal{M}''. \tag{4.1}$$

It can be directly checked that $\langle F \cdot a, G\rangle = a^*\langle F, G\rangle$ for $a \in A$. Besides, for $x, y \in \mathcal{M}$, one has

$$\langle \dot{x}, \dot{y}\rangle = \dot{x}((\dot{y})^{\sim}) = \dot{x}(\widehat{y}) = (\widehat{y}(x))^* = \langle y, x\rangle^* = \langle x, y\rangle.$$

Therefore the inner product defined by formula (4.1) is an extension of the inner product on $\mathcal{M}$. To check the properties of an inner product we require the following construction.

Consider the right A-module $A \times \mathcal{M}$. In addition to the natural direct sum inner product $\langle \cdot, \cdot\rangle_0$ defined by the formula $\langle (A, x), (b, y)\rangle_0 = a^*b + \langle x, y\rangle$, where $a, b \in A$, $x, y \in \mathcal{M}$, let us consider another inner product on $A \times \mathcal{M}$. Take $f \in \mathcal{M}'$, $f \neq 0$, and a number $t > \|f\|$ and put

$$\langle (a, x), (b, y)\rangle_{f,t} := t^2 a^* b + a^* f(y) + f(x)^* b + \langle x, y\rangle. \tag{4.2}$$

Properties (iii) and (iv) of Definition 1.2.1 obviously hold. Let us check properties (i) and (ii). The first one is valid due to the inequality

$$\langle (a,x),(a,x)\rangle_{f,t} = t^2a^*a + a^*f(x) + f(x)^*a + \langle x,x\rangle$$

$$\geq t^2a^*a + a^*f(x) + f(x)^*a + \frac{1}{\|f\|^2} f(x)^*f(x) \tag{4.3}$$

$$\geq t^2a^*a + a^*f(x) + f(x)^*a + \frac{1}{t^2} f(x)^*f(x) \tag{4.4}$$

$$= \left(ta + \frac{1}{t}f(x)\right)^* \left(ta + \frac{1}{t}f(x)\right) \geq 0.$$

Suppose that $\langle (a,x),(a,x)\rangle_{f,t} = 0$. Then equality should be reached at each step in (4.3)–(4.4). Subtracting line (4.3) from line (4.4), we obtain

$$(\|f\|^{-2} - t^{-2})f(x)^*f(x) = 0,$$

hence $f(x) = 0$. Therefore $t^2a^*a + \langle x,x\rangle = 0$ and we can conclude that $a = 0$ and $x = 0$, so we have checked the validity of property (ii). Thus, the module $A \times \mathcal{M}$ with the inner product defined by formula (4.2) is a Hilbert A-module. We denote the Hilbert module $A \times \mathcal{M}$ equipped with this inner product by $(A \times \mathcal{M})_{f,t}$ and let the norm on this module corresponding to this inner product be denoted by $\|\cdot\|_{f,t}$. Note that $\|(0,x)\|_{f,t} = \|x\|$. For $x, y \in \mathcal{M}$, $a \in A$, we have

$$\|(f \cdot a + \widehat{x})(y)\| = \|a^*f(y) + \langle x,y\rangle\| = \|\langle (a,x),(0,y)\rangle_{f,t}\|$$
$$\leq \|(a,x)\|_{f,t} \cdot \|(0,y)\|_{f,t} = \|y\| \cdot \|(a,x)\|_{f,t}\,.$$

Therefore

$$\|(f \cdot a + \widehat{x})\| \leq \|(a,x)\|_{f,t}\,. \tag{4.5}$$

PROPOSITION 4.1.1 ([**101**]). *Let $\mathcal{N} \subset \mathcal{M}'$ be a submodule containing the module $\widehat{\mathcal{M}}$. Then the norm of any functional $\psi \in \mathcal{N}'$ satisfies the equality $\|\psi\| = \|\psi|_{\widehat{\mathcal{M}}}\|$.*

PROOF. Without loss of generality we assume that $\|\psi\| = 1$. Define the functional $f \in \mathcal{M}'$ by the formula $f(x) := \psi(\widehat{x})$, $x \in \mathcal{M}$. Then $\|f\| \leq 1$. It is necessary to prove the inverse inequality, $\|f\| \geq 1$. Take $g \in \mathcal{N}$ such that $\|g\| < 1$ and put $c = \psi(g) \in A$. For $a \in A$ and $x \in \mathcal{M}$ we have

$$\|ca + f(x)\| = \|\psi(g \cdot a + \widehat{x})\| \leq \|g \cdot a + \widehat{x}\| \leq \|(a,x)\|_{g,1}$$

(the last inequality follows from (4.5)), hence the map

$$f_c : A \times \mathcal{M} \longrightarrow A, \qquad (a,x) \longmapsto ca + f(x)$$

is a bounded modular map, $\|f_c\|_{(A\times\mathcal{M})'_{g,1}} \leq 1$. Therefore

$$F_c(a,x)^*f_c(a,x) \leq \langle (a,x),(a,x)\rangle_{g,1} \tag{4.6}$$

for all $a \in A$, $x \in \mathcal{M}$. From the estimate (4.6) we get

$$A^*c^*ca + a^*c^*f(x) + f(x)^*ca + f(x)^*f(x) \leq a^*a + a^*g(x) + g(x)^*a + \langle x,x\rangle.$$

Taking $a = -2g(x)$, we obtain

$$4g(x)^*c^*cg(x) + f(x)^*f(x) \leq \langle x,x\rangle + 2(g(x)^*c^*f(x) + f(x)^*cg(x)).$$

But since

$$G(x)^*c^*f(x) + f(x)^*cg(x) \leq g(x)^*c^*cg(x) + f(x)^*f(x),$$

so that

$$2g(x)^*c^*cg(x) \le \langle x, x\rangle + f(x)^*f(x) \le (1 + \|f\|^2)\langle x, x\rangle,$$

we have $\|g \cdot c^*\| \le \frac{1}{\sqrt{2}}(1 + \|f\|^2)^{1/2}$; thus

$$\|\psi(g \cdot c^*)\| = \|cc^*\| = \|c\|^2 \le \frac{1}{\sqrt{2}}(1 + \|f\|^2)^{1/2}.$$

The last inequality is valid for all $g \in \mathcal{N}$ with $\|g\| < 1$ and, since $\|\psi\| = 1$, the inequality $1 \le \frac{1}{\sqrt{2}}(1 + \|f\|^2)^{1/2}$ should be valid too, whence it follows that $\|f\| \ge 1$. Thus $\|f\| = \left\|\psi|_{\widehat{\mathcal{M}}}\right\| = 1$. □

Note that in the case where $\mathcal{N} = \mathcal{M}'$, Proposition 4.1.1 means that the map $F \longmapsto \widetilde{F}$ is an isometric inclusion $\mathcal{M}'' \subset \mathcal{M}'$.

PROPOSITION 4.1.2 ([**101**]). *For all $F \in \mathcal{M}''$ one has $\langle F, F\rangle \ge 0$ and $\|\langle F, F\rangle\| = \|F\|^2$.*

PROOF. Let $F \in \mathcal{M}''$, $F \ne 0$. Put $c = F(\widetilde{F})$, $D = \|F\|$. Let us show first that $D^2 \in \mathrm{Sp}(c)$. For $t > D$ consider the inner product $\langle\cdot,\cdot\rangle_{\widetilde{F},t}$ on the module $A \times \mathcal{M}$. Since $\|\widetilde{F} \cdot a + \widehat{x}\| \le \|(a, x)\|_{\widetilde{F},t}$ for all $(a, x) \in A \times \mathcal{M}$, the map

$$f_c : A \times \mathcal{M} \longrightarrow A, \qquad (a, x) \longmapsto F(\widetilde{F} \cdot a + \widehat{x}) = ca + \widetilde{F}(x)$$

is bounded and its norm does not exceed D (we mean here the norm defined by the inner product $\langle\cdot,\cdot\rangle_{\widetilde{F},t}$). Therefore

$$(ca + \widetilde{F}(x))^*(ca + \widetilde{F}(x)) \le D^2\langle(a, x), (a, x)\rangle_{\widetilde{F},t}. \tag{4.7}$$

Inequality (4.7) holds for all $t > D$ and, taking the limit $t \to D$, we obtain

$$(ca + \widetilde{F}(x))^*(ca + \widetilde{F}(x)) \le D^2(D^2a^*a + \widetilde{F}(x)^*a + a^*\widetilde{F}(x) + \langle x, x\rangle).$$

Taking $a = -D^{-2}\widetilde{F}(x)$, we get

$$\widetilde{F}(x)^*(D^{-2}c - 1)^*(D^{-2}c - 1)\widetilde{F}(x) \le D^2(-D^{-2}\widetilde{F}(x)^*\widetilde{F}(x) + \langle x, x\rangle),$$

hence

$$\widetilde{F}(x)^*((D^{-2}c - 1)^*(D^{-2}c - 1) + 1)\widetilde{F}(x) \le D^2\langle x, x\rangle.$$

Suppose that $D^2 \notin \mathrm{Sp}(c)$. Then one can find a number $\delta > 0$ such that

$$\widetilde{F}(x)^*(D^{-2}c - 1)^*(D^{-2}c - 1)\widetilde{F}(x) \ge \delta\widetilde{F}(x)^*\widetilde{F}(x)$$

for all $x \in \mathcal{M}$. But then

$$\widetilde{F}(x)^*\widetilde{F}(x) \le \frac{D^2}{1 + \delta}\langle x, x\rangle,$$

whence

$$D^2 = \left\|\widetilde{F}\right\|^2 \le \frac{D^2}{1 + \delta} < D^2.$$

The obtained contradiction shows that $D^2 \in \mathrm{Sp}(c)$. But since

$$\|c\| = \left\|F(\widetilde{F})\right\| \le \|F\| \left\|\widetilde{F}\right\| = \|F\|^2 = D^2,$$

we have $\|c\| = D^2$, hence $\|\langle F, F\rangle\| = \|F\|^2$ and $\|\langle F, F\rangle\| \in \mathrm{Sp}(\langle F, F\rangle)$. For an arbitrary element $a \in A$ we have

$$\|aca\| = \left\|a^*F(\widetilde{F} \cdot a)\right\| = \|\langle F \cdot a, F \cdot a\rangle\| \in \mathrm{Sp}(\langle F \cdot a, F \cdot a\rangle) = \mathrm{Sp}(aca).$$

The following lemma concludes the proof.

LEMMA 4.1.3 ([**101**]). *Let an element* $c \in A$ *be such that the inclusion* $\|aca\| \in \mathrm{Sp}(aca)$ *holds for any* $a \in A$, $a \geq 0$. *Then* $c \geq 0$. □

Proposition 4.1.2 shows that the inner product defined on $\mathcal{M}''$ satisfies conditions (i) and (ii) of Definition 1.2.1. It remains to check condition (iii). Notice that

$$\langle F+G, F+G\rangle \geq 0, \qquad \langle F+iG, F+iG\rangle \geq 0,$$

hence these expressions are selfadjoint. Then

$$(\langle F,G\rangle + \langle G,F\rangle)^* = \langle F,G\rangle + \langle G,F\rangle, \qquad -i(\langle F,G\rangle - \langle G,F\rangle)^* = i(\langle F,G\rangle - \langle G,F\rangle).$$

Therefore $\langle F,G\rangle = \langle G,F\rangle^*$. The module $\mathcal{M}''$ is a Hilbert A-module, since the operator norm on $\mathcal{M}''$ coincides (by Proposition 4.1.2) with the norm defined by the inner product. Thus, we have proved the following theorem.

THEOREM 4.1.4 ([**101**]). *The map* $\langle \cdot,\cdot\rangle : \mathcal{M}'' \times \mathcal{M}'' \longrightarrow A$ *defined by* $\langle F,G\rangle = F(\widetilde{G})$, $F, G \in \mathcal{M}''$, *is an A-valued inner product on* $\mathcal{M}''$. *The norm defined by this inner product coincides with the operator norm on* $\mathcal{M}''$. *The map* $F \longmapsto \widetilde{F}$ *is an isometric inclusion* $\mathcal{M}'' \subset \mathcal{M}'$.

Let us pass now to dual modules of higher order. Let $\Phi \in (\mathcal{M}'')'$. Define the functional $f_\Phi \in \mathcal{M}'$ by the formula

$$F_\Phi(x) := \Phi(\dot{x}), \quad x \in \mathcal{M}.$$

Further, let $f \in \mathcal{M}'$. Define $\Phi_f \in (\mathcal{M}'')'$ by the formula

$$\Phi_f(F) := (F(f))^*, \quad F \in \mathcal{M}''.$$

The maps $\Phi \longmapsto f_\Phi$ and $f \longmapsto \Phi_f$ are A-module morphisms. Consider their composition

$$\mathcal{M}' \longrightarrow \mathcal{M}''' \longrightarrow \mathcal{M}', \qquad F \longmapsto \Phi_f \longmapsto f_{\Phi_f}. \tag{4.8}$$

Since for any $x \in \mathcal{M}$ we have

$$F_{\Phi_f}(x) = \Phi_f(\dot{x}) = (\dot{x}(f))^* = f(x),$$

the composition (4.8) is the identity map, whence the map $\mathcal{M}' \to \mathcal{M}'''$ is an isometric inclusion and the map $\mathcal{M}''' \longrightarrow \mathcal{M}'$ is an epimorphism. Let us show that the last map is also monomorphic. For that purpose, apply Proposition 4.1.1 for the case $\mathcal{N} = \mathcal{M}''$. Let $\Phi \in \mathcal{N}' = \mathcal{M}'''$. Then the functional $f_\Phi \in \mathcal{M}'$ is the restriction on $\widehat{\mathcal{M}}$ of the functional Φ, $f_\Phi = \Phi|_{\widehat{\mathcal{M}}}$. Suppose that $f_\Phi = 0$. Then, by Proposition 4.1.1, we have $\|\Phi\| = \|\Phi|_{\widehat{\mathcal{M}}}\| = 0$, hence the map $\mathcal{M}''' \longrightarrow \mathcal{M}'$, $\Phi \longmapsto f_\Phi$, is monomorphic. Thus this map is an isometric isomorphism.

COROLLARY 4.1.5. *For a Hilbert C^*-module $\mathcal{M}$ one has* $(\mathcal{M}'')' = \mathcal{M}'$ *and* $(\mathcal{M}'')'' = \mathcal{M}''$.

So, the series of dual modules $\mathcal{M}, \mathcal{M}', \dots$ stabilizes on the third step and the inclusions

$$\mathcal{M} \subseteq \mathcal{M}'' = \mathcal{M}'''' \subseteq \mathcal{M}' = \mathcal{M}'''$$

are isometric. Thus $\mathcal{M}$ and $\mathcal{M}''$ are Hilbert C^*-modules unlike the module $\mathcal{M}'$, which is, generally speaking, only a Banach module. Let us illustrate by examples that all possible variants can be realized:

$$\mathcal{M} = \mathcal{M}'' = \mathcal{M}'; \qquad \mathcal{M} \neq \mathcal{M}'' = \mathcal{M}'; \qquad \mathcal{M} = \mathcal{M}'' \neq \mathcal{M}'; \qquad \mathcal{M} \neq \mathcal{M}'' \neq \mathcal{M}'.$$

(i) Let A be a unital C^*-algebra and let $\mathcal{M} = L_n(A)$ be a free A-module with n generators. Then the module $\mathcal{M}$ is self-dual, hence $\mathcal{M} = \mathcal{M}'' = \mathcal{M}'$.

(ii) Let A be a W^*-algebra. By Theorem 3.2.1, for any Hilbert A-module $\mathcal{M}$ its dual module $\mathcal{M}'$ is a self-dual Hilbert module, hence $\mathcal{M} \neq \mathcal{M}'' = \mathcal{M}'$.

(iii) [**44**] Let $A = C_0(0,1]$ be the (nonunital) C^*-algebra of functions on a segment $[0,1]$ vanishing at zero, $\mathcal{M} = A$. Then $\mathcal{M}' = C[0,1]$, $\mathcal{M}'' = C_0(0,1]$ and $\mathcal{M} = \mathcal{M}'' \neq \mathcal{M}'$.

(iv) [**44**] Consider the module $C_0(0,1)$ of functions on the segment $[0,1]$ vanishing at the end points, over the C^*-algebra $A = C_0(0,1]$. In this case one has $\mathcal{M}' = C[0,1]$, $\mathcal{M}'' = C_0(0,1]$, that is, $\mathcal{M} \neq \mathcal{M}'' \neq \mathcal{M}'$.

DEFINITION 4.1.6. A Hilbert C^*-module $\mathcal{M}$ is called *reflexive* if $\mathcal{M}'' = \mathcal{M}$.

Later we shall discuss other examples of reflexive Hilbert C^*-modules.

4.2. Ideals and bidual modules

Let $J \subset A$ be a right ideal in a C^*-algebra A. The ideal J is equipped with the natural structure of a pre-Hilbert A-module with the inner product $\langle x, y\rangle = x^*y$ for $x, y \in J$. In this section we describe the bidual module J''. Note that the inclusion map $J \subset A$ is bounded and A-linear, hence it defines an element of the dual module J'. Denote this element by $f_\iota \in J'$ and put

$$\widetilde{J} := \{F(f_\iota)^* : F \in J''\}.$$

Obviously $\widetilde{J}$ is a linear subspace of A. Since for $a \in A$ and for $F \in J''$ one has

$$F(f_\iota)^*a = (a^*F(f_\iota))^* = ((F \cdot a)(f_\iota))^*,$$

$\widetilde{J}$ is a right ideal in A. Since the equality $\hat{x}(f_\iota)^* = x$ holds for any $x \in J$, we get $J \subseteq \widetilde{J}$.

PROPOSITION 4.2.1 ([**101**]). *If $J \subset A$ is a right ideal in a C^*-algebra A, then*

(i) *the ideal $\widetilde{J}$ is closed;*

(ii) *the map $i : F \longmapsto F(f_\iota)^*$, where $F \in J''$, is an isomorphism of Hilbert A-modules $\widetilde{J}$ and J''.*

PROOF. The map i is obviously a homomorphism of A-modules. This map is also bounded, since $\|f_\iota\|_{J'} = 1$. Since the equality

$$(f_\iota \cdot x)(y) = x^* f_\iota(y) = x^*y = \widehat{x}(y)$$

holds for any $x, y \in J$, one has $f_\iota \cdot x = \widehat{x}$. Therefore

$$\begin{aligned}\|F\|_{J''} &= \|\widetilde{F}\|_{J'} = \sup\{\|\widetilde{F}(x)\| : x \in J, \|x\| \leq 1\} \\ &= \sup\{\|F(f_\iota \cdot x)\| : x \in J, \|x\| \leq 1\} \\ &= \sup\{\|x^*F(f_\iota)\| : x \in J, \|x\| \leq 1\} \leq \|F(f_\iota)\| = \|F(f_\iota)^*\|\end{aligned}$$

for any $F \in J''$. Hence $\|i(F)\| = \|F\|_{J''}$ and the map i is an isometry from J'' onto $\widetilde{J}$. Applying condition (ii) of Theorem 2.1.4 to the map i, we obtain that

$\langle i(F), i(F)\rangle \leq \langle F, F\rangle$. Changing the map i by its inverse, we get $\langle i(F), i(F)\rangle = \langle F, F\rangle$, i.e., $\langle F, F\rangle = \langle F(f_\iota)^*, F(f_\iota)^*\rangle = F(f_\iota)F(f_\iota)^*$. It is easy to derive from here that $\langle F, G\rangle = F(f_\iota)G(f_\iota)^*$ for $F, G \in J''$. Since the module J'' is complete with respect to the norm $\|\cdot\|_{J''}$, $\widetilde{J}$ is closed. □

LEMMA 4.2.2 ([**101**]). *In a C^*-algebra A consider the set of right ideals K, $J \subseteq K \subseteq A$, with the following property: any bounded A-module homomorphism from the ideal J to A admits a unique extension to an A-module homomorphism from K to A. There is a unique maximal element in this set of ideals and it coincides with $\widetilde{J}$.*

PROOF. Let K, $J \subseteq K \subseteq A$, be a right ideal in A such that any element $f \in J'$ can be uniquely extended to a map $\overline{f} \in K'$. For $a \in K$ define $F \in J''$ by the formula $F(f) = \overline{f}(a)^*$. Due to uniqueness, the map $\overline{f}_\iota$ has to be the inclusion map $K \subset A$, hence $F(f_\iota)^* = \overline{f}_\iota(a) = a$. Therefore $K \subseteq \widetilde{J}$. On the other hand, Corollary 4.1.5 shows that any element of J' admits a unique extension to an element in $\widetilde{J}'$. □

Note that $(\widetilde{J})^\sim = \widetilde{J}$ and if $J \subset K$ are two right ideals in a C^*-algebra A, then the inclusion $\widetilde{J} \subseteq \widetilde{K}$ holds. Indeed, for any $F \in J''$ the map $g \longmapsto F(g|_J)$ from K' to A belongs to the module K''.

Consider now the commutative case. Let X be a locally compact Hausdorff space and let $S \subset X$ be its subspace. Denote the C^*-algebra of continuous functions on X vanishing at infinity by $C_0(X)$, the C^*-algebra of bounded continuous functions on S by $C(S)$ and the ideal of functions vanishing on S by $C_0(X, S) \subset C_0(X)$. If $W \subset X$ is an open subset, then it is easy to see that there exists a unique maximal open set $\widetilde{W}$, $W \subseteq \widetilde{W} \subseteq X$, such that any function in $C(W)$ extends uniquely to a function in $C(\widetilde{W})$. There is an obvious inclusion $\widetilde{W} \subseteq \overline{W}$, where $\overline{W}$ denotes the closure of W. If X is a metric space, then one can show that $\widetilde{W} = W$ for any open subset W. This is not true in general. For example, if X is a Stone space and if $W \subset X$ is a dense open subset, then one can show that $\widetilde{W} = X$ [**34**].

Let $E \subset X$ be a closed subspace and let $J = C_0(X, E)$ be an ideal in $A = C_0(X)$. Put $U = X \setminus E$ and denote by Y the space of all bounded functions on X vanishing on E such that their restrictions onto the subset U are continuous on U. The set Y is equipped with the structure of an A-module (with the pointwise multiplication), and products of functions in Y by functions in J belong to J. For $g \in Y$ define $f_g \in J'$ by the formula $f_g(x) = \overline{g}x$, $x \in J$. The following proposition describes the dual module J'.

PROPOSITION 4.2.3 ([**101**]). *The map $g \longmapsto f_g$ is an isometric isomorphism between the Banach A-modules J' and Y.*

PROOF. The map $g \longmapsto f_g$ is obviously an isometric A-homomorphism. We have to show surjectivity of this map, i.e., to show that for any $f \in J'$ there is an element $g \in Y$ such that $f = f_g$. Without loss of generality we can assume that $\|f\|_{J'} \leq 1$. Then $\overline{f(x)}f(x) \leq \overline{x}x$ for all $x \in J$, i.e., $|f(x)(t)| \leq |x(t)|$ holds at each point $t \in X$. For any point $t \in U$ choose a function $x_t \in J$ that satisfies $x_t(t) = 1$. Define the function $h : U \longrightarrow \mathbf{C}$ by $h(t) = f(x_t)(t)$. Note that the estimate $|h(t)| \leq 1$ holds for all $t \in U$. For each $y \in J$ at any point $t \in U$ one has $(x_t y - y)(t) = 0$, hence $f(x_t y - y)(t) = 0$. Therefore

$$f(y)(t) = f(x_t y)(t) = f(x_t)(t)y(t) = h(t)y(t).$$

Since f maps J into A, the function hy should be continuous on the set U for any function $y \in J$. Now one can take $g = \overline{h}$. Then $f = f_{\overline{h}}$. □

After describing the dual module J' we can describe the bidual module $\widetilde{J} = J''$. Let χ_U be the characteristic function of the set U. Then the map f_{χ_U} is the inclusion map $J \subset A$ and $\widetilde{J} = \{F(\chi_U) : F \in Y'\}$.

PROPOSITION 4.2.4 ([**101**]). *The sets $\widetilde{J}$ and $C_0(X, X \setminus \overline{U})$ coincide.*

PROOF. Let $D \subset X$ be the intersection of the zero sets of all functions $F(\chi_U)$, $F \in Y'$. By the closedness of the ideal $\widetilde{J}$ one has $\widetilde{J} = C_0(X, D)$. Since $J \subseteq \widetilde{J}$, one has $D \subseteq E$. Put $V = X \setminus D$. It is easy to see that $V \subseteq \overline{U}$. Let us show that in fact the equality $V = \overline{U}$ holds, i.e., that any function in $C(U)$ extends uniquely to a function in $C(V)$ and that if $W \supset U$ is an open set such that any function in $C(U)$ uniquely extends to a function in $C(W)$, then $W \subseteq V$.

Note that for all $F \in Y'$, $g \in Y$, $t \in U$, the equality $F(g)(t) = F(\chi_U)(t)g(t)$ holds. Indeed, take a function $x \in J$ with $x(t) = 1$. Note that $gx \in A$. Since $gx = \chi_U gx$, one has

$$\begin{aligned} F(g)(t) &= F(g)(t)x(t) = F(gx)(t) = F(\chi_U gx)(t) \\ &= F(\chi_U)(t)(fx)(t) = F(\chi_U)(t)g(t). \end{aligned}$$

Let $h \in C(U)$; let us show that the function h can be extended to a bounded continuous function on V (uniqueness of such an extension follows from $V \subseteq \overline{U}$). Let $g \in Y$ be a function such that its restriction $g|_U$ coincides with h. Let W be a precompact set that lies in V together with its closure. Choose a functional $F \in Y'$ such that $F(\chi_U)|_W = 1$ and $\|F(\chi_U)\| = 1$ (by Proposition 4.2.1, $\|F(\chi_U)\| = \|F\|_{Y'}$). Define the function $\widetilde{h}$ on W by $\widetilde{h} = F(g)(t)$, $t \in W$. The function $\widetilde{h}$ is continuous and $|\widetilde{h}| \leq \|g\| = \|h\|$. For each $t \in W$ and for any net $\{t_\alpha\}$ with $t_\alpha \in U$ and $\lim_\alpha t_\alpha = t$, one has, for sufficiently large α,

$$F(g)(t_\alpha) = F(\chi_U)(t_\alpha)g(t_\alpha) = F(\chi_U)(t_\alpha)h(t_\alpha) = h(t_\alpha),$$

hence one can define the value $\widetilde{h}(t)$ at $t \in W$ as the limit $\widetilde{h}(t) = \lim_\alpha h(t_\alpha)$. Since any point $t \in V$ is contained in some neighborhood of the form W, one can use this argument to define the function $\widetilde{h}$ on the whole V.

Now let $W \supset U$ be an open set with the property that any continuous function in $C(U)$ extends uniquely to a function in $C(W)$. Let us show that then $W \subseteq V$. Take a point $t \in W$ and a function $a \in A$ such that $a(t) = 1$ and $a|_{X \setminus W} = 0$. For a function $g \in Y$, let $\widetilde{g}$ denote the unique bounded continuous extension of the function $g|_U$ to W. Define $F : Y \longrightarrow A$ by

$$F(g)(s) = \begin{cases} \widetilde{g}(s)a(s) & \text{if } s \in W, \\ 0 & \text{if } s \notin W. \end{cases}$$

It is clear that $F \in Y'$. Since $F(\chi_U)(t) = \widetilde{\chi}_U(t)a(t) = 1$, it is easy to see that $t \notin D$, hence $t \in V$, whence $W \subseteq V$ and the proof is finished. □

COROLLARY 4.2.5 ([**101**]). *Let E be a closed subset of a metric locally compact Hausdorff space X. Then the Hilbert $C_0(X)$-module $C_0(X, E)$ is reflexive.*

4.3. Reflexivity of Hilbert modules over $\mathcal{K}^+$

In this section we describe the reflexivity results from [**124**]. Let $\mathcal{K}$ be the C^*-algebra of compact operators on a separable Hilbert space H, and let $\mathcal{K}^+$ be the C^*-algebra of operators of the form $a = \lambda + K$, where $\lambda \in \mathbf{C}$, $K \in \mathcal{K}$.

THEOREM 4.3.1 ([**124**]). *Any countably generated Hilbert $\mathcal{K}^+$-module is reflexive.*

PROOF. By the stabilization Theorem 1.4.2, any countably generated Hilbert module is a direct summand in the standard module $l_2(\mathcal{K}^+)$. Therefore it is sufficient to prove reflexivity of the module $l_2(\mathcal{K}^+)$. Proposition 2.5.5 gives a description of the dual module:

$$l_2(\mathcal{K}^+)' = \left\{ f = (f_i) : f_i \in \mathcal{K}^+, \sup_N \left\| \sum_{i=1}^N f_i^* f_i \right\| < \infty \right\}.$$

LEMMA 4.3.2. *If $f = (f_i) \in l_2(\mathcal{K}^+)'$ and if $K \in \mathcal{K}$, then $f \cdot K = (f_i K) \in l_2(\mathcal{K}^+)$.*

PROOF. Since the operator K can be approximated by finite-dimensional operators, it is sufficient to prove the lemma in the case where K is finite dimensional. Notice that the operator $(f_iK)^*(f_iK) = K^* f_i^* f_i K$ is a positive operator, the kernel of which contains $\operatorname{Ker} K$ and the image of which is contained in $\operatorname{Im} K^*$. Since $\dim \operatorname{Im} K^*$ and $\operatorname{codim} \operatorname{Ker} K$ are finite, the norm convergence of the series $\sum_{i=1}^\infty (f_iK)^*(f_iK)$ follows from its weak convergence. □

Let $F \in l_2(\mathcal{K}^+)''$. Put $F_i = F(\widehat{e}_i)^* \in \mathcal{K}^+$, where $\{e_i\}$ is the standard basis of $l_2(\mathcal{K}^+)$, $\widehat{e}_i \in l_2(\mathcal{K}^+)'$. Since $\mathcal{M}'' \subset \mathcal{M}'$ for any Hilbert module $\mathcal{M}$, the sequence $\widetilde{F} = (F_i)$ is an element of the module $l_2(\mathcal{K}^+)'$. Let us prove that the series $\sum_{i=1}^\infty F_i^* F_i$ is convergent in the C^*-algebra $\mathcal{K}^+$ to the element $F(\widetilde{F}) = \langle F, F \rangle$.

Let $K \in \mathcal{K}$ be a finite-dimensional operator in H. By Lemma 4.3.2,

$$\langle F, F \rangle K = \langle F, F \cdot K \rangle = \sum_{i=1}^\infty F_i^* F_i K,$$

hence

$$K^* \langle F, F \rangle K = \sum_{i=1}^\infty K^* F_i^* F_i K$$

and

$$\sum_{i=1}^\infty (K^* F_i^* F_i K\xi, \xi) = (\langle F, F \rangle K\xi, K\xi)$$

for any $\xi \in H$, where $(\cdot, \cdot)$ is a scalar-valued inner product on H. Let $\eta \in B_1(H)$, where $B_1(H)$ is the unit ball of H. Choose an element $\xi \in H$ and a finite-dimensional operator K' such that $\eta = K'\xi$. Then

$$\sum_{i=1}^\infty (F_i\eta, F_i\eta) = (\langle F, F \rangle \eta, \eta).$$

Therefore

$$\left\| \sum_{i=1}^\infty (F_i^* F_i \eta, \eta) \right\| \leq \|\langle F, F \rangle\| \cdot \|\eta\|^2 \tag{4.9}$$

for any η.

LEMMA 4.3.3. *Let* $f = (k_i) \in l_2(\mathcal{K}^+)'$ *and* $k_i \in \mathcal{K}$ *for all* $i \in \mathbf{N}$. *Then* $F(f) \in \mathcal{K}$.

PROOF. Since F is continuous and the algebra $\mathcal{K}$ is closed, one can assume that all k_i are finite dimensional operators. Denote by $V_i \subset H$ the image of the operator k_i^*, $\dim V_i < \infty$. Let $H = H_1 \oplus H_2$ be an orthogonal decomposition of H into a sum of two closed infinite-dimensional subspaces. Assume first that each of the subspaces V_i lies in one of the subspaces H_1 or H_2. Let $F(f) = K + \lambda$, $K \in \mathcal{K}$, $\lambda \in \mathbf{C}$. Choose a compact operator k with an image $L \subset H_2$ such that $\dim L = \infty$. Replace by zeros those terms in the sequence $(k_1, k_2, \ldots)$, for which $V_i \subset H_1$, and denote the obtained sequence by $(k_1', k_2', \ldots)$. Then $(k_1 k, k_2 k, \ldots) = (k_1' k, k_2' k, \ldots)$ because of the equivalence of the conditions $k_i k = 0$ and $\operatorname{Im} k_i^* \perp \operatorname{Im} k$. Thus

$$F(k_1', k_2', \ldots)k = F(k_1' k, k_2' k, \ldots) = F(k_1 k, k_2 k, \ldots) = F(k_1, k_2, \ldots)k = (K + \lambda)k,$$

i.e.,

$$F(k_1', k_2', \ldots)|_L = (K + \lambda)|_L.$$

Since $\dim L = \infty$, one has $F(k_1', k_2', \ldots) = K' + \lambda$ for some $K' \in \mathcal{K}$. Interchanging the subspaces H_1 and H_2, one can construct a sequence $(k_1'', k_2'', \ldots)$ such that $F(k_1'', k_2'', \ldots) = K'' + \lambda$ for some $K'' \in \mathcal{K}$. Thus

$$(k_1, k_2, \ldots) = (k_1', k_2', \ldots) + (k_1'', k_2'', \ldots),$$

hence

$$F(k_1, k_2, \ldots) = K' + K'' + 2\lambda.$$

Therefore $\lambda = 0$. In the case of arbitrary subspaces V_i one can find finite-dimensional operators l_i, m_i, n_i such that

$$k_i + l_i = m_i + n_i, \quad (l_1, l_2, \ldots) \in l_2(\mathcal{K}^+), \quad (m_1, m_2, \ldots), (n_1, n_2, \ldots) \in l_2(\mathcal{K}^+)',$$

and the image of each of the operators m_i^* and n_i^* lies in one of the subspaces H_1, H_2. □

Put $F_i = K_i + \lambda_i$, $\langle F, F \rangle = K + \lambda$, where $K_i, K \in \mathcal{K}$, $\lambda_i, \lambda \in \mathbf{C}$. Then

$$F_i^* F_i = K_i^* K_i + \lambda_i K_i^* + \overline{\lambda}_i K_i + |\lambda_i|^2.$$

LEMMA 4.3.4. $\sum_{i=1}^{\infty} |\lambda_i|^2 = \lambda$.

PROOF. At first let us show that the series $\sum_{i=1}^{\infty} |\lambda_i|^2$ is convergent. Suppose the contrary. Put $M = \|\langle F, F \rangle\|$ and find a number N such that $\sum_{i=1}^{N} |\lambda_i|^2 > M + 1$. Choose $\varepsilon > 0$ to satisfy the estimate

$$\sum_{i=1}^{N} (1 + 2|\lambda_i|)\varepsilon \le \frac{1}{2}.$$

Choose, further, a vector $\xi \in H$ with $\|\xi\| = 1$ to satisfy the inequalities

$$\|K_i^* K_i \xi\| \le \varepsilon, \quad \|K_i \xi\| \le \varepsilon, \quad \|K_i^* \xi\| \le \varepsilon, \quad i = 1, \ldots, N.$$

Then the inequalities

$$\sum_{i=1}^{\infty} (F_i^* F_i \xi, \xi) \ge \sum_{i=1}^{N} (F_i^* F_i \xi, \xi) \ge M + \frac{1}{2}$$

contradict (4.9). So, $\sum_{i=1}^{\infty} |\lambda_i|^2 < \infty$. But this means that $(\lambda_1, \lambda_2, \dots) \in l_2(\mathcal{K}^+)$. Then

$$\begin{aligned} K + \lambda = \langle F, F\rangle &= F(F_1, F_2, \dots) = F(K_1, K_2, \dots) + F(\lambda_1, \lambda_2, \dots) \\ &= F(K_1, K_2, \dots) + \sum_{i=1}^{\infty} F_i^* \lambda_i = F(K_1, K_2, \dots) + \sum_{i=1}^{\infty} (K_i^* + \overline{\lambda}_i)\lambda_i \\ &= \left(F(K_1, K_2, \dots) + \sum_{i=1}^{\infty} K_i^* \lambda_i \right) + \sum_{i=1}^{\infty} |\lambda_i|^2 . \end{aligned}$$

But since $F(K_1, K_2, \dots) + \sum_{i=1}^{\infty} K_i^* \lambda_i \in \mathcal{K}$, we conclude that $\sum_{i=1}^{\infty} |\lambda_i|^2 = \lambda$. □

LEMMA 4.3.5 ([**124**]). *Let X be a compact Hausdorff space and let f_n, f, g_n, g be real-valued functions on X, $n \in \mathbf{N}$. Assume that the functions f_n, f are continuous, that the functions $f_n + g_n$ and g_n are nonnegative, that the function g is bounded, that the series $\sum_{n=1}^{\infty} g_n$ is uniformly convergent to the function g and that the series $\sum_{n=1}^{\infty}(f_n + g_n)$ is convergent pointwise to the function $f + g$. Then the series $\sum_{n=1}^{\infty}(f_n + g_n)$ is uniformly convergent to the function $f + g$.*

Let $X = B_1(H)$ be the unit ball of H endowed with the weak topology. Take $\xi \in X$ and put

$$f_n(\xi) = ((K_n^* K_n + \lambda_n K_n^* + \overline{\lambda}_n K_n)\xi, \xi), \quad g_n(\xi) = |\lambda_n|^2 \cdot \|\xi\|^2 ,$$
$$f(\xi) = (K\xi, \xi), \quad g(\xi) = \lambda \cdot \|\xi\|^2 .$$

All hypotheses of Lemma 4.3.5 are satisfied, so the series $\sum_{i=1}^{\infty}(F_i^* F_i \xi, \xi)$ uniformly converges on X to the function $(\langle F, F\rangle \xi, \xi)$. Therefore the series $\sum_{i=1}^{\infty} F_i^* F_i$ converges to $\langle F, F\rangle$ in the C^*-algebra $\mathcal{K}^+$, thus $F \in l_2(\mathcal{K}^+)$. □

4.4. Reflexivity of modules over $C(X)$

In this section we present results of the papers [**91, 125**].

DEFINITION 4.4.1 ([**125**]). A compact Hausdorff space X is called an L-space if, for any sequence $f_1, f_2, \dots$ of continuous functions on X converging pointwise to some bounded function f, the set of the continuity points of the function f is dense in X.

Examples of L-spaces are compact subsets of a finite-dimensional Euclidean space [**75**]. Infinite Stone spaces are not L-spaces.

Definition of L-spaces allows us to give a description of the bidual Hilbert modules over C^*-algebras of functions on such spaces.

THEOREM 4.4.2 ([**91, 125**]). *Let $A = C(X)$, where X is an L-space. Then any countably generated Hilbert A-module is reflexive.*

PROOF. By the stabilization Theorem 1.4.2, any countably generated Hilbert module is a direct summand in the standard module H_A. Therefore it is sufficient to prove reflexivity of the module H_A. Proposition 2.5.5 gives a description of the dual module,

$$H_A' = \left\{ f = (f_i(t)) : f_i(t) \in C(X), \sup_N \left\| \sum_{i=1}^{N} |f_i(t)|^2 \right\| < \infty \right\}.$$

Since the sequence of partial sums $\sum_{i=1}^N |f_i(t)|^2$ is monotone and bounded at each point $t \in X$, the corresponding series is convergent pointwise to a bounded function. By assumption, the set of the continuity points for the limit function $\sum_{i=1}^\infty |f_i(t)|^2$ is dense in X. Let us fix a continuity point $t_0 \in X$ and let $\lambda(t)$ be a continuous function on X equal to 1 at t_0. For $F \in H''_A$ we have

$$F(f\lambda) = F\left(\sum_{i=1}^N e_i f_i \cdot \lambda\right) + F\left(\left(f - \sum_{i=1}^N e_i f_i\right)\lambda\right), \tag{4.10}$$

where $\{e_i\}$ is the standard basis in $H_A \subset H'_A$. The element $\sum_{i=1}^N e_i f_i \lambda$ belongs to the module H_A, hence

$$F\left(\sum_{i=1}^N e_i f_i \cdot \lambda\right) = \sum_{i=1}^N F_i^* f_i \cdot \lambda,$$

where by $F_i = F_i(t)$ we denote $F(\widehat{e}_i)^*$. Let $\omega(f)$ be the least upper bound of oscillation of the function $\sum_{i=1}^N |f_i(t)|^2$. Then, obviously,

$$\omega(f) = \lim_{N\to\infty} \sup_{t\in X} \sum_{i=N+1}^\infty |f_i(t)|^2 = \lim_{N\to\infty} \left\| f - \sum_{i=1}^N e_i f_i \right\|^2.$$

Let us choose an arbitrary $\varepsilon > 0$ and a function $\lambda(t)$ such that the oscillation $\omega(f)$ on the support $\operatorname{Supp}\lambda(t)$ is less than ε^2. Then let us find N such that the inequality

$$\left\| \left(f - \sum_{i=1}^{N'} e_i f_i\right)\lambda \right\|^2 < 2\varepsilon^2 \tag{4.11}$$

holds for all $N' > N$. It follows from the inequalities (4.10), (4.11) that at the point t_0 (where $\lambda(t_0) = 1$) we have

$$|F(f)(t_0) - \sum_{i=1}^{N'} F_i^*(t_0) f_i(t_0)| \leq \|F\| \cdot 2\varepsilon$$

for all $N' > N$. Thus, the series $\sum_{i=1}^\infty F_i^*(t) f_i(t)$ converges to a continuous function $F(f)(t)$ at all the continuity points t_0 of the sum $\sum_{i=1}^\infty |f_i(t)|^2$. Moreover, we shall see soon that this series is uniformly convergent, hence it converges to a continuous function, which should coincide with the function $F(f)(t)$ due to the property L of X. The uniform convergence of this series would follow from that of the series

$$\sum_{i=1}^\infty F_i^*(t) F_i(t) \tag{4.12}$$

because of the Cauchy-Bunyakovskii inequality (Proposition 1.2.4). So we have to prove the uniform convergence of (4.12) on X.

Since $H''_A \subset H'_A$, the series (4.12) converges at each point and its sum coincides with the continuous function $\langle F, F\rangle(t) = F(\widetilde{F})(t)$ at each point of continuity. Denote by $E \subset X$ the set of the continuity points for the series (4.12). Now let t_0 be some point of discontinuity for the series (4.12). Without loss of generality (multiplying by some continuous function, if necessary) one can assume that there is a sequence of points $t_n \in E$ converging to the point t_0 and satisfying the conditions

$\langle F, F\rangle(t_n) = 1$ and $\sum_{i=1}^{\infty} |F_i(t_0)|^2 < 1$. Choose functions $h_i(t) \in C(X)$ to satisfy the conditions

(i) $h_i(t_0) = F_i(t_0)$; $|h_i(t)| \le |F_i(t)|$;
(ii) $h = (h_i) \in H_A$, i.e., the series $\sum_{i=1}^{\infty} |h_i(t)|^2$ uniformly converges.

Define the function $\lambda(t)$ on X by the following conditions:

(i) $0 \le \lambda(t) \le 1$; $\lambda(t_{2n}) = 0$, $\lambda(t_{2n+1}) = 1$;
(ii) outside the point t_0 the function $\lambda(t)$ is continuous.

Consider the element

$$f_i(t) = h_i(t) + \lambda(t)(F_i(t) - h_i(t)).$$

It can be easily checked that the functions $f_i(t)$ are continuous on the whole X. Moreover, since $|f_i(t)| \le |F_i(t)|$, we have $f = (f_i) \in H'_A$. Then

$$F(f)(t) = \langle F, f\rangle(t) = \sum_{i=1}^{\infty} F_i^*(t)h_i(t) + \lambda(t)\sum_{i=1}^{\infty} F_i^*(t)(F_i(t) - h_i(t)).$$

The series

$$\sum_{i=1}^{\infty} F_i^*(t)h_i(t) \tag{4.13}$$

is uniformly convergent since $h \in H_A$. The series

$$\sum_{i=1}^{\infty} F_i^*(t)(F_i(t) - h_i(t))$$

uniformly converges on the set $E \subset X$ to the function $\langle F, F - h\rangle(t)$, which is continuous on the whole X. Then, for $t \in E$, one has

$$\langle F, f\rangle(t) = \langle F, h\rangle(t) + \lambda(t)(\langle F, F\rangle(t) - \langle F, h\rangle(t))$$

and, for each t_n,

$$\langle F, f\rangle(t_n) = \langle F, h\rangle(t_n) + \lambda(t_n)(1 - \langle F, h\rangle(t_n)). \tag{4.14}$$

The functions $\langle F, f\rangle(t)$ and $\langle F, h\rangle(t)$ are continuous and, since the series (4.13) is uniformly convergent,

$$\langle F, h\rangle(t_0) = \sum_{i=1}^{\infty} |F_i(t_0)|^2 < 1.$$

But this contradicts the continuity of the function $\langle F, f\rangle(t)$, because, due to our choice of the function $\lambda(t)$, the limits of the right-hand side and the left-hand side of the equality (4.14) are different for even and for odd n. □

4.5. Hilbert modules related to conditional expectations of finite index

In this section we describe results of the paper [**47**]. Let $E : A \to B \subseteq A$ be a faithful conditional expectation on a C^*-algebra A, i.e., a projection of norm one onto a C^*-subalgebra B such that the condition $E(x^*x) = 0$, $x \in A$, implies $x = 0$.

A conditional expectation $E : A \to B \subseteq A$ is called a conditional expectation of *algebraically finite index* if there exists a set of elements $\{u_1, \dots, u_n\} \subset A$ such that $x = \sum_{i=1}^{n} u_i E(u_i^* x)$ for any $x \in A$. Then the element $\mathrm{Ind}(E) = \sum_{i=1}^{n} u_i u_i^*$ of the center of the C^*-algebra A is called the *index* of E. $\mathrm{Ind}(E)$ does not depend on a choice of the elements $\{u_1, \dots, u_n\} \subset A$, it is positive and $\mathrm{Ind}(E) \ge 1_A$ [**133, 5**]. It is shown in [**5**] that the algebraic finiteness of the index is equivalent to the property of A to be a projective finitely generated C^*-Hilbert module over the C^*-algebra B.

It is also interesting to consider another class of conditional expectations $E : A \to B \subseteq A$, for which there exists a number $K \geq 1$ such that the map $(K \cdot E - \mathrm{id}_A)$ is a positive operator on the C^*-algebra A (cf. [**45**]). Such conditional expectations are called conditional expectations of *finite index* and they have the following property:

PROPOSITION 4.5.1. *Let A be a C^*-algebra and let $E : A \to B \subseteq A$ be a conditional expectation, for which the set of fixed points coincides with B. Then there exists a finite number $K \geq 1$ such that the map $(K \cdot E - \mathrm{Id}_A)$ is positive iff E is faithful and the (right) pre-Hilbert B-module $\{A, E(\langle \cdot, \cdot \rangle_A)\}$ is complete with respect to the norm $\|E(\langle \cdot, \cdot \rangle_A)\|_B^{1/2}$ (where $\langle a, b \rangle_A = a^* b$ for $a, b \in A$).*

PROOF. If the algebra A is complete with respect to the norm $\|E(\langle \cdot, \cdot \rangle_A)\|_B^{1/2}$, then there exists a number K such that the inequality $K\|E(x^*x)\| \geq \|x^*x\|$ holds for each $x \in A$. For $a \in A$, $\varepsilon > 0$, put $x = a(\varepsilon + E(a^*a))^{-1/2}$. Note that

$$(\varepsilon + E(a^*a))^{-1/2} \cdot E(a^*a) \cdot (\varepsilon + E(a^*a))^{-1/2} \leq 1_A \,,$$

whence the inequality $K \cdot 1_A \geq (\varepsilon + E(a^*a))^{-1/2} \cdot a^*a \cdot (\varepsilon + E(a^*a))^{-1/2}$ follows. Multiplying both sides of this by $(\varepsilon + E(a^*a))^{1/2}$, we conclude that $K \cdot (\varepsilon + E(a^*a)) \geq a^*a$ for all $a \in A$, $\varepsilon > 0$. The inverse statement easily follows from the inequality $\|E(x)\| \geq K^{-1/2}\|x\|$, which is valid for any $x \in A$. □

Notice that, unlike algebraic finiteness of the index, in the case of conditional expectation of a finite index the Hilbert C^*-module $\{A, E(\langle \cdot, \cdot \rangle_A)\}$ can be infinitely generated.

Put

$$K(E) = \inf\{K \text{ such that } (K \cdot E - \mathrm{id}_A) \text{ is positive on } A\} \,.$$

We call $K(E)$ the *characteristic number* of the conditional expectation E.

Let X be a compact Hausdorff space with an action of a group G. Denote by $C^G(X)$ the C^*-algebra of G-invariant continuous functions on X and by $G_x = \{g \in G : gx = x\}$ the stabilizer of a point $x \in X$.

DEFINITION 4.5.2. A continuous action of a group G on X is called *uniformly continuous* if for each point $x \in X$ and for each neighborhood U_x of x there is a neighborhood V_x of the point x such that $g(V_x) \subseteq U_x$ for each $g \in G_x$.

Note that the continuous action of a compact group satisfies this definition.

DEFINITION 4.5.3. Let a group G act uniformly continuously on a compact Hausdorff space X in such a way that the cardinality of each orbit $\#Gx$ does not exceed some number $k \in \mathbf{N}$. Define the conditional expectation $E_G : C(X) \to E_G(C(X)) \subseteq C(X)$ by the formula

$$E_G(f)(x) = \frac{1}{\#(G/G_x)} \cdot \sum_{g_a \in G/G_x} f(g_a x), \qquad x \in X.$$

LEMMA 4.5.4 ([**47**]). *The conditional expectation E_G is well defined.*

THEOREM 4.5.5 ([**47**]). *Let a group G act uniformly continuously on a locally compact Hausdorff space X so that $k := \max\{\#(Gx) : x \in X\} < +\infty$. Then the characteristic number of the conditional expectation E_G satisfies the equality*

$$K(E_G) = k := \max_{x \in X} \#(Gx).$$

PROOF. Let $x \in X$ be an arbitrary point and let $k_x = \#Gx$. Then

$$KE_G(f)x = k\frac{1}{k_x} \cdot \sum_{g_a \in G/G_x} f(g_a x) \geq f(x),$$

where f is an arbitrary nonnegative function in $C(X)$. Assume that $K(E_G) < k$ and choose a point x such that $k_x > K(E_G)$. Then one can choose a sufficiently small neighborhood U_x of the point x such that $g_i U_x \cap g_j U_x = \emptyset$ $(i \neq j)$ for the set $\{g_1 = 1, g_2, \ldots, g_m\} = G/G_x$. Let f be a continuous nonnegative function with support inside U_x. Then

$$K(E_G)E_G(f)x = K(E_G)\frac{1}{k_x} \cdot \sum_{g_a \in G/G_x} F(g_a x) < f(x).$$

Contradiction with the definition of $K(E_G)$ completes the proof. □

THEOREM 4.5.6 (combination of Proposition 4.5.1, Lemma 4.5.4, and Theorem 4.5.5). *Suppose a discrete group G acts on a locally compact Hausdorff space X in such a way that $k := \max\{\#(Gx) : x \in X\} < +\infty$. Then E_G is well defined.*

If X is a normal space, then $K(E_G) = k$. Hence, by Proposition 4.5.1, $C(X)$ is a Hilbert C^*-module over $C_G(X)$.

THEOREM 4.5.7. *Let X be a compact Hausdorff space and let G be a group acting uniformly continuously on X. If all orbits of the action of G have the same finite number of points, then the conditional expectation*

$$E(f)(x) = \frac{1}{\#(Gx)} \sum_{g_i \in (G/G_x)_i} f(g_i x)$$

is well defined on $C(X)$, and the Hilbert $C^G(X)$-module $\{C(X), E(\langle \cdot, \cdot \rangle)\}$ is finitely generated and projective.

PROOF. The idea of the proof is contained in [**133**] and requires two technical lemmas.

LEMMA 4.5.8. *Let X be a compact Hausdorff space with a uniformly continuous action of a group G and let all orbits contain an equal finite number of points. Then, for any point $x \in X$ and any element $g \in G_x$, one can find an open neighborhood U_x of a point x, on which g acts identically.*

PROOF. Let us denote the cardinality of orbits $\#(Gx)$ by n. Let $x_1, \ldots, x_n \in X$ be the orbit of the point x and let $h_i \in G$ be elements such that $h_i x = x_i$. Choose $g_0 \in G_x$ and assume that each neighborhood U_x of the point x contains some point $y \in U_x$ such that $g_0 y \neq y$. Fix neighborhoods U_{x_i} of the points x_i satisfying the condition $U_{x_i} \cap U_{x_j} = \emptyset$ for $i \neq j$. Then we can find a neighborhood $V_x \subseteq U_x$ of the point x such that $h_i(V_x) \subseteq U_{x_i}$. Since the group G acts uniformly continuously, one can find a neighborhood $W_x \subseteq V_x$ of the point x such that $g(W_x) \subseteq V_x$ for each $g \in G_x$. If $y \in W_x$ and $g_0 y \neq y$, then the orbit Gy of this point contains at least $n+1$ different points $\{h_i y \in U_{x_i} : i = 1, \dots, n\} \cup \{y, g_0 y \in V_x\}$. The obtained contradiction proves the lemma. $\square$

LEMMA 4.5.9. *Under the hypotheses of Lemma 4.5.8, for any point $x \in X$ one can find a neighborhood V_x of this point such that the action of the subgroup G_x on V_x is trivial.*

PROOF. We should show that a neighborhood U_x from Lemma 4.5.8 can be chosen for all $g \in G_x$ simultaneously. For each $g \in G_x$ put

$$U_x(g) = \{y \in X : gy = y\}.$$

Suppose the contrary, i.e., that the set $\bigcap_{g \in G_x} U_x(g)$ does not contain a neighborhood of x. It means that any neighborhood U_x of x contains some point z such that for some $g_z \in G_x$ we have $g_z z \neq z$. Consider a neighborhood V_x of x and neighborhoods $\{U_{h_i x}\}$ for fixed representatives $\{h_i \neq e\} \in G$ of cosets in G/G_x such that their intersections are empty pairwise and $h_i(V_x) \subseteq U_{h_i x}$. As in the proof of Lemma 4.5.8, we find a neighborhood $W_x \subseteq V_x$ of the point x such that $g(W_x) \subseteq V_x$ for each $g \in G_x$. Put $U_x = \bigcup_{g \in G_x} G(W_x) \subseteq V_x$. It is a G_x-invariant open neighborhood of the point $x \in X$. The assumption $g_z z \neq z$ for some $z \in U_x$, $g_z \in G_x$ means that the orbit of the point z consists of at least $n+1$ points. $\square$

Let $U_x \subset V_x$ be a neighborhood of the point x such that the action of G_x on V_x is trivial. Then one can find a function $f_x \in C(X)$ such that $\operatorname{supp} f_x \subset V_x$ and $f_x|_{U_x} = 1$. For each $g \in G$ one has either $(gV_x) \cap V_x = \emptyset$ or $gV_x \equiv V_x$. Therefore

$$\alpha_g(f_x) \cdot f_x = \begin{cases} \alpha_g(f_x)^2 & \text{if } gx = x, \\ 0 & \text{if } gx \neq x, \end{cases}$$

where α denotes the action of G on functions, $\alpha_g(f)(x) = f(g^{-1}x)$. Consider a finite covering $\{U_{x_1}, \dots, U_{x_k}\}$ of the space X by sets of the above form. Put

$$v = \sum_{i=1}^{k} f_{x_i} \geq 1, \quad U_i = v^{-1/2}(f_{x_i})^{1/2} \in C(X).$$

Note that if we take one element g_i in each coset G/G_x, then, by Lemma 4.5.4, the map

$$E_G(f)(x) = \frac{1}{\#(Gx)} \sum_{i=1}^{n} f(g_i x)$$

is well defined for all $x \in X$, $f \in C(X)$ and it is a conditional expectation on $C(X)$. Moreover, one has

$$\sum_{i=1}^{k} u_i \cdot E_G(u_i^* f) = \sum_{i=1}^{k} \left(\frac{1}{n} \sum_{j=1}^{n}, U_i(x) u_i^*(g_j x) f(g_j x) \right) = f, \quad E_G(u_i^* u_j) = \delta_{i,j},$$

for each function $f \in C(X)$; hence the set $\{u_1, \dots, u_n\}$ is a basis of the Hilbert $C^G(X)$-module $\{C(X), E_G(\langle \cdot, \cdot \rangle)\}$. Therefore this Hilbert module is finitely generated and projective. $\square$

Theorem 4.5.7 generalizes results of [**133**] and shows that if all orbits consist of an equal finite number of points, then the corresponding conditional expectation is of algebraically finite index. Being finitely generated and projective, the Hilbert module $A = \{C(X), E(\langle \cdot, \cdot \rangle)\}$ is self-dual. In the case of a finite index (when

$K(E_G) < \infty$ and the pre-Hilbert module A is complete) we cannot expect this module to be self-dual. However, sometimes this module is reflexive, i.e., $A'' = A$, where A' is the dual Banach $C^G(X)$-module of bounded $C^G(X)$-homomorphisms from A into $C^G(X)$.

THEOREM 4.5.10 ([**47**]). *Let the group G act uniformly continuously on a compact Hausdorff space X. Suppose that all orbits consist of not more than n points, and that the number of points for which the length of their orbit is less than n is finite. Then the Hilbert $C^G(X)$-module $\{C(X), E_G(\langle\cdot,\cdot\rangle)\}$ is reflexive.*

PROOF. Describe first the dual Banach $C^G(X)$-module A'. Let $x_1, \ldots, x_m$ be the points with cardinality of orbits smaller than n. One can choose open neighborhoods $U_1, \ldots, U_m$ of these points in such a way that each neighborhood U_i is invariant with respect to the action of the subgroup G_{x_i} and if, for some $h \in G$, one has $hx_i = x_j$, then $hU_i = U_j$. Denote by Y the G-invariant compact set $X \setminus (U_1 \cup \cdots \cup U_m)$. Let $F \in A'$ be a $C^G(X)$-valued functional on the module A. Consider its restriction on the Hilbert $C^G(Y)$-module $\{C(Y), E_G(\langle\cdot,\cdot\rangle)\}$. For a function $g \in C(Y)$ we take its extension $\widetilde{g} \in C(X)$ and define $F|_Y$ by the formula $F|_Y(g) = F(\widetilde{g})|_Y$. This definition does not depend on the choice of an extension $\widetilde{g}$. If $Y' \supset Y$ is another compact G-invariant subspace not containing the points $x_1, \ldots, x_m$, then $(F|_{Y'})|_Y = F|_Y$. Since the orbit of each point of the set Y has the same cardinality, by Theorem 4.5.7, the $C^G(Y)$-module $\{C(Y), E_G(\langle\cdot,\cdot\rangle)\}$ is finitely generated and projective, hence self-dual. Denote by $C(X \setminus \{x_1, \ldots, x_m\})$ the set of continuous functions on the *noncompact* space $X \setminus \{x_1, \ldots, x_m\}$. The restriction onto this space defines the map

$$A' \longrightarrow C(X \setminus \{x_1, \ldots, x_m\}). \tag{4.15}$$

It is easy to verify that the map (4.15) is injective.

Let us study local properties of functionals from A' in a neighborhood of the points $x_1, \ldots, x_m$. Let x_0 be one of these points. It has a neighborhood U_{x_0} such that if $gx_0 = x_0$, then $gU_{x_0} = U_{x_0}$. The group G_{x_0} contains a normal subgroup G_0 of the elements that do not move points from the neighborhood U_{x_0}. Choose a representative g_i in each coset G/G_{x_0}. Then outside the point x_0 the action of the functional $F \in A'$ can be written as

$$F(f)(x) = \frac{1}{n} \sum_{i=1}^{n} F^*(g_i x) \cdot f(g_i x) \tag{4.16}$$

and this action can be continuously extended to the point x_0. Consider the orbit $x_0 = x^0, x^1, \ldots, x^{k-1}$ of the point x_0. Then one can write the sum (4.16) in the form

$$F(f)(x) = \frac{1}{n} \sum_{j=0}^{k-1} \left(\sum_{i: g_i x_0 = x^j} F^*(g_i x) \cdot f(g_i x) \right).$$

Passing to the limit (which exists by assumption), we obtain

$$F(f)(x_0) = \lim_{x \to x_0} \frac{1}{n} \sum_{j=0}^{k-1} \left(\sum_{i: g_i x_0 = x^j} F^*(g_i x) \cdot f(x^j) \right),$$

hence there exists (for $f \equiv 1 \in C(X)$) the limit

$$\frac{1}{n} \lim_{x \to x_0} \sum_{i: g_i x_0 = x^j} F(g_i x)$$

for any $x \in X \setminus \{x_1, \dots, x_m\}$. Recall that the function $F(x)$ is defined only outside the point x_0. If we would like the action F on the Hilbert $C^G(Y)$-module $\{C(Y), E_G(\langle \cdot, \cdot \rangle)\}$ to be of the form (4.16) on the whole X, it would be necessary to define the function $F(x)$ at the point x_0 by

$$F(x_0) = \frac{1}{n} \lim_{x \to x_0} \sum_{i: g_i x_0 = x^j} F(g_i x). \tag{4.17}$$

To complete the proof we need the following lemma.

LEMMA 4.5.11. *The module A' is isomorphic to the module of all bounded functions $F(x)$ on X that are continuous on $X \setminus \{x_1, \dots, x_m\}$ and satisfy the condition* (4.17).

PROOF. We have to show that the image of the monomorphism (4.15) consists of bounded functions. Suppose the contrary. Then there exists a point $\bar{x}$ such that $|F(\bar{x})| > n \cdot \|F\|$, where $\|F\|$ is the norm of F in the dual Banach $C^G(X)$-module A'. Moreover, one can choose a neighborhood $U_{\bar{x}}$ of the point $\bar{x}$ such that $U_{\bar{x}} \cap g_i U_{\bar{x}} = \emptyset$ for those elements of the group G, for which $g_i \bar{x} \neq \bar{x}$. Consider a function $f \in C(X)$ such that $f(\bar{x}) = 1$ and supp $f \subset U_{\bar{x}}$. Then it follows from the equality (4.16) that

$$F(f)(\bar{x}) = \frac{1}{n} F^*(\bar{x}) \cdot f(\bar{x}),$$

and the inequality

$$|F(f)(\bar{x})| = \frac{1}{n} |F^*(\bar{x})| \cdot |f(\bar{x})| = \frac{1}{n} |F(\bar{x})| > \|F\|$$

gives a contradiction. □

Having the above description of the dual module A', we can describe the bidual module A''. Since one has the canonical inclusion $A'' \subset A'$, and since the inner product on A can be naturally extended to an inner product on A'' (see Theorem 4.1.4) making it a Hilbert C^*-module, it is sufficient to verify to which functions in A' one can extend the $C^G(X)$-valued inner product. Consider a function F from $A' \cap A''$. Adding to it (if necessary) a continuous function from A, we can assume that $F(x_0) = 0$. Then the $C^G(X)$-valued inner product of F by itself is an element of $C^G(X)$ of the form

$$\langle F, F \rangle (x) = E(|F(x)|^2) = \frac{1}{n} \sum_{i: g_i x_0 = x^j} |F(g_i x)|^2 \tag{4.18}$$

for all $x \in X$. But since $\langle F, F \rangle (x_0) = 0$, it follows from the assumption $F \in A''$ that

$$\lim_{x \to x_0} F(g_i x) = 0$$

for each summand of the equality (4.18). Therefore we obtain from (4.17) that the function F is continuous at the point x_0, hence the module A is reflexive. □

Later in this section we will prove a final version of Theorem 4.5.10 due to V. Seregin.

CONSTRUCTION 4.5.12. Consider a uniformly continuous action of G on a Hausdorff compact space X with bounded cardinality of orbits. Let us develop the construction used in the proof of Theorem 4.5.10. For any point $x^0 \in X$ let U_{x^0} be a neighborhood of x that is invariant under the action of the stabilizer G_{x^0} of the given point and is disjoint from its own images under the action of the group elements not belonging to G_{x^0}. Such a neighborhood exists due to uniform continuity: indeed, take a neighborhood U'_{x^0} with all these properties except that of invariance; then choose, by the definition of uniform continuity, a neighborhood W_{x^0} with $G_{x^0}W_{x^0} \subset U'_{x^0}$ and put $U_{x^0} := G_{x^0}W_{x^0}$.

In this case one can take representatives g_j ($j = 1, \dots, \#Gx^0$) for the cosets for the subgroup G_{x^0} of G and the neighborhoods that appear in this construction are of the form $U_{x_j^0} = g_j U_{x^0}$, $x_j^0 = g_j(x^0)$, where $\{x_j^0\}$ is the orbit of x^0. Let $\varphi_{x^0} : X \to [0,1]$, $\varphi_{x^0} \in H$, be a continuous function such that

$$\varphi_{x_0}(x) = \begin{cases} 0, & x \notin U_{x^0}, \\ 1, & x = x^0. \end{cases}$$

Since the cardinality of orbits is bounded, E is well defined and one can assume that φ_{x^0} is G_{x^0}-invariant. Suppose $\varphi_j(x) = \varphi_{x^0}(g_j^{-1}x)$, so $\operatorname{supp}\varphi_j \subset U_{x_j^0}$. Then $\varphi_U := \sum_j \varphi_j$ is invariant, equals 1 on the orbit x^0 and is supported in $\bigcup_j U_{x_j^0}$. All these properties, except the invariance, are obvious. If $g \in G$ and $y \in U_{x_j^0}$, then $y = g_j z$, where $z \in U_{x^0}$, and $gg_j = g_k h$ for some g_k and $h \in G_{x^0}$, so

$$\varphi_U(gy) = \varphi_k(g_k hz) = \varphi_{x_0}(hz) = \varphi_{x_0}(z) = \varphi_j(y) = \varphi_U(y).$$

Similarly, we can choose a neighborhood $V_{x_j^0}$ with the same properties as $U_{x_j^0}$ and contained in $U_{x_j^0}$ together with its closure. Then, taking a function $\varphi^V_{x_0}$ instead of φ_{x_0} such that it equals 1 on V_{x^0} and has support in U_{x^0}, we obtain the corresponding function φ^V_U.

THEOREM 4.5.13 ([**118**]). *Let X be a compact Hausdorff space and let G be a discrete group acting uniformly continuously on X. If the cardinalities of all orbits are uniformly bounded by some number N, then the Hilbert $C_G(X)$-module $H = \{C(X), E_G(\langle \cdot, \cdot \rangle)\}$ is reflexive.*

PROOF. Denote by $B(X)$ the space of all bounded functions on X and by $\widetilde{H}$ the space of all maps from H to $B(X)$.

Choose U_x and φ_x for any $x \in X$ as in Construction 4.5.12 and define the maps $\xi : H' \to B(X)$ and $\eta : B(X) \to \widetilde{H}$ as follows:

$$\xi(F)(x) = \#Gx \cdot (F(\varphi_x))(x) \tag{4.19}$$

and

$$\eta(f)(h)(x) = \frac{1}{\#Gx} \sum_{x_i \in Gx} f(x_i)h(x_i), \tag{4.20}$$

where $F \in H'$ and $\#Gx$ denotes the number of points in the orbit of x, as before. Let us verify that ξ is well defined, i.e., that the function $f = \xi(F)$ is bounded: $|f(x)| \le N|F(\varphi_x)(x)| \le N\|F\|$.

Let us prove that $\eta(\xi(F)) = F$. Take an element $h \in C(X)$ and a point $x \in X$. Then

$$\begin{aligned}
F(h)(x) &= F\left(h - \sum_{x_i \in Gx} h(x_i)\varphi_{x_i} + \sum_{x_i \in Gx} h(x_i)\varphi_{x_i}\right)(x) \\
&= F\left(h - \sum_{x_i \in Gx} h(x_i)\varphi_{x_i}\right)(x) + F\left(\sum_{x_i \in Gx} h(x_i)\varphi_{x_i}\right)(x) \\
&= F(h_1)(x) + \sum_{x_i \in Gx} h(x_i) F(\varphi_{x_i})(x) \\
&= F(h_1)(x) + \sum_{x_i \in Gx} h(x_i) F(\varphi_{x_i})(x_i) \\
&= F(h_1)(x) + \sum_{x_i \in Gx} h(x_i) \frac{1}{\#Gx} f(x_i) F(h_1)(x) + \eta(\xi(F))(h)(x),
\end{aligned}$$

where $h_1 = h - \sum_{x_i \in Gx} h(x_i)\varphi_{x_i}$. Here we have taken into account that $F(\varphi_{x_i}) \in C^G(X)$. Note that the function h_1 is identically 0 on the orbit of x. Then $F(h_1)$ also is identically 0 on the orbit of x. Indeed, choose (as in Construction 4.5.12) the neighborhoods U_{x_j} so small that $|h_1| < \varepsilon$ holds on their union. Then $\|h_1 \varphi_U\| < \varepsilon$ for the corresponding invariant function φ_U and for arbitrary $\varepsilon > 0$ we have

$$\|F(h_1)(x)\| = \|F(h_1)\varphi_U(x)\| = \|F(h_1\varphi_U)(x)\| < \varepsilon \|F\|.$$

Since x is an arbitrary point, this implies $\eta \circ \xi = \mathrm{Id}$ and $\eta(\xi(H')) = H'$.

Let us prove now that the map η is injective. Suppose that $f \neq 0 \in B(X)$, so we have $f(x^0) \neq 0$ for some point $x^0 \in X$. Choose a continuous function $h \in H$ satisfying the following conditions:

$$h(x^0) = 1, \qquad h(x_i^0) = 0 \quad (x_i^0 \in Gx^0,\, x_i^0 \neq x^0).$$

Then $\eta(f)(h)(x^0) = \frac{1}{\#Gx_0} f(x^0) \neq 0$, hence $\eta(f) \neq 0$ and $\operatorname{Ker} \eta = 0$. Therefore

$$\xi : H' \cong \operatorname{Im} \xi \subset B(X), \qquad \eta : \operatorname{Im} \xi \cong H' \subset \widetilde{H}. \tag{4.21}$$

LEMMA 4.5.14. *The image of ξ consists of functions $f \in B(X)$ such that for any point $x_0 \in X$ there exists a neighborhood U_{x_0} of x_0 such that*

$$f(x_0) = \lim_{x \to x_0} \frac{\#Gx_0}{\#Gx} \sum_{x_i \in Gx \cap U_{x_0}} f(x_i). \tag{4.22}$$

PROOF. Let f satisfy condition (4.22). To prove that $f \in \operatorname{Im} \xi$ it suffices to show that, in this case, $\eta(f) \in H'$. Then (4.21) immediately implies that $\xi\eta(f) = f$.

The property $\eta(f) \in H'$ means that $\eta(f)(h)$ is a continuous invariant function for any function $h \in H$ and that the correspondence $h \mapsto \eta(f)(h)$ is bounded and linear over $C^G(X)$. Invariance follows at once from (4.20) and continuity follows from the calculation below, where we denote the points of the orbit of x (resp. x_0) by x_i (resp. by x_i^0) and we suppose that $x \in U_{x^0}$ (see Construction 4.5.12), so $Gx \subset \bigcup_i U_{x_i^0}$ and one can take $\varepsilon_{x_0}^j := \max_{x_i \in U_{x_j^0}} |h(x_i) - h(x_j^0)|$ $(j = 1, \dots, \#Gx^0)$.

Note that $\varepsilon^j_{x_0} \to 0$ as the neighborhood U_{x^0} decreases (as well as $U_{x^0_j}$). Then

$$|\eta(f)(h)(x) - \eta(f)(h)(x^0)| = \left| \frac{1}{\#Gx} \sum_i f(x_i)h(x_i) - \frac{1}{\#Gx^0} \sum_j f(x^0_j)h(x^0_j) \right|$$

$$= \left| \frac{1}{\#Gx} \sum_j \left(\sum_{x_i \in Gx \cap U_{x^0_j}} f(x_i)h(x_i) \right) - \frac{1}{\#Gx^0} \sum_j f(x^0_j)h(x^0_j) \right|$$

$$\leq \sup |f(y)| \cdot \max |\varepsilon^j_{x^0}|$$

$$+ \left| \frac{1}{\#Gx} \sum_j \left(\sum_{x_i \in Gx \cap U_{x^0_j}} f(x_i)h(x^0_j) \right) - \frac{1}{\#Gx^0} \sum_j f(x^0_j)h(x^0_j) \right|$$

$$\leq \|f\| \cdot \max |\varepsilon^j_{x^0}| + \left| \frac{1}{\#Gx^0} \sum_j h(x^0_j) \left(\left[\frac{\#Gx^0}{\#Gx} \sum_{x_i \in Gx \cap U_{x^0_j}} f(x_i) \right] - f(x^0_j) \right) \right|.$$

Hence, by (4.22) (with x^0_j instead of x^0),

$$\lim_{x \to x^0} |\eta(f)(h)(x) - \eta(f)(h)(x^0)| = 0.$$

Additivity of $\eta(f)$ follows immediately from (4.20). Now let $r \in C^G(X)$ be an invariant function. Then

$$\eta(f)(hr)(x) = \frac{1}{\#Gx} \sum_{x_i \in Gx} f(x_i)h(x_i)r(x_i) = \frac{1}{\#Gx} \sum_{x_i \in Gx} f(x_i)h(x_i)r(x)$$
$$= \eta(f)(h)(x)r(x)$$

and $\eta(f)$ is a $C^G(X)$-functional.

For any $h \in H$ one has

$$\eta(f)(h)(x) = \frac{1}{\#Gx} \sum_{x_i \in Gx} f(x_i)h(x_i) \leq \frac{1}{\#Gx} \sum_{x_i \in Gx} |f(x_i)| \cdot |h(x_i)| \leq \|f\| \cdot \|h\|,$$

so $\eta(f)$ is a continuous functional. Hence, $\eta(f) \in H'$.

Conversely, let us prove that condition (4.22) is satisfied for any function f from the image of ξ. For f, there exists a functional $F \in H'$ such that $f = \xi(F)$. Namely, $F = \eta(f)$. Consider an arbitrary point x^0 and choose for it a neighborhood $V_{x^0} \subset U_{x^0}$ (see 4.5.12). Take the function $\varphi^V_{x^0}$ (see 4.5.12) and evaluate F at this function. Then, for $x \in V_{x^0}$, the function defined by the formula

$$F(\varphi^V_{x^0})(x) = \eta(f)(\varphi^V_{x^0}) \frac{1}{\#Gx} \sum_i f(x_i)\varphi^V_{x^0}(x_i) = \frac{1}{\#Gx} \sum_{x_i \in Gx \cap V_{x^0}} f(x_i)$$

is continuous because it lies in the image of the functional F. This is equivalent to the condition

$$\lim_{x \to x^0} \frac{1}{\#Gx} \sum_{x_i \in Gx \cap V_{x^0}} f(x_i) = \frac{1}{\#Gx^0} \sum_{z \in Gx^0 \cap V_{x^0}} f(z) = \frac{1}{\#Gx^0} f(x^0).$$

Thus, (4.22) holds for f and Lemma 4.5.14 is proved. □

Let us continue the proof of the theorem and let us give a functional description of H''. Define the maps $\nu : H'' \to B(X)$ and $\mu : B(X) \to \widetilde{\mathbf{H}}$, where $\widetilde{\mathbf{H}}$ is the space of all maps from H'' to $B(X)$, by the formulas

$$\nu(\mathbf{F})(x) = \#Gx \cdot (\mathbf{F}(\mathbf{f}_x))(x)$$

and

$$\mu(f)(\mathbf{h})(x) = \frac{1}{\#Gx} \sum_{x_i \in Gx} f(x_i)h(x_i),$$

where $\mathbf{f}_x = \eta(\varphi_x) \in H'$, $x \in X$, $h = \xi\mathbf{h}$, so $\mathbf{h} = \eta(h)$. Note that

$$\|h\| = \sup_x |\xi(\mathbf{h})(x)| = \sup_x \#Gx \cdot |((\mathbf{h})(\varphi_x))(x)| \le N \cdot \|\mathbf{h}\|. \tag{4.23}$$

Then

$$\begin{aligned}
\mathbf{F}(\mathbf{h})(x) &= \mathbf{F}\left(\mathbf{h} - \sum_{x_i \in Gx} h(x_i)\mathbf{f}_{x_i} + \sum_{x_i \in Gx} h(x_i)\mathbf{f}_{x_i}\right)(x) \\
&= \mathbf{F}\left(\mathbf{h} - \sum_{x_i \in Gx} h(x_i)\mathbf{f}_{x_i}\right)(x) + \mathbf{F}\left(\sum_{x_i \in Gx} h(x_i)\mathbf{f}_{x_i}\right)(x) \\
&= \mathbf{F}(\mathbf{h}_1)(x) + \sum_{x_i \in Gx} h(x_i)\mathbf{F}(\mathbf{f}_{x_i})(x) \\
&= \mathbf{F}(\mathbf{h}_1)(x) + \sum_{x_i \in Gx} h(x_i)\mathbf{F}(\mathbf{f}_{x_i})(x_i) \\
&= \mathbf{F}(\mathbf{h}_1)(x) + \sum_{x_i \in Gx} h(x_i)\frac{1}{\#Gx}\nu(\mathbf{F})(x_i) = \mathbf{F}(\mathbf{h}_1)(x) + \mu\nu(\mathbf{F})(\mathbf{h})(x),
\end{aligned}$$

where $\mathbf{h}_1 = \mathbf{h} - \sum_{x_i \in Gx} h(x_i)\mathbf{f}_{x_i}$ and invariance of the functions $\mathbf{F}(\mathbf{f}_{x_i})$ was used. Further, for any point $y \in Gx$, one has

$$\begin{aligned}
h_1(y) := \xi\mathbf{h}_1(y) &= h(y) - \sum_{x_i \in Gx} h(x_i)\xi(\mathbf{f}_{x_i})(y) \\
&= h(y) - \sum_{x_i \in Gx} h(x_i)\varphi_{x_i}(y) \\
&= h(y) - h(y) = 0,
\end{aligned}$$

so the function $\mathbf{F}(\mathbf{h}_1)$ equals 0 on this orbit.

Indeed, choose, as in Construction 4.5.12, the neighborhoods U_{x_j} so small that $|h_1| < \varepsilon$ holds on their union. Then $\|h_1\varphi_U\| < \varepsilon$ for the corresponding invariant function φ_U and

$$|[\mathbf{h}_1\varphi_U](f)(x)| = |(\eta(h_1\varphi_U))(f)(x)| = \left|\frac{1}{\#Gx}\sum_{x_i \in Gx}(h_1\varphi_U)(x_i)f(x_i)\right| < \varepsilon\|f\|.$$

So we obtain

$$\|\mathbf{F}(\mathbf{h}_1)(x)\| = \|\mathbf{F}(\mathbf{h}_1)\varphi_U(x)\| = \|\mathbf{F}(\mathbf{h}_1\varphi_U)(x)\| < \varepsilon\|\mathbf{F}\|$$

for arbitrary $\varepsilon > 0$, hence

$$\mu\nu = \mathrm{Id}\,. \tag{4.24}$$

Let us prove now that μ is injective. Suppose that $f \neq 0 \in B(X)$, so, for some point $x^0 \in X$, we have $f(x^0) \neq 0$. Then

$$\begin{aligned}\mu(f)(\mathbf{f}_{x^0})(x^0) &= \frac{1}{\#Gx_0} \sum_{x_j^0 \in Gx^0} \xi(\mathbf{f}_{x^0})(x_j^0) f(x_j^0) \\ &= \frac{1}{\#Gx_0} \sum_{x_j^0 \in Gx^0} \varphi_{x^0}(x_j^0) f(x_j^0) = \frac{1}{\#Gx_0} f(x^0) \neq 0,\end{aligned}$$

hence $\mu(f) \neq 0$ and $\operatorname{Ker} \mu = 0$.

Suppose that $f \in \operatorname{Im} \nu$. Then there is a functional $\mathbf{F} \in H''$ such that $f = \nu(\mathbf{F})$. Since μ is injective, by (4.24), one has $\mathbf{F} = \mu(f)$. For an arbitrary point x choose V_x and φ_x^V (see Construction 4.5.12). The function $\mathbf{F}(\eta(\varphi_x^V))$ is continuous and one has

$$\begin{aligned}\mathbf{F}(\eta(\varphi_x^V))(y) = \mu(f)(\eta(\varphi_x^V))(y) &= \frac{1}{\#Gy} \sum_{y_i \in Gy} f(y_i) \xi \eta(\varphi_x^V)(y_i) \\ &= \frac{1}{\#Gy} \sum_{y_i \in Gy} f(y_i)(\varphi_x^V)(y_i) = \frac{1}{\#Gy} \sum_{y_i \in Gy \cap V_x} f(y_i)\end{aligned}$$

in the neighborhood V_x of x. As was already shown, condition (4.22) is satisfied for f. Thus f is bounded due to the estimate

$$|f(x)| \leq \frac{N}{\#Gx} |f(x)| = N |\mathbf{F}(\mathbf{f}_x)(x)| \leq N \|\mathbf{F}\|.$$

Hence $f \in \xi(H')$. Moreover, this implies the continuity of $\nu : H' \to \nu(H') \subset B(X)$, when the last space is equipped with the uniform norm. On the other hand, taking into account (4.23), one has

$$\begin{aligned}\|\mathbf{F}\| = \sup_{\|\mathbf{h}\|=1} \|\mathbf{F}(\mathbf{h})\| = \sup_{x, \|\mathbf{h}\|=1} \|\mu(f)(\mathbf{h})(x)\| = \sup_{x, \|\mathbf{h}\|=1} \left\| \frac{1}{\#Gx} \sum_{x_i \in Gx} f(x_i) h(x_i) \right\| \\ \leq \sup_{\|\mathbf{h}\|=1} N \cdot \|f\| \cdot \|h\| \leq N^2 \|\mathbf{h}\|.\end{aligned}$$

Hence, H' is isomorphic to the image of $\nu(H')$ equipped with the uniform norm.

Let us prove that each $f \in \operatorname{Im} \nu$ is continuous. Let $x \in X$. One can assume that $f = 0$ on Gx. Indeed, it is clear from the definition of ν that f is invariant and $\nu\mu(1) = 1$ for the identity element, since otherwise we can change f by a constant. It follows from above that one can define $\mathbf{F} := \mu(f) \in H''$ and $F = \eta(\overline{f}) \in H'$. Then the function

$$\mathbf{F}(F)(x) = \frac{1}{\#Gx} \sum_{x_j \in Gx} |f(x_j)|^2$$

is continuous and invariant.

Let x^a be a directed set of points of X that converges to x^0. Then

$$\lim_a \frac{1}{\#Gx^a} \sum_{x_j^a \in Gx^a} |f(x_j^a)|^2 = \frac{1}{\#Gx^0} \sum_{x_i^0 \in Gx^0} |f(x_i^0)|^2 = 0,$$

hence

$$\lim_a \sum_{x_j^a \in Gx^a} |f(x_j^a)|^2 = 0.$$

In particular, $\lim_a f(x^a) = 0$. Therefore the function f is continuous at an arbitrary point x. Hence $f \in C(X)$ and $H = H''$. $\square$

Let us complete this section with some statements, which are not directly related to reflexivity. Details can be found in [**131**].

THEOREM 4.5.15. *Let G be a discrete group acting uniformly continuously on a compact separable Hausdorff connected topological space X. Suppose that either the averaging over orbits defines the structure of a self-dual Hilbert $C_G(X)$-module on $C(X)$ and $C_G(X)$ is* MI *(cf.* 2.5.8) *or $C(X)$ is a finitely generated projective module over $C_G(X)$. Then all the orbits have the same finite cardinality.*

DEFINITION 4.5.16. An action of a group G on a metric space X is called (Lyapunov) *stable* if for any $\varepsilon > 0$ and any $x \in X$ there exists $\delta > 0$ such that

$$\rho(gx, gy) < \varepsilon \quad \text{for each} \quad g \in G \quad \text{provided} \quad \rho(x, y) < \delta.$$

THEOREM 4.5.17. *Let a discrete group G act on a compact metric space X. If the action is stable, then E_G is well defined.*

CHAPTER 5

Multipliers of A-Compact Operators. Structure Results

5.1. Extension of a Hilbert C^*-module by the enveloping W^*-algebra

CONSTRUCTION 5.1.1. Let A be a C^*-algebra, A^{**} the enveloping W^*-algebra of A and $\mathcal{M}$ a Hilbert A-module. Consider the algebraic tensor product $\mathcal{M} \otimes A^{**}$ (over $\mathbf{C}$). One can equip this tensor product with the structure of a right A^{**}-module by the formula $(x \otimes a) \cdot b := x \otimes ab$, $x \in \mathcal{M}$, $a, b \in A^{**}$. Define the inner product

$$[\cdot, \cdot] : \mathcal{M} \otimes A^{**} \times \mathcal{M} \otimes A^{**} \longrightarrow A^{**}$$

by the formula

$$\left[\sum_{i=1}^{n} x_i \otimes a_i, \sum_{j=1}^{m} y_j \otimes b_j\right] = \sum_{i,j} a_i^* \langle x_i, y_j \rangle b_j,$$

where $x_i, y_j \in \mathcal{M}$, $a_i, b_j \in A^{**}$. The sesquilinearity and the properties $[z, w] = [w, z]^*$ and $[z, w \cdot a] = [z, w]a$ clearly hold. To verify that this inner product is positive, we use the following well-known statement.

LEMMA 5.1.2 ([**123**, Lemma IV.3.2]). *Let B be a C^*-algebra, $c_{ij} \in B$, $i, j = 1, \dots, n$. A matrix $[c_{ij}] \in M_n(B)$ is positive iff $\sum_{i,j} b_i^* c_{ij} b_j \geq 0$ for any $b_1, \dots, b_n \in B$.*

Since for any $a_1, \dots, a_n \in A$, one has

$$\sum_{i,j} a_i^* \langle x_i, x_j \rangle a_j = \left\langle \sum_{i=1}^{n} x_i \cdot a_i, \sum_{i=1}^{n} x_i \cdot a_i \right\rangle \geq 0,$$

the matrix $[\langle x_i, x_j \rangle] \in M_n(A)$ is positive. Therefore the element $\sum_{i,j} b_i^* \langle x_i, x_j \rangle b_j$ is positive for any $b_i \in A^{**}$, hence $[z, z] \geq 0$ for any $z \in \mathcal{M} \otimes A^{**}$. Set

$$\mathcal{N} = \{z \in \mathcal{M} \otimes A^{**} : [z, z] = 0\}.$$

Then $\mathcal{N}$ is an A^{**}-submodule in $\mathcal{M} \otimes A^{**}$ and the quotient module $\mathcal{M} \otimes A^{**}/\mathcal{N}$ is a pre-Hilbert A^{**}-module. Denote by $\mathcal{M}^{\#}$ the Hilbert A^{**}-module obtained by completion of $\mathcal{M} \otimes A^{**}/\mathcal{N}$ with respect to the norm given by the inner product $[\cdot, \cdot]$. We call this module the *extension* of the module $\mathcal{M}$ by the algebra A^{**}. The W^*-algebra A^{**} always contains the unit element and for any $x \in \mathcal{M}$, $a \in A$, we have $(x \cdot a) \otimes 1 - x \otimes a \in \mathcal{N}$. Therefore the A-module map $x \longmapsto x \otimes 1 + \mathcal{N}$, $\mathcal{M} \longrightarrow \mathcal{M}^{\#}$, is well defined. Since $[x \otimes 1 + \mathcal{N}, y \otimes 1 + \mathcal{N}] = \langle x, y \rangle$, this map is an isometric inclusion.

Denote by $\operatorname{Hom}_A(\mathcal{M}, A^{**})$ the set of all bounded A-linear maps from $\mathcal{M}$ to A^{**}. Introduce on this set the structure of a vector space over $\mathbf{C}$ by the formula

$(\lambda\phi)(x) := \overline{\lambda}\phi(x)$, where $\lambda \in \mathbf{C}$, $x \in \mathcal{M}$, $\phi \in \mathrm{Hom}_A(\mathcal{M}, A^{**})$, and also with the structure of a right A^{**}-module by the formula $(\phi \cdot b)(x) := b^*\phi(x)$, $b \in A^{**}$. For a functional $f \in (\mathcal{M}^{\#})'$ we can define the map $f_R \in \mathrm{Hom}_A(\mathcal{M}, A^{**})$ as the restriction of f onto $\mathcal{M}$, namely, $f_R(x) := f(x \otimes 1 + \mathcal{N})$. Obviously $\|f_R\| \leq \|f\|$.

THEOREM 5.1.3 ([**100**]). *For any C^*-algebra A and for any Hilbert A-module $\mathcal{M}$, the map $f \longmapsto f_R$ is an isometry of $(\mathcal{M}^{\#})'$ onto* $\mathrm{Hom}_A(\mathcal{M}, A^{**})$.

PROOF. Let a matrix $[c_{ij}] \in M_n(A^{**})$ satisfy the condition $\sum_{i,j} a_i^* c_{ij} a_j \geq 0$ for any $a_1, \dots, a_n \in A$. Let us show that it is sufficient to prove positivity of the matrix $[c_{ij}]$. For this purpose, it is sufficient to show that

$$\sum_{i,j} b_i^* c_{ij} b_j \geq 0 \tag{5.1}$$

for any $b_1, \dots, b_n \in A^{**}$. Without loss of generality one can assume that the elements b_i are in the unit ball $B_1(A^{**})$ of the W^*-algebra A^{**}. Since the unit ball $B_1(A)$ of the C^*-algebra A is dense in $B_1(A^{**})$ with respect to the strong* topology, one can find nets $a_{i;\lambda} \in A$, $\lambda \in \Lambda$, converging to the elements $b_i \in A^{**}$ with respect to the strong* topology. Then the net $\sum_{ij} a_{i;\lambda}^* c_{ij} a_{j;\lambda}$ converges to $\sum_{i,j} b_i^* c_{ij} b_j$ with respect to the weak topology (see [**123**, §II.2]), whence the inequality (5.1) follows.

We have to show that any map $\phi \in \mathrm{Hom}_A(\mathcal{M}, A^{**})$ with $\|\phi\| \leq 1$ can be extended up to a unique functional $f \in (\mathcal{M}^{\#})'$ with $\|f\| \leq 1$. Consider the functional $f_0 : \mathcal{M} \otimes A^{**} \longrightarrow A^{**}$ given by the formula

$$f_0\left(\sum_{i=1}^n x_i \otimes a_i\right) = \sum_{i=1}^n \phi(x_i) a_i.$$

Obviously f_0 is an A^{**}-module map. For $a_i \in A$, one has

$$\begin{aligned}\sum_{i,j} a_i^* \phi(x_i)^* \phi(x_j) a_j &= \sum_{i,j} \phi(x_i \cdot a_i)^* \phi(x_j \cdot a_j) \\ &= \left(\phi\left(\sum_{i=1}^n x_i \cdot a_i\right)\right)^* \phi\left(\sum_{i=1}^n x_i \cdot a_i\right) \\ &\leq \left\langle \sum_{i=1}^n x_i \cdot a_i, \sum_{i=1}^n x_i \cdot a_i \right\rangle = \sum_{i,j} a_i^* \langle x_i, x_j \rangle a_j.\end{aligned}$$

Therefore for any $b_i \in A^{**}$, the following inequality holds:

$$\sum_{i,j} b_i^* \phi(x_i)^* \phi(x_j) b_j \leq \sum_{i,j} b_i^* \langle x_i, x_j \rangle b_j,$$

i.e.,

$$f_0(z)^* f_0(z) \leq [z, z]$$

for any $z \in \mathcal{M} \otimes A^{**}$. Therefore the functional $f : \mathcal{M}^{\#} \longrightarrow A^{**}$ is well defined by the formula

$$f\left(\sum_{i=1}^n x_i \otimes a_i + \mathcal{N}\right) := \sum_{i=1}^n \phi(x_i) a_i$$

and satisfies the inequality $f(y)^* f(y) \leq [y, y]$ for all $y \in \mathcal{M}^{\#}$, whence $\|f\| \leq 1$. Thus $f \in (\mathcal{M}^{\#})'$. It follows from the equality $f(x \otimes 1 + \mathcal{N}) = \phi(x)$ that f is an extension of ϕ. $\square$

COROLLARY 5.1.4. *Let A be a C^*-algebra and let $\mathcal{M}$ be a Hilbert A-module. Then an A-valued inner product on $\mathcal{M}$ can be extended up to an A^{**}-valued inner product on the set* $\mathrm{Hom}_A(\mathcal{M}, A^{**})$ *making this set a self-dual Hilbert A^{**}-module.*

COROLLARY 5.1.5. *Let A be a C^*-algebra and let $\mathcal{M}$ be a self-dual Hilbert A-module. Then the Hilbert A^{**}-module $\mathcal{M}^{\#}$ is self-dual too.*

As one more corollary we shall present the following characterization of self-dual Hilbert modules.

THEOREM 5.1.6 ([**42, 40**]). *Let A be a C^*-algebra. Then the following statements are equivalent:*

(i) *the Hilbert A-module H_A is self-dual;*
(ii) *the C^*-algebra A is finite-dimensional.*

PROOF. Note that each condition of the theorem implies the existence of a unit in the C^*-algebra A. Indeed, if A is finite-dimensional, then $1 \in A$. If the module H_A is self-dual, then the bounded A-module map $f : H_A \longrightarrow A$ defined by the formula $f(a) = a_1$, where $a = (a_i)$, $a_i \in A$, $i \in \mathbf{N}$, has to be an element of the module H_A itself. This means in particular that $\mathrm{End}_A(A) = \mathrm{End}^*_A(A)$ and so the identity mapping $A \longrightarrow A$ is identified with the unit element of A.

In what follows, let us use the description of the dual module H'_A as the set of all sequences $b = (b_i)$, $b_i \in A$, such that the partial sums $c_n = \sum_{i=1}^n b_i^* b_i$ of the series $\sum_i b_i^* b_i$ are uniformly bounded. Self-duality means that for any increasing sequence (c_n) of positive elements of the C^*-algebra A, its boundedness is equivalent to its norm-convergence. But, for finite-dimensional C^*-algebras, the monotone bounded sequences are convergent, and this proves the implication (ii) $\Rightarrow$ (i). To prove the opposite implication we pass to the Hilbert module $H_A^{\#} = H_{A^{**}}$ over the enveloping W^*-algebra A^{**}. This module is self-dual by Corollary 5.1.5. Since any monotone bounded sequence in a W^*-algebra is convergent, any positive linear functional on A^{**} has to be normal, i.e., $(A^{**})_* = (A^{**})^*$. Suppose that the W^*-algebra A^{**} is infinite-dimensional. Then it contains an infinite collection of mutually orthogonal projections $p_k \in A^{**}$ such that $\sum_{k=1}^{\infty} p_k = 1$. Hence there exists an inclusion of the commutative W^*-algebra of bounded sequences l_∞ into A^{**}. Let $\varphi \in (l_\infty)^*$ be a positive linear functional on the algebra l_∞. Let us extend it to a positive linear functional $\overline{\varphi}$ on the larger algebra A^{**}. According to the assumption, $\overline{\varphi}$ is normal. Therefore its restriction $\overline{\varphi}|_{l_\infty} = \varphi$ to the algebra l_∞ is normal too. Hence we have obtained a false conclusion $(l_\infty)_* = (l_\infty)^*$. This contradiction shows that the W^*-algebra A^{**} is finite-dimensional, hence the C^*-algebra A is finite-dimensional as well and this proves the implication (i) $\Rightarrow$ (ii). □

An interesting general theory of extensions of Hilbert C^*-modules can be found in [**6**].

5.2. Multipliers and centralizers

While writing this section we extensively used results of [**104, 134**]. Let $\mathcal{B}(H)$ be the algebra of all bounded operators on a Hilbert space H and let A be a C^*-algebra.

DEFINITION 5.2.1. A two-sided closed ideal $J \subset A$ is called *essential* if $J \cap J' \neq \emptyset$ for any nonzero ideal $J' \subset A$.

REMARK 5.2.2. An ideal $J \subset A$ is essential if and only if

$$J^{\perp} := \{a \in A \mid aJ = 0\} = 0.$$

DEFINITION 5.2.3. A representation $\rho : A \to \mathcal{B}(H)$ is called *nondegenerate* if for any $h \in H$ there exists an element $a \in A$ such that $\rho(a)h \neq 0$.

REMARK 5.2.4. For an arbitrary representation ρ, we can consider its restriction onto the orthogonal complement H' to the invariant subspace $H^0_\rho := \{h \in H \mid \rho(A)h = 0\}$, which is invariant as well. The new representation $\rho' : A \to \mathcal{B}(H')$ is then nondegenerate. Thus, roughly speaking, we lose nothing working only with nondegenerate representations.

LEMMA 5.2.5. *A representation is nondegenerate if and only if $\rho(A)(H)$ is dense in H.*

PROOF. Let a representation ρ be nondegenerate and let $h \perp \rho(A)(H)$, i.e., for any $f \in H$ and any $a \in A$,

$$0 = \langle h, \rho(a)f \rangle = \langle \rho(a^*)h, f \rangle$$

holds. Then $\rho(b)h = 0$ for any $b \in A$, hence $h = 0$.

Conversely, let $\overline{\rho(A)(H)} = H$ and let $h \in H$ be an arbitrary nonzero vector. Without loss of generality one can assume that $\|h\| = 1$. Since $\rho(A)(H)$ is dense in H, one can find $g \in H$ and $a \in A$ such that $\|h - \rho(a)g\| < 1/2$. Then $\|\rho(a)g\| > 1/2$,

$$1/4 > \langle h - \rho(a)g, h - \rho(a)g \rangle = 1 - \langle g, \rho(a^*)h \rangle - \langle \rho(a^*)h, g \rangle + 1/4,$$
$$\langle g, \rho(a^*)h \rangle + \langle \rho(a^*)h, g \rangle > 1, \qquad \rho(a^*)h \neq 0.$$

□

DEFINITION 5.2.6. Let $\rho : A \to \mathcal{B}(H)$ be a faithful nondegenerate representation, so we can assume $A \subset \mathcal{B}(H)$. An operator $x \in \mathcal{B}(H)$ is called a (two-sided) *multiplier* of A if

$$xa \in A, \qquad ax \in A$$

for each $a \in A$. Denote by $\mathbf{M}(A)$ the set of all multipliers. It is easy to see that $\mathbf{M}(A)$ is an involutive unital algebra.

REMARK 5.2.7. It seems that the definition of multiplier depends on the choice of a (nondegenerate faithful) representation. In Theorem 5.2.11 we shall see that this is not the case.

PROPOSITION 5.2.8. *The set $\mathbf{M}(A)$ is a unital C^*-algebra,*

$$A \subset \mathbf{M}(A) \subset A^{**}$$

and A is an essential ideal in $\mathbf{M}(A)$. If A has no unit, then $A^+ \subset \mathbf{M}(A)$.

PROOF. Three statements are nontrivial: 1) that $\mathbf{M}(A)$ is norm-closed, 2) that the ideal A is essential, and 3) that there exists an inclusion into the bidual space.

1) Let $x_\alpha \to x$ with respect to the norm, $x_\alpha \in \mathbf{M}(A)$, $x \in \mathcal{B}(H)$. Then $x_\alpha a \to xa$ and $ax_\alpha \to ax$ for any a. Since A is closed, $xa \in A$ and $ax \in A$ for any a, i.e., $x \in \mathbf{M}(A)$ by definition.

2) Let J be an ideal in $\mathbf{M}(A)$, $J \cap A = 0$, and let $x \in J$ be an arbitrary element. Then $xa \in A$ (since x is a multiplier) and $xa \in J$ (since $a \in A \subset \mathbf{M}(A)$ and J is an ideal) for any $a \in A$. Therefore $xa \in J \cap A = 0$, $xa = 0$ for any $a \in A$. Then $x = 0$ by Lemma 5.2.5.

3) Since $A^{!!} \hookrightarrow A^{!!}_u \cong A^{**}$ (cf. remark after Theorem 3.1.3), it is sufficient to prove that $\mathbf{M}(A) \subset A^{!!}$. For this purpose, note that for any $x \in \mathbf{M}(A)$ and for any weakly convergent net $a_\lambda \in A$, we have

$$x \operatorname*{w-lim}_{\lambda\in\Lambda} a_\lambda = \operatorname*{w-lim}_{\lambda\in\Lambda}(xa_\lambda) \in [A]_w,$$

since $xa_\lambda \in A$, where $[A]_w$ is the weak closure of A in $\mathcal{B}(H)$ and where w-lim denotes the limit with respect to the weak topology. One has $[A]_w = A^{!!}$ due to the nondegeneracy of the representation, hence $xA^{!!} = x[A]_w \subset [A]_w = A^{!!}$. Since $1 \in A^{!!}$, one has $x \in A^{!!}$. □

The next definition historically precedes Definition 5.2.6.

DEFINITION 5.2.9. A pair (L, R) of maps

$$L : A \longrightarrow A, \qquad R : A \longrightarrow A, \qquad R(a)b = aL(b) \text{ for each } a, b \in A,$$

is called a *double centralizer* of A. Let us denote the set of all double centralizers of A by $\mathbf{DC}(A)$.

PROPOSITION 5.2.10. *Let* $(L, R) \in \mathbf{DC}(A)$. *Then*

(i) $L(ab) = L(a)b$ *and* $R(ab) = aR(b)$;
(ii) L *and* R *are linear;*
(iii) L *and* R *are bounded and* $\|L\| = \|R\|$.

The set $\mathbf{DC}(A)$ *with the operations*

$$\begin{aligned}(L_1, R_1) + (L_2, R_2) &:= (L_1 + L_2, R_1 + R_2), \qquad z(L, R) = (zL, zR), \qquad z \in \mathbf{C},\\ (L_1, R_1)(L_2, R_2) &:= (L_1L_2, R_2R_1),\\ (L, R)^* &:= (R^*, L^*),\\ L^*(a) &:= (L(a^*))^*, \qquad R^*(a) := (R(a^*))^*, \qquad a \in A,\end{aligned}$$

is a normed involutive algebra with respect to the norm

$$\|(L, R)\| := \|L\| = \|R\|.$$

PROOF. 1) Let $a, b \in A$, $z \in \mathbf{C}$ and let e_α $(\alpha \in \mathcal{A})$ be an approximate unit of A. Then

$$\begin{aligned}e_\alpha L(ab) &= R(e_\alpha)ab = e_\alpha L(a)b, \qquad L(ab) = L(a)b,\\ e_\alpha L(za + zb) &= R(e_\alpha)(za + zb) = R(e_\alpha)za + R(e_\alpha)zb = zR(e_\alpha)a + zR(e_\alpha)b\\ &= ze_\alpha L(a) + ze_\alpha L(b) = e_\alpha(z(L(a) + L(b))),\\ L(za + zb) &= z(L(a) + L(b)).\end{aligned}$$

2) Thus L is a linear operator on the Banach space A and, to prove its continuity, it is sufficient to check that its graph is closed. Let $a_n \longrightarrow a$ and $L(a_n) \longrightarrow b$. Then

$$\begin{aligned}\|v(L(a) - b)\| &\le \|vL(a) - vL(a_n)\| + \|vL(a_n) - vb\|\\ &= \|R(v)(a - a_n)\| + \|vL(a_n) - vb\|\\ &\le \|R(v)\| \cdot \|a - a_n\| + \|v\| \cdot \|L(a_n) - b\| \longrightarrow 0\end{aligned}$$

for any $v \in A$, hence $vL(a) = vb$, whence $b = L(a)$, since v was taken arbitrarily. Thus the graph of L is closed, so L is continuous. Properties of R can be verified in the same way.

3) Let us compare $\|L\|$ and $\|R\|$:

$$\|L\|^2 = \sup_{\|a\|=1} \|L(a)\|^2 = \sup_{\|a\|=1} \|L(a)^*L(a)\| = \sup_{\|a\|=1} \|R(L(a)^*)a\|$$
$$\leq \sup_{\|a\|=1} \|R\| \cdot \|L(a)^*\| \cdot \|a\| \leq \sup_{\|a\|=1} \|R\| \cdot \|L\| \cdot \|a\|^2 = \|R\| \cdot \|L\|,$$

whence $\|L\| \leq \|R\|$. A similar calculation gives the opposite inequality.

The remaining statements are obvious, one has only to verify that

$$R_2(R_1(a))b = R_1(a)L_2(b) = aL_1(L_2(b)).$$

□

THEOREM 5.2.11. *The map*

$$\mu : \mathbf{M}(A) \to \mathbf{DC}(A), \qquad x \mapsto (L_x, R_x), \qquad L_x(a) = xa, \quad R_x(a) = ax,$$

is an isometric $*$-isomorphism between $\mathbf{M}(A)$ and $\mathbf{DC}(A)$. Therefore $\mathbf{DC}(A)$ is a C^*-algebra and $\mathbf{M}(A)$ does not depend on our choice of a nondegenerate representation.

PROOF. First of all one has $aL_x(b) = axb = R_x(a)b$, so the pair (L_x, R_x) does belong to $\mathbf{DC}(A)$. The linearity of the map μ is clear. Also,

$$L_{xy}(a) = (xy)a = x(ya) = L_x(ya) = L_x(L_y(a)),$$
$$R_{xy}(a) = a(xy) = (ax)y = R_y(ax) = R_y(R_x(a)),$$
$$(L_{xy}, R_{xy}) = (L_xL_y, R_yR_x) = (L_x, R_x)(L_y, R_y),$$

thus μ is a homomorphism of algebras. It is involutive,

$$(L_x)^*(a) = (L_xa^*)^* = (xa^*)^* = ax^* = R_{x^*}(a),$$
$$(R_x)^*(a) = (R_xa^*)^* = (a^*x)^* = x^*a = L_{x^*}(a).$$

Since $\|xa\| \leq \|x\|\,\|a\|$, one has $\|L_x\| \leq \|x\|$. Conversely, since the representation is nondegenerate, $xe_\alpha \longrightarrow x$ with respect to the strong topology, where e_α is an approximate unit of A. Indeed, for any $a \in A$, the net $e_\alpha a \longrightarrow a$ is norm-convergent, whence $xe_\alpha a \longrightarrow xa$ with respect to the norm. From the nondegeneracy we obtain the strong convergence $xe_\alpha \longrightarrow x$ on the dense set AH, and, due to the boundedness, $\|xe_\alpha\| \leq \|x\|$, the strong convergence takes place everywhere on H. Let $\varepsilon > 0$ be taken arbitrarily. Choose $h \in H$ such that $\|h\| = 1$ and $\|x\| \leq \|xh\| + \varepsilon/2$. Since $x = \text{s-lim}_\alpha\, xe_\alpha$, $xe_\alpha h \longrightarrow xh$ and one can find α such that $\|xh - xe_\alpha h\| < \varepsilon/2$. Then $\|x\| \leq \|xe_\alpha\| + \varepsilon$. Therefore

$$\|L_x\| \geq \|L_x(e_\alpha)\| = \|xe_\alpha\| \geq \|x\| - \varepsilon$$

and $\|L_x\| \geq \|x\|$, since ε is arbitrary. Hence μ is an isometry.

It remains to verify that $\operatorname{Im}\mu = \mathbf{DC}(A)$. Consider an arbitrary element $(L, R) \in \mathbf{DC}(A)$. Then the sets $L(e_\alpha)$ and $R(e_\alpha)$ are bounded and by the weak compactness of the unit ball $\mathcal{B}(H)$, they have points of accumulation $x_L \in A^{!!}$ and $x_R \in A^{!!}$, respectively, with respect to the weak topology. Passing, if necessary, to sub-nets, we can suppose without loss of generality that

$$x_L = \operatorname*{w-lim}_{\mathcal{A}} L(e_\alpha), \qquad x_R = \operatorname*{w-lim}_{\mathcal{A}} R(e_\alpha).$$

Then $x_R = x_L$. Indeed, for any $a, b \in A$ we have

$$ax_L b = a \text{ w-}\lim_{\mathcal{A}} L(e_\alpha)b = a \text{ w-}\lim_{\mathcal{A}} L(e_\alpha b) = aL(b) = R(a)b,$$
$$ax_R b = a \text{ w-}\lim_{\mathcal{A}} R(e_\alpha)b = \text{w-}\lim_{\mathcal{A}} R(ae_\alpha)b = R(a)b,$$

whence $x_L = x_R$ (cf. the proof of the second part of Proposition 5.2.8). Let us denote $x := x_L = x_R$. Then

$$L(a) = \text{w-}\lim_{\mathcal{A}} L(e_\alpha a) = xa = L_x(a),$$
$$R(a) = \text{w-}\lim_{\mathcal{A}} R(ae_\alpha) = ax = R_x(a),$$

in particular, $x \in \mathbf{M}(A)$. The above equalities show that $\mu(x) = (L, R)$. □

EXAMPLE 5.2.12. 1) The equality $\mathbf{M}(A) = A$ holds if and only if A is unital.

2) For a commutative algebra $A = C_0(X)$ one has

$$\mathbf{M}(C_0(X)) = C_b(X) \cong C(\beta X),$$

where $C_b(X)$ is the C^*-algebra of all bounded functions with the sup norm and βX is the Stone-Čech compactification of X.

3) For the algebra $\mathcal{K} = \mathcal{K}(H)$ of compact operators one has $\mathbf{M}(\mathcal{K}(H)) = \mathcal{B}(H)$.

The proof of 1) and 2) can be found, for example, in [**134**] and 3) will be proved below in a more general situation (see Theorem 5.3.1).

DEFINITION 5.2.13. Let $\rho : A \to \mathcal{B}(H)$ be a faithful nondegenerate representation, so we can assume $A \subset \mathcal{B}(H)$. An operator $x \in \mathcal{B}(H)$ is called a *left multiplier* of A if

$$xa \in A$$

for any $a \in A$. Denote by $\mathbf{LM}(A)$ the set of all left multipliers. It is clear that they form a unital algebra. Similarly one can define *right multipliers*, $\mathbf{RM}(A)$.

An operator $x \in \mathcal{B}(H)$ is called a *quasi-multiplier* of A if

$$axb \in A$$

for any a, $b \in A$. Denote by $\mathbf{QM}(A)$ the set of all quasi-multipliers. It is clear that they form an involutive linear space.

DEFINITION 5.2.14. A linear map $\lambda : A \to A$ is called a *left centralizer* if

$$\lambda(ab) = \lambda(a)b, \qquad \text{for each } a,\ b \in A.$$

Similarly one defines a *right centralizer.* Denote the spaces of left and right centralizers by $\mathbf{LC}(A)$ and $\mathbf{RC}(A)$ respectively.

DEFINITION 5.2.15. A linear map $q : A \times A \to A$ is called a *quasi-centralizer* if

$$\lambda_a \in \mathbf{LC}(A), \text{ where } \lambda_a : b \mapsto q(a, b), \qquad \text{for any } a,\ b \in A,$$

and

$$\rho_b \in \mathbf{RC}(A), \text{ where } \rho_b : a \mapsto q(a, b), \qquad \text{for any } a,\ b \in A.$$

In other words,

$$q(ca, bd) = cq(a, b)d, \qquad \text{for any } a,\ b,\ c,\ d \in A.$$

LEMMA 5.2.16. *If* $\rho \in \mathbf{RC}(A)$*, then* $\rho^* \in \mathbf{LC}(A)$.

PROOF. Recall that ρ^* is defined by $\rho^*(a) := (\rho(a^*))^*$. Then

$$\rho^*(ab) = (\rho((ab)^*))^* = (\rho(b^*a^*))^* = (b^*\rho(a^*))^* = \rho^*(a)b.$$

□

LEMMA 5.2.17 ([**104**, Lemma 3.12.2]). *Each right centralizer, each left centralizer and each quasi-centralizer is bounded.*

PROOF. Let $\rho \in \mathbf{RC}(A)$. Suppose that it is unbounded, i.e., there exists a sequence $x_n \in A$ such that $\|x_n\| < 1/n$ and $\|\rho(x_n)\| > n$. Then the element $a := \sum_n x_n^* x_n$ is well defined. By Proposition 1.1.5, for each x_n there exists an element $u_n \in A$ such that $\|u_n\| \leq \|a^{1/6}\|$ and $x_n = u_n a^{1/3}$. Then the equality

$$\|\rho(x_n)\| = \|u_n \rho(a^{1/3})\| \leq \|\rho(a^{1/3})\| \cdot \|a^{1/6}\|$$

gives a contradiction, hence ρ is bounded. In a similar way one can prove that any left centralizer is bounded too. Thus, $q \in \mathbf{QC}(A)$ is continuous separately in each variable as a map $A \times A \to A$. By the uniform boundedness principle, such an operator is jointly continuous (see [**35**]). □

PROPOSITION 5.2.18 ([**104**, Prop. 3.12.3]). *Let $A \to \mathcal{B}(H)$ be a nondegenerate faithful representation. Then there exists a one-to-one isometric linear correspondence between left, right and quasi-multipliers and left, right and quasi-centralizers, respectively. In the first two cases this correspondence is a homomorphism of algebras, in the third one it is a homomorphism of involutive spaces.*

PROOF. The correspondence for the left and right multipliers, together with its properties, were already described in Theorem 5.2.11. Let $q \in \mathbf{QC}(A)$ and let $x \in A^{!!}$ be an accumulation point (with respect to the weak topology) of the bounded directed net $\{q(e_\alpha, e_\alpha)\}$, where $\{e_\alpha\}$ is an approximate unit for A. Passing, if necessary, to a sub-net, we can, as before, assume that $x = \text{w-lim}_\alpha\, q(e_\alpha, e_\alpha)$. Then

$$A \ni q(a,b) = \lim_\alpha q(ae_\alpha, e_\alpha b) = \lim_\alpha aq(e_\alpha, e_\alpha)b = axy$$

for any a, $b \in A$, so $x \in \mathbf{QM}(A) \subset A^{!!}$. The required properties can be verified exactly as the similar properties in 5.2.11. □

PROPOSITION 5.2.19. *Let A be a (closed two-sided $*$-) ideal of a C^*-algebra B. Then there exists a unique homomorphism $\gamma : B \to \mathbf{M}(A)$ such that its restriction to A is the identity map.*

PROOF. Put $\gamma(b) := (L_b, R_b)$, i.e., $L_b(a) = ba$, $R_b(a) = ab$, where we identify $\mathbf{M}(A) = \mathbf{DC}(A)$. Since $A \subset B$ is an ideal, $ba \in A$ and $ab \in A$, so $\gamma(b) \in \mathbf{DC}(A)$. Thus, obviously $\gamma|_A : A \hookrightarrow \mathbf{M}(A)$.

Now assume that in addition to γ, there exists a homomorphism $\delta : B \to \mathbf{M}(A)$ that coincides with γ on A. Then

$$\delta(b)a = \delta(b)\delta(a) = \delta(ba) = ba, \qquad \gamma(b)a = \gamma(b)\gamma(a) = \gamma(ba) = ba$$

for any $b \in B$ and $a \in A$, i.e., $\gamma(b)$ and $\delta(b)$ coincide as multipliers of A, hence $\delta = \gamma$. □

COROLLARY 5.2.20. *Let $\rho : A \to \mathcal{B}(H)$ be a faithful representation of A and let $A \subset B$ be an ideal. Then there exists a representation of B that extends ρ.*

PROPOSITION 5.2.21. *Let A and B be C^*-algebras and let $\varphi : A \to B$ be a* surjective *morphism. Then φ can be extended to a morphism $\varphi'' : \mathbf{M}(A) \to \mathbf{M}(B)$ and it induces a morphism $\overline{\varphi} : \mathbf{M}(A)/A \to \mathbf{M}(B)/B$, which makes the diagram*

$$\begin{array}{ccccc} A & \longrightarrow & \mathbf{M}(A) & \xrightarrow{\pi} & \mathbf{M}(A)/A \\ \downarrow{\scriptstyle\varphi} & & \downarrow{\scriptstyle\varphi''} & & \downarrow{\scriptstyle\overline{\varphi}} \\ B & \longrightarrow & \mathbf{M}(B) & \xrightarrow{\pi} & \mathbf{M}(B)/B \end{array}$$

commutative. If φ is an isomorphism, then φ'' and $\overline{\varphi}$ are isomorphisms too.

PROOF. Let $(L,R) \in \mathbf{DC}(A)$. Define $\widehat{L}, \widehat{R} : B \to B$ by

$$\widehat{L}(b) := \varphi(L(a)), \qquad \widehat{R}(b) := \varphi(R(a)), \qquad b \in B,\ b = \varphi(a).$$

Let us show that these maps are well defined. Let e_α be an approximate unit of the C^*-algebra A and let $b = \varphi(a) = \varphi(a')$. Then

$$\varphi(L(a) - L(a')) = \lim_\alpha \varphi(e_\alpha L(a) - e_\alpha L(a')) = \lim_\alpha \varphi(R(e_\alpha))\,\varphi(a - a') = 0.$$

A similar equality holds for right multipliers. Since for $b_1 = \varphi(a_1)$ and $b_2 = \varphi(a_2)$ one has

$$\begin{aligned} \widehat{R}(b_1)b_2 &= \varphi(R(a_1))\,\varphi(a_2) = \varphi(R(a_1)a_2) = \varphi(a_1 L(a_2)) \\ &= \varphi(a_1)\,\varphi(L(a_2)) = \beta_1 \widehat{L}(b_2), \end{aligned}$$

$(\widehat{L}, \widehat{R}) \in \mathbf{DC}(A)$. Define $\varphi'' : \mathbf{DC}(A) \to \mathbf{DC}(B)$ by $\varphi''(L,R) = (\widehat{L}, \widehat{R})$. Then it is a $*$-morphism of C^*-algebras that extends φ. This map induces a map for quotients. Indeed, if $(x - y) \in A$, $x, y \in \mathbf{M}(A)$, then $\varphi''(x - y) = \varphi(x - y) \in B$. We have obtained the desired commutative diagram.

If now φ is an isomorphism, then $\varphi''(L,R) = (\varphi \circ L \circ \varphi^{-1}, \varphi \circ R \circ \varphi^{-1})$. Hence, φ'' is an isomorphism and the inverse map is defined by

$$(\widehat{L}, \widehat{R}) \mapsto (\varphi^{-1} \circ \widehat{L} \circ \varphi, \varphi^{-1} \circ \widehat{R} \circ \varphi).$$

By the five-lemma, $\overline{\varphi}$ is also an isomorphism. □

REMARK 5.2.22. The homomorphism φ'' coincides with the canonical extension of φ to $A^{!!}$ restricted onto $\mathbf{M}(A)$.

5.3. Multipliers of A-compact operators

THEOREM 5.3.1 ([**63**]). *Let $\mathcal{M}$ be a Hilbert A-module. Define the map*

$$\phi : \operatorname{End}^*_A(\mathcal{M}) \to \mathbf{DC}(\mathcal{K}(\mathcal{M})), \qquad T \mapsto (T_1, T_2),$$

$$T_1(\theta_{x,y}) := \theta_{Tx,y}, \quad T_2(\theta_{x,y}) := \theta_{x,T^*y}.$$

Then ϕ is an isomorphism.

PROOF. First of all, note that

$$\theta_{Tx,y}z = Tx\langle y, z\rangle = T \circ \theta_{x,y}(z), \qquad \theta_{x,T^*y}z = x\langle T^*y, z\rangle = x\langle y, Tz\rangle = \theta_{xy} \circ T(z),$$

so that T_1 and T_2 can be defined in an equivalent way (and for all compact operators simultaneously) by the formulas

$$T_1(k) := T \circ k, \qquad T_2(k) := k \circ T, \qquad k \in \mathcal{K}(\mathcal{M}).$$

From these equalities we obtain at once that T_1 and T_2 are well defined as maps $\mathcal{K}(\mathcal{M}) \to \mathcal{K}(\mathcal{M})$ (since $\mathcal{K}(\mathcal{M}) \subset \mathrm{End}^*_A(\mathcal{M})$ is a two-sided ideal) and are bounded by $\|T\|$. Since

$$T_2(k_1)k_2 = k_1Tk_2 = k_1T_1(k_2), \qquad k_1,\, k_2 \in \mathcal{K}(\mathcal{M}),$$

$(T_1, T_2) \in \mathbf{DC}(\mathcal{K}(\mathcal{M}))$. Since

$$(TS)_1(k) = TSk = T_1(S_1k), \qquad (TS)_2(k) = kTS = S_2(T_2k),$$

ϕ is a homomorphism of algebras. It also respects the involution:

$$T_1^*(\theta_{x,y}) = (T_1(\theta^*_{x,y}))^* = (T\theta_{y,x})^* = \theta_{x,y}T^*, \quad T_1^* = (T^*)_2,$$

$$T_2^*(\theta_{x,y}) = (T_2(\theta^*_{x,y}))^* = (\theta_{y,x}T)^* = T^*\theta_{x,y}, \quad T_2^* = (T^*)_1.$$

The map ϕ is algebraically injective. Indeed, let $T_1 = 0$ and $T_2 = 0$. Then for any $x \in \mathcal{M}$, $0 = T_1(\theta_{x,Tx})(Tx) = Tx\langle Tx, Tx\rangle$ holds, whence $\langle Tx, Tx\rangle^3 = 0$ and $Tx = 0$. Hence $T = 0$.

To prove the surjectivity of ϕ, we construct an inverse continuous map ψ. Let (T_1, T_2) be an element of $\mathbf{DC}(\mathcal{K}(\mathcal{M}))$ and let $x \in \mathcal{M}$. Consider the limits

$$T(x) := \lim_{n\to\infty} T_n(x), \qquad T_n(x) := T_1(\theta_{x,x})(x)[\langle x, x\rangle + 1/n]^{-1}, \tag{5.2}$$

$$T^*(x) = \lim_{n\to\infty} T_n^*(x), \qquad T_n^*(x) := [T_2(\theta_{x,x})]^*(x)[\langle x, x\rangle + 1/n]^{-1}. \tag{5.3}$$

Let us prove that these limits exist. By Theorem 5.2.11, $(T_1, T_2) = (L_F, R_F)$, where $F \in \mathbf{M}(\mathcal{K}(\mathcal{M}))$. Then

$$\begin{aligned}(T_1(k))^*T_1(k) &= (Fk)^*Fk = k^*F^*Fk \le \|F\|^2k^*k = \|T_1\|^2k^*k,\\ T_2(k)(T_2(k))^* &= kF(kF)^* = kFF^*k^* \le \|F\|^2kk^* = \|T_2\|^2kk^*,\end{aligned}$$

where the inequalities are the operator inequalities in $\mathcal{K}(\mathcal{M})^{**}$. Then

$$\begin{aligned}
&\langle T_n(x) - T_m(x), T_n(x) - T_m(x)\rangle\\
&= \{[\langle x, x\rangle + 1/n]^{-1} - [\langle x, x\rangle + 1/m]^{-1}\}\langle T_1(\theta_{x,x})(x), T_1(\theta_{x,x})(x)\rangle\\
&\qquad \times \{[\langle x, x\rangle + 1/n]^{-1} - [\langle x, x\rangle + 1/m]^{-1}\}\\
&\le \{[\langle x, x\rangle + 1/n]^{-1} - [\langle x, x\rangle + 1/m]^{-1}\}\langle (T_1(\theta_{x,x}))^*T_1(\theta_{x,x})(x), x\rangle\\
&\qquad \times \{[\langle x, x\rangle + 1/n]^{-1} - [\langle x, x\rangle + 1/m]^{-1}\}\\
&\le \{[\langle x, x\rangle + 1/n]^{-1} - [\langle x, x\rangle + 1/m]^{-1}\}\|T_1\|^2\langle (\theta_{x,x}\theta_{x,x})(x), x\rangle\\
&\qquad \times \{[\langle x, x\rangle + 1/n]^{-1} - [\langle x, x\rangle + 1/m]^{-1}\}\\
&= \|T_1\|^2\{[\langle x, x\rangle + 1/n]^{-1} - [\langle x, x\rangle + 1/m]^{-1}\}\langle \theta_{x,x}x\langle x, x\rangle, x\rangle\\
&\qquad \times \{[\langle x, x\rangle + 1/n]^{-1} - [\langle x, x\rangle + 1/m]^{-1}\}\\
&= \|T_1\|^2\langle x, x\rangle^3\{[\langle x, x\rangle + 1/n]^{-1} - [\langle x, x\rangle + 1/m]^{-1}\}^2.
\end{aligned}$$

Thus, the Cauchy criterion of convergence for (5.2) is the same as for the limit from Lemma 1.3.9. Therefore the convergence is proved. The convergence of (5.3) can be proved similarly. We obtained the maps T and T^* defined everywhere on $\mathcal{M}$.

Also, by Lemma 1.3.9,

$$\begin{aligned}\langle x, T^*y\rangle &= \lim_{n\to\infty} \langle x, [T_2(\theta_{y,y})]^*(y)\cdot[\langle y,y\rangle + 1/n]^{-1}\rangle \\ &= \lim_{n\to\infty} \langle T_2(\theta_{y,y})\theta_{x,x}(x)[\langle x,x\rangle + 1/n]^{-1}, y\cdot[\langle y,y\rangle + 1/n]^{-1}\rangle \\ &= \lim_{n\to\infty} \langle \theta_{y,y}T_1(\theta_{x,x})(x)[\langle x,x\rangle + 1/n]^{-1}, y\cdot[\langle y,y\rangle + 1/n]^{-1}\rangle \\ &= \lim_{n\to\infty} \langle T_1(\theta_{x,x})(x)[\langle x,x\rangle + 1/n]^{-1}, \theta_{y,y}(y)\cdot[\langle y,y\rangle + 1/n]^{-1}\rangle = \langle Tx, y\rangle.\end{aligned}$$

Hence, by Lemma 2.1.1, T, $T^* \in \mathrm{End}^*_A(\mathcal{M})$. It remains to verify that $\phi(T) = (T_1, T_2)$. Put

$$x_n := \left(\langle x,x\rangle + \frac{1}{n}\right)^{-1}$$

and note that, by Lemma 1.3.9,

$$\theta_{x,y}(z) = x\langle y,z\rangle = \lim_{n\to\infty} x\langle x,x\rangle x_n\langle y,z\rangle = \lim_{n\to\infty} \theta_{x,x}\theta_{xx_n,y}(z). \tag{5.4}$$

Therefore

$$\begin{aligned}T_1(\theta_{x,y})(z) &= \lim_{n\to\infty} T_1(\theta_{x,x}\theta_{xx_n,y})(z) = \lim_{n\to\infty} T_1(\theta_{x,x})\theta_{xx_n,y}(z) \\ &= \lim_{n\to\infty} T_1(\theta_{x,x})xx_n\langle y,z\rangle = T(x)\langle y,z\rangle = T\,\theta_{x,y}(z).\end{aligned}$$

A similar argument for T_2 completes the proof. □

It is easy to obtain the following extension of this theorem.

THEOREM 5.3.2 ([**73**, Theorem 1.5]). *There exists an isometric isomorphism of Banach algebras*

$$\phi : \mathrm{End}_A(\mathcal{M}) \longrightarrow \mathbf{LM}(\mathcal{K}(\mathcal{M}))$$

that extends the homomorphism ϕ from Theorem 5.3.1.

PROOF. As usual, unitalizing, if necessary, we can assume that the C^*-algebra A is already unital. As before, define ϕ by the formula

$$\phi(T)(k) = Tk, \qquad k \in \mathcal{K}(\mathcal{M}),$$

so that it extends ϕ from Theorem 5.3.1. Then the calculations presented in the proof of 5.3.1 for T_1 show that ϕ is an algebraically injective homomorphism of algebras and $\|\phi\| \le 1$. To prove that it is surjective, define the continuous inverse map ψ for ϕ similarly to 5.3.1 by

$$\psi(S)(x) := \lim_{n\to\infty} T_n(x), \qquad T_n(x) := S(\theta_{x,x})(x)[\langle x,x\rangle + 1/n]^{-1},$$

$$S \in \mathbf{LM}(\mathcal{K}(\mathcal{M})), \quad x \in \mathcal{M}.$$

By the same argument as in Theorem 5.3.1, the above limit exists. It defines an A-linear map. The boundedness of the operator $\psi(S) : \mathcal{M} \to \mathcal{M}$ and the continuity of ψ can be verified as follows. For any $x \in \mathcal{M}$, $\|x\| \le 1$, we have (see [**100**, 3.11] and Proposition 3.4.1) $x = u\langle x,x\rangle^{1/2}$, where $u \in (\mathcal{M}^\#)'$, and one has $u\langle x,x\rangle^\beta \in \mathcal{M}$ for any $\beta > 0$. Put

$$y := u\langle x,x\rangle^{3\alpha - 1/2} \in \mathcal{M}, \qquad z := u\langle x,x\rangle^{1/2-\alpha} \in \mathcal{M}, \qquad \frac{1}{4} < \alpha < \frac{1}{2}.$$

Then

$$\langle x,y\rangle = \langle x,x\rangle^{3\alpha}, \qquad \langle z,z\rangle = \langle x,x\rangle^{1-2\alpha}$$

and

$$\theta_{z,z}\theta_{z,y}(v) = z\langle z,z\rangle\langle y,v\rangle = u\langle x,x\rangle^{1/2-\alpha}\langle x,x\rangle^{1-2\alpha+3\alpha-1}\langle x,v\rangle = \theta_{x,x}(v).$$

For any $n = 1, 2, \dots$, one has

$$\begin{aligned}\langle T_n x, T_n x\rangle &= [\langle x,x\rangle + 1/n]^{-1}\,\langle S(\theta_{x,x})(x), S(\theta_{x,x})(x)\rangle\,[\langle x,x\rangle + 1/n]^{-1}\\ &= [\langle x,x\rangle + 1/n]^{-1}\,\langle S(\theta_{z,z})\theta_{z,y}(x), S(\theta_{z,z})\theta_{z,y}(x)\rangle\,[\langle x,x\rangle + 1/n]^{-1}\\ &= [\langle x,x\rangle + 1/n]^{-1}\,\langle y,x\rangle^*\langle S(\theta_{z,z})z, S(\theta_{z,z})z\rangle\langle y,x\rangle\,[\langle x,x\rangle + 1/n]^{-1}\\ &\le \|S(\theta_{z,z})\|^2[\langle x,x\rangle + 1/n]^{-1}\,\langle y,x\rangle^*\langle z,z\rangle\langle y,x\rangle\,[\langle x,x\rangle + 1/n]^{-1}\\ &= \|S(\theta_{z,z})\|^2[\langle x,x\rangle + 1/n]^{-1}\,\langle x,x\rangle^{4\alpha+1}\,[\langle x,x\rangle + 1/n]^{-1},\end{aligned}$$

whence, as $n \longrightarrow \infty$, we obtain that

$$\langle \psi(S)x, \psi(S)x\rangle \le \|S\|^2\langle x,x\rangle^{4\alpha-1}, \qquad 1/4 < \alpha < 1/2.$$

Passing here to the limit as $\alpha \longrightarrow 1/2$, we get the estimate

$$\langle \psi(S)x, \psi(S)x\rangle \le \|S\|^2\langle x,x\rangle.$$

Thus, $\|\psi(S)\| \le \|S\|$ and, by 2.1.4, $\psi(S)$ is an A-homomorphism. Hence $\psi(S) \in \mathrm{End}_A(\mathcal{M})$ and $\|\psi\| \le 1$.

Let us show that $\phi\psi = \mathrm{Id}_{\mathbf{LM}}$. For this purpose it is sufficient to verify that $S(\theta_{x,y}) = \psi(S)\circ\theta_{x,y} = \theta_{\psi(S)x,y}$. Using (5.4) and the notation from it, we obtain

$$\begin{aligned}S(\theta_{x,y})(z) &= \lim_{n\to\infty} S(\theta_{x,x}\theta_{xx_n,y})(z) = \lim_{n\to\infty} S(\theta_{x,x})\theta_{xx_n,y}(z)\\ &= \lim_{n\to\infty} S(\theta_{x,x})xx_n\langle y,z\rangle = \psi(S)(x)\langle y,z\rangle = \psi(S)\circ\theta_{x,y}(z).\end{aligned}$$

Since the norm of ϕ and ψ does not exceed 1, ϕ is an isometry. □

5.4. Quasi-multipliers of A-compact operators

In this section we present a modified proof of Theorem 1.6 of [**73**]. Note that this theorem and similar statements about left and double multipliers can be deduced from some general results about multipliers (see [**103**]).

THEOREM 5.4.1. *Let $\mathcal{M}$ be a Hilbert A-module. Then the map ϕ from Theorem 5.3.2 can be extended up to an isometric involutive isomorphism*

$$\phi : \mathrm{End}_A(\mathcal{M},\mathcal{M}') \longrightarrow \mathbf{QM}(\mathcal{K}(\mathcal{M})).$$

PROOF. The formula

$$\phi(T)(\theta_{x',y'},\theta_{x,y}) := \theta_{x',y\cdot(T(x)(y'))}, \qquad x,\, y,\, x',\, y' \in \mathcal{M}, \quad T \in \mathrm{End}_A(\mathcal{M},\mathcal{M}'),$$

obviously is bilinear, thus, it defines a bilinear map on a dense subset in $\mathcal{K}(\mathcal{M}) \times \mathcal{K}(\mathcal{M})$ with values in $\mathcal{K}(\mathcal{M})$.

Let us estimate the norm of ϕ. Let $x = u\langle x,x\rangle^{1/2}$ be the polar decomposition of x in $(\mathcal{M}^\#)'$. Let $w_\varepsilon := u\langle x,x\rangle^\varepsilon$, where $0 < \varepsilon < 1/2$. Recall that the structure of a right module on $\mathcal{M}'$ is defined by $(\varphi a)(y) = a^*\varphi(y)$ for $a \in A$, $\varphi \in \mathcal{M}'$ and

$y \in \mathcal{M}$. By 2.1.4, we obtain that

$$\begin{aligned}
\|\theta_{x',y(T(x)(y'))}(z)\|^2 &= \|\langle z,y\rangle (T(x)(y'))\langle x',x'\rangle (T(x)(y'))^*\langle y,z\rangle\| \\
&= \|\langle z,y\rangle \langle x,x\rangle^{\frac{1}{2}-\varepsilon}[T(w_\varepsilon)(y')]\langle x',x'\rangle [T(w_\varepsilon)(y')]^*\langle x,x\rangle^{\frac{1}{2}-\varepsilon}\langle y,z\rangle\| \\
&\le \|\langle x',x'\rangle^{1/2}[T(w_\varepsilon)(y')]^*\|^2\cdot \|\langle x,x\rangle^{\frac{1}{2}-\varepsilon}\langle y,z\rangle\|^2 \\
&= \|\langle x',x'\rangle^{1/2}[T(w_\varepsilon)(y')]^*[T(w_\varepsilon)(y')]\langle x',x'\rangle^{1/2}\|\cdot \|\langle x,x\rangle^{\frac{1}{2}-\varepsilon}\langle y,z\rangle\|^2 \\
&\le \|T(w_\varepsilon)\|^2\cdot \|\langle x',x'\rangle^{1/2}\langle y',y'\rangle \langle x',x'\rangle^{1/2}\|\cdot \|\langle x,x\rangle^{\frac{1}{2}-\varepsilon}\langle y,z\rangle\langle z,y\rangle\langle x,x\rangle^{\frac{1}{2}-\varepsilon}\| \\
&\le \|T(w_\varepsilon)\|^2\cdot \|\langle x',x'\rangle^{1/2}\langle y',y'\rangle^{1/2}\|^2\cdot \|\langle x,x\rangle^{\frac{1}{2}-\varepsilon}\langle y,y\rangle^{1/2}\|\cdot\|z\|^2.
\end{aligned}$$

Passing to the limit $\varepsilon \longrightarrow 0$, we obtain $\|w_\varepsilon\| \longrightarrow 1$ and

$$\|\theta_{x',y(T(x)(y'))}(z)\| \le \|T\|\cdot \|\langle x',x'\rangle^{1/2}\langle y',y'\rangle^{1/2}\|\cdot \|\langle x,x\rangle^{1/2}\langle y,y\rangle^{1/2}\|\cdot\|z\|.$$

Therefore $\|\phi(T)(\theta_{x',y'},\theta_{x,y})\| \le \|T\|\cdot\|\theta_{x',y'}\|\cdot\|\theta_{x,y}\|$. Thus, $\phi(T):\mathcal{K}(\mathcal{M})\times\mathcal{K}(\mathcal{M})\to \mathcal{K}(\mathcal{M})$. Since

$$\begin{aligned}
\phi(T)(k'\theta_{x',y'},\theta_{x,y}k) &= \phi(T)(\theta_{k'x',y'},\theta_{x,k^*y}) \\
&= \theta_{k'x',k^*y\cdot(T(x)(y'))} = k'\theta_{x',y\cdot(T(x)(y'))}k,
\end{aligned} \tag{5.5}$$

we have $\phi(T) \in \mathbf{QC}(\mathcal{K}(\mathcal{M}))$. Moreover, the previous calculation shows that ϕ is continuous: $\|\phi\| \le 1$.

Let us show the algebraic injectivity of ϕ, i.e., that $\operatorname{Ker}\phi = 0$. Let $T \ne 0$. This means that there exists a vector $x \in \mathcal{M}$ such that $T(x)$ is a nonzero functional, $T(x)(y') \ne 0$ for some (nonzero) $y' \in \mathcal{M}$. Then

$$(T(x)(y'))^*T(x)(y') \le \|T(x)\|^2\,\langle y',y'\rangle, \qquad \langle y',y'\rangle \ge \|T(x)\|^{-2}(T(x)(y'))^*T(x)(y').$$

Therefore

$$\begin{aligned}
&\langle y'\cdot(T(x)(y'))^*(T(x)(y')), y'\cdot(T(x)(y'))^*(T(x)(y'))\rangle \\
&\quad = (T(x)(y'))^*(T(x)(y'))\langle y',y'\rangle(T(x)(y'))^*(T(x)(y')) \\
&\quad \ge \|T(x)\|^{-2}\{\,(T(x)(y'))^*T(x)(y')\,\}^3 \ne 0
\end{aligned}$$

and

$$\begin{aligned}
&\langle \phi(T)(\theta_{y'\cdot(T(x)(y'))^*(T(x)(y')),y'},\theta_{x,y'\cdot(T(x)(y'))^*})[y'\cdot(T(x)(y'))^*(T(x)(y'))], \\
&\quad \phi(T)(\theta_{y'\cdot(T(x)(y'))^*(T(x)(y')),y'},\theta_{x,y'\cdot(T(x)(y'))^*})[y'\cdot(T(x)(y'))^*(T(x)(y'))]\rangle \\
&= \langle \theta_{y'\cdot(T(x)(y'))^*(T(x)(y')),y'\cdot(T(x)(y'))^*(T(x)(y'))}[y'\cdot(T(x)(y'))^*(T(x)(y'))], \\
&\quad \theta_{y'\cdot(T(x)(y'))^*(T(x)(y')),y'\cdot(T(x)(y'))^*(T(x)(y'))}[y'\cdot(T(x)(y'))^*(T(x)(y'))]\rangle \\
&= \langle y'\cdot(T(x)(y'))^*(T(x)(y'))\cdot\langle y'\cdot(T(x)(y'))^*(T(x)(y')), y'\cdot(T(x)(y'))^*(T(x)(y'))\rangle, \\
&\quad y'\cdot(T(x)(y'))^*(T(x)(y'))\cdot\langle y'\cdot(T(x)(y'))^*(T(x)(y')), y'\cdot(T(x)(y'))^*(T(x)(y'))\rangle\rangle \\
&= \langle y'\cdot(T(x)(y'))^*(T(x)(y')), y'\cdot(T(x)(y'))^*(T(x)(y'))\rangle^3 \ne 0.
\end{aligned}$$

Thus, $\phi(T) \ne 0$ and the algebraic injectivity of ϕ is proved.

To prove that the map ϕ is surjective and isometric, it is sufficient to define, as in the previous theorem, a map

$$\psi : \mathbf{QC}(\mathcal{K}(\mathcal{M})) \longrightarrow \operatorname{End}(\mathcal{M},\mathcal{M}'), \qquad \phi\psi(S) = S, \qquad \|\psi\| \le 1.$$

For any $S \in \mathbf{QC}(\mathcal{K}(\mathcal{M}))$ and any $k \in \mathcal{K}(\mathcal{M})$, the map $S(k,\cdot\,):\mathcal{K}(\mathcal{M})\to\mathcal{K}(\mathcal{M})$ is a left centralizer, or, in terms of multipliers, for any $S \in \mathbf{QM}(\mathcal{K}(\mathcal{M}))$ and any $k \in \mathcal{K}(\mathcal{M})$, the element $kS \in (K(\mathcal{M}))^{!!}$ is a left multiplier. Then the map $\psi : \mathbf{LM}(\mathcal{K}(\mathcal{M})) \to \mathrm{End}(\mathcal{M})$ from the previous theorem is applicable to it. To distinguish between different maps, we denote the maps constructed in the previous theorem by ϕ' and ψ'. To define ψ, put, for each $x, y \in \mathcal{M}$,

$$(\psi(S)(x))(y) := \lim_{n\to\infty} T_n(x, y),$$

$$T_n(x, y) := \langle \psi'(\theta_{y,y}S)(x), y\rangle \left(\langle y, y\rangle + \frac{1}{n}\right)^{-1}. \tag{5.6}$$

We have to verify the following properties:

1: existence of the limit in (5.6);
2: the linearity (over A and $\mathbf{C}$) of this expression in x and y;
3: the estimate $\|(\psi(S)(x))(y)\| \le \|S\|\,\|x\|\,\|y\|$;
4: the identity $\phi\psi(S) = S$.

Let $y = u \cdot \langle y, y\rangle^{1/2}$, $u \in (\mathcal{M}^{\#})'$. Put

$$z_1 = u \cdot \langle y, y\rangle^{3\alpha - \frac{1}{2}}, \qquad z_2 = u \cdot \langle y, y\rangle^{\frac{1}{2}-\alpha}, \qquad \frac{1}{4} < \alpha < \frac{1}{2}.$$

Then $y = z_1 \cdot \langle y, y\rangle^{1-3\alpha}$ and $y = z_2 \cdot \langle y, y\rangle^{\alpha}$, hence $z_1 \in \mathcal{M}$ and $z_2 \in \mathcal{M}$, since $1 - 3\alpha < 1/4 < 1/2$ and $\alpha < 1/2$. For $n = 1, 2, \dots,$ we have (similarly to the proof of Theorem 5.3.2)

$$\begin{aligned}
\langle \psi'(\theta_{y,y}S)(x), y\rangle \left[\langle y, y\rangle + \frac{1}{n}\right]^{-1} &= \langle \psi'(\theta_{z_1,z_2}\theta_{z_2,z_2}S)(x), y\rangle \left[\langle y, y\rangle + \frac{1}{n}\right]^{-1}\\
&= \langle \psi'(\theta_{z_1,z_2})\psi'(\theta_{z_2,z_2}S)(x), y\rangle \left[\langle y, y\rangle + \frac{1}{n}\right]^{-1}\\
&= \langle \theta_{z_1,z_2}\psi'(\theta_{z_2,z_2}S)(x), y\rangle \left[\langle y, y\rangle + \frac{1}{n}\right]^{-1}\\
&= \langle \psi'(\theta_{z_2,z_2}S)(x), z_2\rangle\, \langle z_1, y\rangle \left[\langle y, y\rangle + \frac{1}{n}\right]^{-1}\\
&= \langle \psi'(\theta_{z_2,z_2}S)(x), z_2\rangle\, \langle y, y\rangle^{3\alpha} \left[\langle y, y\rangle + \frac{1}{n}\right]^{-1}.
\end{aligned}$$

One can deduce three corollaries from this equality. At first, for any y with $\|y\| \le 1$ one has

$$\begin{aligned}
&\langle T_n(x, y), T_n(x, y)\rangle\\
&\quad\le \|\psi'(\theta_{z_2,z_2}S)(x)\|^2 \left[\langle y, y\rangle + \frac{1}{n}\right]^{-1} \langle y, y\rangle^{3\alpha}\langle z_2, z_2\rangle\, \langle y, y\rangle^{3\alpha} \left[\langle y, y\rangle + \frac{1}{n}\right]^{-1}\\
&\quad\le \|\theta_{z_2,z_2}\|^2\|S\|^2\|x\|^2\langle y, y\rangle^{6\alpha+(1-2\alpha)} \left[\langle y, y\rangle + \frac{1}{n}\right]^{-2}\\
&\quad\le \|y\|^{4(1-2\alpha)}\|S\|^2\|x\|^2\langle y, y\rangle^{1+4\alpha} \left[\langle y, y\rangle + \frac{1}{n}\right]^{-2}\\
&\quad\le \|y\|^{4(1-2\alpha)}\|S\|^2\|x\|^2\langle y, y\rangle^{2} \left[\langle y, y\rangle + \frac{1}{n}\right]^{-2} \le (\|S\|\,\|x\|)^2
\end{aligned}$$

and statement **3** is proved. Secondly, by fixing some α, we have

$$\langle T_n(x,y) - T_m(x,y), T_n(x,y) - T_m(x,y)\rangle$$
$$\leq \|y\|^{4(1-2\alpha)}\|S\|^2\|x\|^2\langle y,y\rangle^{1+4\alpha}\left[\left(\langle y,y\rangle + \frac{1}{n}\right)^{-1} - \left(\langle y,y\rangle + \frac{1}{m}\right)^{-1}\right]^2 \longrightarrow 0,$$

since $4\alpha > 1$. So statement **1** is proved. Also,

$$\langle T_n(x,y), T_n(x,y)\rangle \leq \|y\|^{4(1-2\alpha)}\|S\|^2\|x\|^2\langle y,y\rangle^{1+4\alpha}\left[\langle y,y\rangle + \frac{1}{n}\right]^{-2}$$
$$\leq \|y\|^{4(1-2\alpha)}\|S\|^2\|x\|^2\langle y,y\rangle^{4\alpha-1} \longrightarrow \|S\|^2\|x\|^2\langle y,y\rangle \qquad (\alpha \longrightarrow \frac{1}{2}),$$

and this proves **2** by 2.1.4, as the linearity in x is clear.

To prove **4** it is sufficient to verify, for elementary compact operators, that

$$S(\theta_{x',y'}, \theta_{x,y}) = \theta_{x',y\cdot(\psi(S)(x)(y'))}, \qquad S \in \mathbf{QC}(\mathcal{K}(\mathcal{M})), \quad x,\ y,\ x',\ y' \in \mathcal{M}.$$

Let

$$x = u\cdot\langle x,x\rangle^{1/2}, \quad y' = u'\cdot\langle y',y'\rangle^{1/2}, \quad v := u\cdot\langle x,x\rangle^{1/6} \in \mathcal{M},$$

$$v' := u'\cdot\langle y',y'\rangle^{1/6} \in \mathcal{M}, \quad u,\ u' \in (\mathcal{M}^{\#})'.$$

Then for $w := u\langle x,x\rangle^{1/3} \in \mathcal{M}$ and any $z \in \mathcal{M}$, we have

$$\theta_{v,v}\theta_{w,w}(z) = v\cdot\langle x,x\rangle^{\frac{1}{6}+\frac{1}{3}-\frac{1}{6}}\langle u\cdot\langle x,x\rangle^{1/2}, z\rangle = x\cdot\langle x,z\rangle = \theta_{x,x}(z), \tag{5.7}$$

while $\theta_{w,w}(x) = x\cdot\langle x,x\rangle^{-\frac{1}{6}-\frac{1}{6}+1} = x\cdot\langle x,x\rangle^{2/3}$. Therefore, if $T \in \mathbf{LM}(\mathcal{K}(\mathcal{M}))$, then for $w' := u\cdot\langle x,x\rangle^{1/12}$

$$\begin{aligned}
\psi'(T)(x) &= \lim_{n\to\infty}(T(\theta_{x,x}))(x)\left[\langle x,x\rangle + \frac{1}{n}\right]^{-1} \\
&= \lim_{n\to\infty}(T(\theta_{v,v}\theta_{w,w}))(x)\left[\langle x,x\rangle + \frac{1}{n}\right]^{-1} \\
&= \lim_{n\to\infty}(T(\theta_{v,v}))x\cdot\langle x,x\rangle^{2/3}\left[\langle x,x\rangle + \frac{1}{n}\right]^{-1} \\
&= \lim_{n\to\infty}(T(\theta_{v,v}))w'\cdot\langle x,x\rangle^{\frac{5}{12}+\frac{2}{3}}\left[\langle x,x\rangle + \frac{1}{n}\right]^{-1} \\
&= (T(\theta_{v,v}))w'\cdot\langle x,x\rangle^{\frac{13}{12}-1} = (T(\theta_{v,v}))v.
\end{aligned}$$

Similarly to (5.7), we obtain that

$$\theta_{y',y'} = \theta_{w'',w''}\theta_{v',v'}, \qquad w'' := v'\langle y',y'\rangle^{1/3} \in \mathcal{M}.$$

Then, after putting $w_* := v'\langle y', y'\rangle^{1/12} \in \mathcal{M}$, we have

$$\begin{aligned}
&\lim_{n\to\infty} \langle \psi'(\theta_{y',y'}S)(x), y'\rangle \, [\langle y', y'\rangle + 1/n]^{-1} \\
&\quad = \lim_{n\to\infty} \langle \psi'(\theta_{w'',w''}\theta_{v',v'}S)(x), y'\rangle \, [\langle y', y'\rangle + 1/n]^{-1} \\
&\quad = \lim_{n\to\infty} \langle \theta_{w'',w''}\psi'(\theta_{v',v'}S)(x), y'\rangle \, [\langle y', y'\rangle + 1/n]^{-1} \\
&\quad = \lim_{n\to\infty} \langle w'' \cdot \langle w'', \psi'(\theta_{v',v'}S)(x)\rangle, y'\rangle \, [\langle y', y'\rangle + 1/n]^{-1} \\
&\quad = \lim_{n\to\infty} \langle \psi'(\theta_{v',v'}S)(x), w''\rangle \, \langle w'', y'\rangle \, [\langle y', y'\rangle + 1/n]^{-1} \\
&\quad = \lim_{n\to\infty} \langle \psi'(\theta_{v',v'}S)(x), w_*\rangle \, \langle y', y'\rangle^{\frac{1}{3}-\frac{1}{12}-\frac{1}{6}+1} \, [\langle y', y'\rangle + 1/n]^{-1} \\
&\quad = \lim_{n\to\infty} \langle \psi'(\theta_{v',v'}S)(x), w_*\rangle \, \langle y', y'\rangle^{\frac{13}{12}} \, [\langle y', y'\rangle + 1/n]^{-1} \\
&\quad = \langle \psi'(\theta_{v',v'}S)(x), w_*\rangle \, \langle y', y'\rangle^{\frac{1}{12}} = \langle \psi'(\theta_{v',v'}S)(x), v'\rangle,
\end{aligned}$$

whence

$$\begin{aligned}
\theta_{x',y\cdot(\psi(S)(x)(y'))} &= \lim_{n\to\infty} \theta_{x',y\cdot\langle\psi'(\theta_{y',y'}S)(x),(y')\rangle\,[\langle y',y'\rangle+1/n]^{-1}} \\
&= \theta_{x',y\cdot\langle\psi'(\theta_{v',v'}S)(x),v'\rangle} = \theta_{x',y\cdot\langle(\theta_{v',v'}S\theta_{v,v})(v),v'\rangle} \\
&= \theta_{x',y\cdot\langle S(\theta_{v',v'},\theta_{v,v})(v),v'\rangle}.
\end{aligned}$$

In the last expression we used the presentation of quasi-multipliers in the form of quasi-centralizers. On the other hand, $y' = v' \cdot \langle v', v'\rangle$, $x = v \cdot \langle v, v\rangle$ and

$$\begin{aligned}
S(\theta_{x',y'}, \theta_{x,y}) &= S(\theta_{x',v'}\theta_{v',v'}, \theta_{v,v}\theta_{v,y}) = \theta_{x',v'}S(\theta_{v',v'}, \theta_{v,v})\theta_{v,y} \\
&= \theta_{x',v'}\theta_{S(\theta_{v',v'},\theta_{v,v})v,y} = \theta_{x',y\cdot\langle S(\theta_{v',v'},\theta_{v,v})v,v'\rangle}.
\end{aligned}$$

Statement **4** is proved, and this completes the proof of the theorem. □

5.5. Strict topology

DEFINITION 5.5.1. Let $A \hookrightarrow \mathcal{B}(H)$ be a nondegenerate faithful representation of a C^*-algebra A. The *strict topology* on $\mathcal{B}(H)$ is the topology satisfying one of the following (obviously, equivalent) conditions:

(i) it is the weakest topology, for which the maps

$$\begin{aligned}
&r_a : \mathcal{B}(H) \to \mathcal{B}(H), \qquad r_a : x \mapsto xa, \\
&l_a : \mathcal{B}(H) \to \mathcal{B}(H), \qquad l_a : x \mapsto ax, \qquad x \in \mathcal{B}(H),\ a \in A,
\end{aligned}$$

are continuous;

(ii) it is the topology generated by the system of seminorms

$$\{\nu_a^R, \nu_a^L\}_{a\in A}, \qquad \nu_a^R(x) := \|xa\|, \qquad \nu_a^L(x) := \|ax\|.$$

Usually this topology is denoted by β analogously with the Stone-Čech compactification (cf. 5.2.12). For example, we denote by $[X]_\beta$ the space of the maximal ideals for the closure of the algebra $C(X)$ in $\mathcal{B}(H)$ with respect to the strict topology, and we denote by β-lim the limit with respect to the strict topology.

PROPOSITION 5.5.2. *The set* $\mathbf{M}(A)$ *is strictly closed,*

$$[A]_\beta \subset [\mathbf{M}(A)]_\beta = \mathbf{M}(A).$$

PROOF. Let the net $\{x_\alpha\}_{\alpha\in\mathcal{A}} \subset \mathbf{M}(A)$ be strictly convergent to $x \in \mathcal{B}(H)$. Then for any $a \in A$ there exist the norm-limits

$$L(a) := \lim_\alpha x_\alpha a, \qquad R(a) := \lim_\alpha a x_\alpha$$

defining the maps $L,\ R : A \to A$. Since

$$aL(b) = a\ \lim_\alpha(x_\alpha b) = \lim_\alpha(a x_\alpha b) = \Big(\lim_\alpha(a x_\alpha)\Big)\, b = R(a)b,$$

the pair (L, R) is an element of $\mathbf{DC}(A)$, i.e., a double centralizer. Identifying double centralizers with multipliers, by Theorem 5.2.11 we obtain that $y \in \mathbf{M}(A)$. Then $x_\alpha \xrightarrow{\beta} y$. Indeed,

$$ya - x_\alpha a = L(a) - x_\alpha a \longrightarrow 0, \qquad ay - a x_\alpha = R(a) - a x_\alpha \longrightarrow 0$$

with respect to the norm topology. So, $\mathbf{M}(A)$ is β-closed. □

LEMMA 5.5.3. *The net $\{e_\alpha\}_{\alpha\in\mathcal{A}}$ is an approximate unit for A if and only if $\mathbf{M}(A) \ni 1 = \beta\text{-}\lim_\alpha e_\alpha$.*

PROOF. The result follows immediately from the definition. □

PROPOSITION 5.5.4. (i) *The conjugation in $\mathbf{M}(A)$ is β-continuous.*
(ii) *Multiplication in $\mathbf{M}(A)$ by a fixed element is β-continuous.*
(iii) *Multiplication in $\mathbf{M}(A)$ is jointly β-continuous on bounded sets.*

PROOF. Let $x \in \mathbf{M}(A)$ and let $x_\alpha \xrightarrow{\beta} x$. Then

$$x_\alpha a \xrightarrow{\|\cdot\|} xa, \qquad a x_\alpha \xrightarrow{\|\cdot\|} ax \quad \text{for any } a \in A,$$

whence, after conjugation,

$$b x_\alpha^* \xrightarrow{\|\cdot\|} b x^*, \qquad x_\alpha^* b \xrightarrow{\|\cdot\|} x^* b \quad \text{for any } b \in A \qquad (b = a^*),$$

i.e., $x_\alpha^* \xrightarrow{\beta} x^*$.

Now let $y \in \mathbf{M}(A)$ be a fixed element and $x_\alpha \xrightarrow{\beta} x$. Then

$$\begin{aligned} \|(x_\alpha y)a - (xy)a\| &= \|x_\alpha(ya) - x(ya)\| \longrightarrow 0, \\ \|a(x_\alpha y) - a(xy)\| &\le \|a x_\alpha - ax\| \cdot \|y\| \longrightarrow 0 \end{aligned}$$

for any $a \in A$. This means that $x_\alpha y \xrightarrow{\beta} xy$.

Now let $x_\alpha \xrightarrow{\beta} x$, $\|x_\alpha\| < c_x$ for any $\alpha \in \mathcal{A}$ and $y_\gamma \xrightarrow{\beta} y$, $\|y_\gamma\| < c_y$ for any $\gamma \in \Gamma$. Then, for any $a \in A$ and for any $\varepsilon > 0$, there exists a pair (α_0, γ_0) such that for any pair $(\alpha, \gamma) > (\alpha_0, \gamma_0) \in \mathcal{A} \times \Gamma$ (i.e., for $\alpha > \alpha_0$ and $\gamma > \gamma_0$) one has

$$\begin{aligned} \|(x_\alpha y_\gamma)a - (xy)a\| &\le \|x_\alpha y_\gamma a - x_\alpha y a\| + \|x_\alpha y a - x y a\| \\ &\le c_x \cdot \|y_\gamma a - ya\| + \|x_\alpha(ya) - x(ya)\| < \varepsilon, \\ \|a(x_\alpha y_\gamma) - a(xy)\| &\le \|a x_\alpha y_\gamma - a x y_\gamma\| + \|a x y_\gamma - a x y\| \\ &\le \|a x_\alpha - ax\| \cdot c_y + \|(ax)y_\gamma - (ax)y\| < \varepsilon. \end{aligned}$$

This means that $(xy) = \beta\text{-}\lim_{(\alpha,\gamma)\in\mathcal{A}\times\Gamma}(x_\alpha y_\gamma)$. □

THEOREM 5.5.5. *The algebra of multipliers $\mathbf{M}(A)$ coincides with the β-closure of A in $\mathcal{B}(H)$,*

$$\mathbf{M}(A) = [A]_\beta.$$

PROOF. By Proposition 5.5.2, it is sufficient to prove that $\mathbf{M}(A) \subset [A]_\beta$. Let $\{e_\alpha\}_{\alpha\in\mathcal{A}}$ be an approximate unit for A. By Lemma 5.5.3, $e_\alpha \xrightarrow{\beta} 1 \in \mathbf{M}(A)$. Let $x \in \mathbf{M}(A)$ be an arbitrary element. Consider the net xe_α. Since x is a multiplier, the elements of this net belong to A and, by Proposition 5.5.4, the net is β-convergent. Hence, $y = \beta\text{-}\lim_\alpha(xe_\alpha) \in [A]_\beta$. But $y = x$. Indeed, for any $a \in A$, one has

$$ay = a\,\beta\text{-}\lim_\alpha(xe_\alpha) = \lim_\alpha(axe_\alpha) = ax,$$
$$yb = (\beta\text{-}\lim_\alpha(xe_\alpha))b = \lim_\alpha(xe_\alpha b) = xb,$$

so x is the same multiplier as y. □

DEFINITION 5.5.6. Let $A \hookrightarrow \mathcal{B}(H)$ be a nondegenerate faithful representation of a C^*-algebra A. The *left strict* topology on $\mathcal{B}(H)$ is the topology satisfying one of the following equivalent conditions:

(i) it is the weakest topology for which the maps
$$r_a : \mathcal{B}(H) \to \mathcal{B}(H), \quad r_a : x \mapsto xa, \qquad x \in \mathcal{B}(H),\ a \in A,$$
are continuous;

(ii) it is the topology generated by the system of seminorms
$$\{\nu_a^R\}_{a\in A}, \qquad \nu_a^R(x) := \|xa\|.$$

For the left strict topology, one can prove the following analog of Theorem 5.5.5.

THEOREM 5.5.7. *The algebra of left multipliers* $\mathbf{LM}(A)$ *coincides with the closure of* A *in* $A^{!!}$ *with respect to the left strict topology.*

PROOF. This statement can be obtained by the same method as in 5.5.5, it is sufficient to use only the "left half" of the argument. □

For Hilbert modules, it is natural to consider the following two topologies on the space of bounded homomorphisms.

DEFINITION 5.5.8. Let $\mathcal{M}$ be a Hilbert A-module. The *strong module* topology on $\mathrm{End}(\mathcal{M})$ is the topology generated by the system of seminorms
$$\{s^h\}_{h\in\mathcal{M}}, \qquad s^h(x) := \|x(h)\|, \quad x \in \mathrm{End}(\mathcal{M}),$$
and the $*$*-strong module* topology on $\mathrm{End}^*(\mathcal{M})$ is the topology generated by the system of seminorms
$$\{s^h; s_*^h\}_{h\in\mathcal{M}}, \qquad s^h(x) := \|x(h)\|, \quad s_*^h(x) := \|x^*(h)\|, \quad x \in \mathrm{End}^*(\mathcal{M}).$$

PROPOSITION 5.5.9. *The strong topology is not weaker than the $*$-strong module topology on bounded sets in* $\mathcal{K}(\mathcal{M})$ *(hence, by Theorems* 5.5.5 *and* 5.3.1, *everywhere on* $\mathrm{End}^*(\mathcal{M})$*).*

The left strong topology is not weaker than the strong module topology on $\mathcal{K}(\mathcal{M})$ *(hence, by Theorems* 5.5.7 *and* 5.3.2, *everywhere on* $\mathrm{End}(\mathcal{M})$*).*

The corresponding topologies coincide on bounded sets in $\mathcal{K}(\mathcal{M})$.

PROOF. We will verify the equivalence of the appropriate seminorms. We have
$$s^h(x) = \|x(h)\| = \|xk(g)\| \le \|xk\| \cdot \|g\| = \|g\| \cdot \nu_k^R(x)$$
for some $k \in \mathcal{K}(\mathcal{M})$, $g \in \mathcal{M}$ (see Lemma 2.2.3) and
$$s_*^h(x) = \|x^*(h)\| = \|x^*k(g)\| \le \|x^*k\| \cdot \|g\| = \|k^*x\| \cdot \|g\| = \|g\| \cdot \nu_{k^*}^L(x)$$

for some $k \in \mathcal{K}(\mathcal{M})$, $g \in \mathcal{M}$. Conversely, let $k \in \mathcal{K}(\mathcal{M})$ be an arbitrary element and let $x_\alpha \to 0$ with respect to the strong module topology. Assume that the net x_α is bounded: $\|x_\alpha\| < c$. Then for any $\varepsilon > 0$ there exist vectors $h_1, \ldots, h_n$ and $g_1, \ldots, g_n$ from $\mathcal{M}$ such that

$$\left\| k - \sum_{i=1}^{n} \theta_{h_i, g_i} \right\| < \frac{\varepsilon}{c}$$

and there is α_0 such that

$$\|x_\alpha(h_i)\| < \frac{\varepsilon}{n \cdot \|g_i\|}, \qquad i = 1, \ldots, n,$$

for any $\alpha > \alpha_0$. Then, for these α,

$$\begin{aligned} \nu_k^R(x_\alpha) = \|x_\alpha k\| &\leq \frac{\varepsilon}{c} \cdot \|x_\alpha\| + \sum_{i=1}^{n} \|x_\alpha \theta_{h_i, g_i}\| \\ &\leq c \cdot \frac{\varepsilon}{c} + \sum_{i=1}^{n} \|x_\alpha h_i\| \cdot \|g_i\| \leq \varepsilon + n \cdot \frac{\varepsilon}{n \cdot \|g_i\|} \cdot \|g_i\| < 2\varepsilon. \end{aligned}$$

Similarly, if $x_\alpha \to 0$ with respect to the $*$-strong module topology, then we can additionally require that

$$\|x_\alpha^*(g_i)\| < \frac{\varepsilon}{n \cdot \|h_i\|}, \qquad i = 1, \ldots, n,$$

for $\alpha > \alpha_0$. Then

$$\begin{aligned} \nu_k^L(x_\alpha) = \|k x_\alpha\| &\leq \frac{\varepsilon}{c} \cdot \|x_\alpha\| + \sum_{i=1}^{n} \|\theta_{h_i, g_i} x_\alpha\| \\ &\leq c \cdot \frac{\varepsilon}{c} + \sum_{i=1}^{n} \|h_i\| \cdot \|x_\alpha^* g_i\| \leq \varepsilon + \sum_{i=1}^{n} \frac{\varepsilon}{n \cdot \|g_i\|} \|g_i\| = 2\varepsilon. \end{aligned}$$

□

PROPOSITION 5.5.10. *Let A and B be C^*-algebras and let $\psi : \mathbf{M}(A) \to \mathbf{M}(B)$ be a morphism such that $B \subset \psi(A)$. Then ψ is strictly continuous.*

PROOF. Let $x_\alpha \xrightarrow{\beta} x$ in $\mathbf{M}(A)$ and let b be an arbitrary element of B. Then $b = \psi(a)$ for some $a \in A$. The nets $x_\alpha a$ and $a x_\alpha$ are norm-convergent in A. The map ψ, being a homomorphism of C^*-algebras, does not increase the norm. Therefore $\psi(\chi_\alpha a) = \psi(x_\alpha) b$ and $\psi(a x_\alpha) = b \psi(x_\alpha)$ are Cauchy nets, so they are norm-convergent. Since b was arbitrary, this means that $\psi(x_\alpha) \xrightarrow{\beta} y \in \mathbf{M}(B)$. Thus, for any $b' \in B$, $b' = \psi(a')$, one has

$$y b' = \beta\text{-}\lim_\alpha \psi(x_\alpha) b' = \lim_\alpha \psi(x_\alpha) b' = \lim_\alpha \psi(x_\alpha a') = \psi(x a') = \psi(x) b'.$$

Thus, $y = \psi(x)$. □

REMARK 5.5.11. In particular, the map φ'' from Proposition 5.2.21 is the extension by continuity of the morphism φ, thereby, is unique.

A similar theory can be developed for quasi-multipliers, if one uses the following definition.

DEFINITION 5.5.12. Let $A \hookrightarrow \mathcal{B}(H)$ be a nondegenerate faithful representation of a C^*-algebra A. The *quasi-strict topology* on $\mathcal{B}(H)$ is the topology satisfying one of the following equivalent conditions:

(i) it is the weakest topology for which the maps

$$Q_{ab} : \mathcal{B}(H) \to \mathcal{B}(H), \quad Q_{ab} : x \mapsto axb, \qquad x \in \mathcal{B}(H),\ a,\, b \in A,$$

are continuous;

(ii) it is the topology generated by the system of seminorms

$$\{\nu_{ab}\}_{a,\, b \in A}, \qquad \nu_{ab}(x) := \|axb\|.$$

5.6. Multipliers and Hilbert modules. The commutative case

The following results describing the modules $l_2(C_0(X, A))$ and spaces of operators on them in terms of spaces of maps are obtained by combination and small modification of [**41, 1**].

DEFINITION 5.6.1. Denote by $C_0(X, \mathcal{M})$ (resp. by $A_0(X)$) the space of continuous maps $X \to \mathcal{M}$ (resp. $X \to A$) vanishing at infinity.

Note that $A_0(X)$ is a C^*-algebra with respect to the sup-norm and $C_0(X, \mathcal{M})$ is a Hilbert $A_0(X)$-module with the inner product given by $\langle f, g\rangle = \langle f(x), g(x)\rangle_{\mathcal{M}}$, where $f, g \in C_0(X, \mathcal{M})$, $x \in X$.

DEFINITION 5.6.2. Let X be a locally compact Hausdorff topological space and let $\mathcal{M}$ be a Hilbert A-module. Let us call the pair $(X, \mathcal{M})$ *compatible* if the following conditions are satisfied:

(i) the map

$$j : \mathcal{M} \otimes_{\mathbf{C}} C_0(X) \to C_0(X, \mathcal{M}), \quad j(m \otimes f)(x) := f(x)m, \quad m \in \mathcal{M},\ f \in C_0(X),$$

is an isometric isomorphism;

(ii) let $\varphi \in C_0(X, \mathcal{M})$ be such that $\varphi(x_0) = 0$ for some $x_0 \in X$ and let $F \in \mathrm{End}_{A_0(X)}(C_0(X, \mathcal{M}))$ be an arbitrary operator; then $(F\varphi)(x_0) = 0$.

REMARK 5.6.3. Here and further, by a tensor product we always mean the projective tensor product. In the case of C^*-algebras it is also called spatial or minimal. However in (i) (e.g. if $\mathcal{M} = A$) we use the tensor product with a commutative algebra, which is nuclear, so all C^*-norms on this tensor product coincide. For more details see [**94**, §6.3].

REMARK 5.6.4. Generalizing [**41**], we shall prove in the following two lemmas that the pair $(X, l_2(A))$ is compatible (in [**41**] the case of compact X is considered). Note that a weaker compatibility condition for an arbitrary pair (namely, if in (ii) we replace End by End*) follows from [**1**]. Thus, the second part (the one concerning adjointable operators) of Theorem 5.6.8 [**41**] below can be deduced from [**1**] using identification of $\mathrm{End}^*_A(\mathcal{M})$ with multipliers for $\mathcal{K}(\mathcal{M})$.

LEMMA 5.6.5. *The map*

$$j : l_2(A_0(X)) \to C_0(X, l_2(A)), \qquad j(f)(x) := (f_1(x), f_2(x), \dots),$$
$$f = (f_1, f_2, \dots) \in l_2(A_0(X)),$$

is an isometry.

PROOF. It is clear that j is an isometric inclusion. Let us show that it is surjective. Let $\varphi \in C_0(X, l_2(A))$ be an arbitrary element. Then, since

$$\|(\varphi(x))_i\| \leq \|\varphi(x)\|, \qquad \|(\varphi(x))_i - (\varphi(y))_i\| \leq \|\varphi(x) - \varphi(y)\|,$$

we have $(\varphi(x))_i \in A_0(X)$. It is necessary to verify that the series $\sum_i (\varphi(x))_i^*(\varphi(x))_i$ is norm-convergent. Take an arbitrary $\varepsilon > 0$ and let $K \subset X$ be a compact set such that the inequality $\|\varphi(y)\| < \varepsilon$ holds for any $y \in Y := X \setminus K$. For each point $x \in K$ choose a number $n(x)$ such that

$$\sum_{i=n(x)}^{\infty} (\varphi(x))_i^*(\varphi(x))_i < \frac{\varepsilon}{2}.$$

Since the map

$$X \xrightarrow{\varphi} l_2(A) \xrightarrow{1-p_n} L_n^{\perp}$$

is continuous, for each $x \in K$ one can find an open neighborhood U_x in K such that

$$\sum_{i=n(x)}^{\infty} (\varphi(z))_i^*(\varphi(z))_i < \varepsilon$$

for each $z \in U_x$. Due to the compactness of K one can choose a finite subcovering $U_{x_1}, \ldots, U_{x_s}$. Take $n := \max\{n_{x_1}, \ldots, n_{x_s}\}$. Then, for any $m > n$,

$$\begin{aligned}
&\sup_{x \in X} \left[\sum_{i=n}^{m} (\varphi(x))_i^*(\varphi(x))_i \right] \\
&\quad \leq \max \left\{ \sup_{x \in Y} \left[\sum_{i=n}^{m} (\varphi(x))_i^*(\varphi(x))_i \right], \sup_{x \in K} \left[\sum_{i=n}^{m} (\varphi(x))_i^*(\varphi(x))_i \right] \right\} \\
&\quad \leq \max \left\{ \sup_{x \in Y} \|\varphi(x)\|, \max_{j=1}^{s} \sup_{x \in U_j} \left[\sum_{i=n(x_j)}^{m} (\varphi(x))_i^*(\varphi(x))_i \right] \right\} \\
&\quad \leq \max \left\{ \varepsilon, \max_{j=1}^{s} \varepsilon \right\} = \varepsilon.
\end{aligned}$$

By the Cauchy criterion the series is convergent. □

LEMMA 5.6.6. *Let $\varphi \in C_0(X, l_2(A))$ and let $F \in \mathrm{End}_{A_0(X)}(C_0(X, l_2(A)))$ be an arbitrary operator. Suppose that $\varphi(x_0) = 0$ for some $x_0 \in X$. Then $(F\varphi)(x_0) = 0$.*

PROOF. By 2.1.4, one has $\langle F\varphi, F\varphi \rangle \leq \|F\|^2 \langle \varphi, \varphi \rangle$, so

$$\langle F\varphi(x_0), F\varphi(x_0) \rangle = \langle F\varphi, F\varphi \rangle(x_0) \leq \|F\|^2 \langle \varphi, \varphi \rangle(x_0) = \|F\|^2 \langle \varphi(x_0), \varphi(x_0) \rangle = 0.$$

□

DEFINITION 5.6.7. Denote by $\mathcal{B}(X, \mathrm{End}_A(\mathcal{M}))$ (resp. $\mathcal{B}^*(X, \mathrm{End}_A^*(\mathcal{M}))$) the algebra of bounded continuous maps from X to $\mathrm{End}_A(\mathcal{M}) = \mathbf{LM}(\mathcal{K}(\mathcal{M}))$ equipped with the left strict topology (resp. the algebra of bounded continuous maps from X to $\mathrm{End}_A^*(\mathcal{M}) = \mathbf{M}(\mathcal{K}(\mathcal{M}))$ equipped with the strict topology). By Proposition 5.5.9, one can consider the strong module (resp., $*$-strong module) topology instead of the left strict (resp., strict) topology. We equip these algebras, $\mathcal{B}(X, \mathrm{End}_A(\mathcal{M}))$ and $\mathcal{B}^*(X, \mathrm{End}_A^*(\mathcal{M}))$, with the sup-norm, making the first one a Banach algebra and the second one a C^*-algebra.

THEOREM 5.6.8. *Let $(X, \mathcal{M})$ be a compatible pair. Then*
(i) *the Banach algebras $\operatorname{End}_{A_0(X)}(\mathcal{M} \otimes C_0(X))$ and $\mathcal{B}(X, \operatorname{End}_A(\mathcal{M})))$ are naturally isomorphic;*
(ii) *the C^*-algebras $\operatorname{End}^*_{A_0(X)}(\mathcal{M} \otimes C_0(X))$ and $\mathcal{B}^*(X, \operatorname{End}^*_A(\mathcal{M})))$ are naturally isomorphic.*

PROOF. By condition (i) of Definition 5.6.2, we can identify $\mathcal{M} \otimes_{\mathbf{C}} C_0(X) = C_0(X, \mathcal{M})$ and define the map

$$J : \mathcal{B}(X, \operatorname{End}_A(\mathcal{M}))) \to \operatorname{End}_{A_0(X)}(C_0(X, \mathcal{M})),$$
$$(J(D)\varphi)(x) := D(x)(\varphi(x)), \quad x \in X, \quad \varphi \in C_0(X, \mathcal{M}).$$

First of all, let us show that $J(D)\varphi \in C_0(X, \mathcal{M})$. For this purpose we should verify that it is continuous and vanishes at infinity. Since $\|D(x)\|$ is bounded by some constant C, we can choose a compact set $K \subset X$ such that $\|\varphi(x)\| < \varepsilon/C$ outside K. We obtain that

$$\|(J(D)\varphi)(x)\| = \|D(x)(\varphi(x))\| < C \cdot \frac{\varepsilon}{C} = \varepsilon \qquad \text{outside} \quad K.$$

Vanishing at infinity is verified. To verify continuity, we choose an arbitrary $x \in X$ and $\varepsilon > 0$. Let us find an open neighborhood V_1 of the point x in X such that the estimate $\|\varphi(x) - \varphi(y)\| < \varepsilon/C$ holds for any $y \in V_1$. By the definition of continuity with respect to the strong module topology, for (a fixed element) $\varphi(x) \in \mathcal{M}$, there exists an open neighborhood V_2 of the point x in X such that the inequality

$$\|D(x)(\varphi(x)) - D(y)(\varphi(x))\|_{\mathcal{M}} < \varepsilon$$

holds for any $y \in V_2$. Then, for any $y \in U := V_1 \cap V_2$, one has

$$\begin{aligned}
&\|(J(D)\varphi)(x) - (J(D)\varphi)(y)\|_{\mathcal{M}} = \|D(x)(\varphi(x)) - D(y)(\varphi(y))\|_{\mathcal{M}} \\
&\le \|D(x)(\varphi(x)) - D(y)(\varphi(x))\|_{\mathcal{M}} + \|D(y)(\varphi(x)) - D(y)(\varphi(y))\|_{\mathcal{M}} \\
&< \varepsilon + C \cdot \frac{\varepsilon}{C} = 2\varepsilon,
\end{aligned}$$

so the continuity is proved.

The linearity over $\mathbf{C}$ and $A_0(X)$ of the operator $J(D)$ is clear. Since

$$\begin{aligned}
\|J(D)\varphi\|_{C_0(X,\mathcal{M})} &= \sup_{x \in X} \|D(x)(\varphi(x))\|_{\mathcal{M}} \\
&\le \sup_{x \in X} \|D(x)\|_{\operatorname{End}\mathcal{M}} \cdot \sup_{x \in X} \|\varphi(x)\|_{\mathcal{M}} \le C \cdot \|\varphi\|_{C_0(X,\mathcal{M})},
\end{aligned}$$

the operator $J(D)$ is bounded. Thereby the map J is well defined. It is clear that it is $\mathbf{C}$-linear and

$$\begin{aligned}
(J(DC)\varphi)(x) &= DC(x)(\varphi(x)) = D(x)[C(x)(\varphi(x))] \\
&= D(x)[(J(C)\varphi)(x)] = (J(D)J(C)\varphi)(x),
\end{aligned}$$

so that J is a homomorphism of algebras. Let us check the (algebraic) injectivity of J. Suppose that the relation $(J(D)\varphi)(x) = 0$, i.e., $D(x)\varphi(x) = 0$, holds for any $\varphi \in C_0(X, \mathcal{M})$ and any $x \in X$. Note that passing to the one-point compactification X^+ of the space X, we can define a continuous function $\chi^{x_0} : X^+ \to [0, 1]$ on the normal space X^+ that equals 1 at x_0 and vanishes at ∞. Then, for any $m \in \mathcal{M}$, the function $\chi^{x_0}_m(x) := m\chi^{x_0}(x)$ tends to zero at infinity. Since $0 = D(x_0)\chi^{x_0}_m(x_0) = D(x_0)m$ and since m and x_0 are arbitrary, we have $D(x_0) = 0$ for any x_0, i.e., $D = 0$.

Note that the estimate above, which proves the continuity of $J(D)$, also gives the inequality $\|J\| \leq 1$. Thus, the first part of the theorem will be proved if we can manage to define an inverse linear map

$$S : \operatorname{End}_{A_0(X)} C_0(X, \mathcal{M}) \to \mathcal{B}(X, \operatorname{End}_A(\mathcal{M})), \qquad \|S\| \leq 1.$$

Put

$$(S(T)(x))(m) := (T\varphi)(x), \tag{5.8}$$

where φ is a (nonuniquely defined) map $\varphi \in C_0(X, \mathcal{M})$, satisfying the condition $\varphi(x) = m$. Let us verify the independence of the definition from this nonunique choice of φ. Let $\varphi(x) = 0$. Then, by property (ii), one has $(T\varphi)(x) = 0$ for any operator T and we have proved that (5.8) is well defined. Once more, the linearity of $S(T)(x)$ is clear, so we should verify its boundedness. We have

$$\begin{aligned} \|(S(T)(x))(m)\| &= \|(T\chi^x_m)(x)\| \\ &\leq \|T\|_{\operatorname{End}(C_0(X,\mathcal{M}))} \cdot \|\chi^x_m\|_{C_0(X,\mathcal{M})} = \|T\|_{\operatorname{End}(C_0(X,\mathcal{M}))} \cdot \|m\|_{\mathcal{M}}. \end{aligned}$$

This estimate shows that $S(T)(x)$ is bounded, that the bound does not depend on x, and that $\|S\| \leq 1$. To complete the proof of the first part of the theorem, the continuity of $S(T)(x)$ in x with respect to the strong module topology should be verified. Let $x \in X$ be an arbitrary point, $\varepsilon > 0$ an arbitrary number and $m \in \mathcal{M}$ an arbitrary (fixed) element. One can find an open neighborhood U of the point x such that the inequality

$$\|(T\chi^x_m)(x) - (T\chi^x_m)(y)\| < \varepsilon$$

holds for any $y \in U$ and for some map of the form χ^x_m. Then one has

$$\|S(T)(x)m - S(T)(y)m\| = \|(T\chi^x_m)(x) - (T\chi^x_m)(y)\| < \varepsilon$$

for any $y \in U$. The first statement is proved.

Now let $J' = J|_{\mathcal{B}^*(X,\operatorname{End}^*_A(\mathcal{M}))}$. Then, for any $D \in \mathcal{B}^*(X, \operatorname{End}^*_A(\mathcal{M}))$, one has

$$\begin{aligned} \langle (J'(D)\varphi), \psi\rangle(x) &= \langle (J'(D)\varphi)(x), \psi(x)\rangle = \langle D(x)(\varphi(x)), \psi(x)\rangle \\ &= \langle \varphi(x), (D(x))^*(\psi(x))\rangle = \langle \varphi(x), D^*(x)(\psi(x))\rangle \\ &= \langle \varphi(x), [J'(D^*)(\psi)](x)\rangle = \langle \varphi, J'(D^*)(\psi)\rangle(x), \end{aligned}$$

so that $\operatorname{Im} J' \subset \operatorname{End}^*_{A_0(X)} C_0(X, \mathcal{M})$.

To complete the proof of the second part, we have to verify, firstly, that

$$\operatorname{Im} S|_{\operatorname{End}^*_{A_0(X)} C_0(X,\mathcal{M})} \subset \mathcal{B}^*(X, \operatorname{End}^*_A(\mathcal{M})),$$

i.e., that the operator $S(T)(x)$ is adjointable for each $x \in X$, and, secondly, that $S(T)(x)$ is continuous in the $*$-strong module topology.

The first assertion follows from the calculation

$$\begin{aligned} \langle S(T)(x)m, m'\rangle &= \langle (T\chi^x_m)(x), \chi^x_{m'}(x)\rangle = \langle (T\chi^x_m), \chi^x_{m'}\rangle(x) = \langle \chi^x_m, T^*\chi^x_{m'}\rangle(x) \\ &= \langle \chi^x_m(x), (T^*\chi^x_{m'})(x)\rangle = \langle m, S(T^*)(x)m'\rangle, \qquad m, m' \in \mathcal{M}. \end{aligned}$$

Moreover, it implies that S is involutive. Now let $x \in X$ be an arbitrary point, $\varepsilon > 0$ an arbitrary number and $m \in \mathcal{M}$ an arbitrary (fixed) element. One can find an open neighborhood U of the point x such that

$$\|(T^*\chi^x_m)(x) - (T^*\chi^x_m)(y)\| < \varepsilon$$

holds for any $y \in U$ and for some map of the form χ_m^x. Then one has

$$\|S(T)^*(x)m - S(T)^*(y)m\| = \|(T^*\chi_m^x)(x) - (T^*\chi_m^x)(y)\| < \varepsilon$$

for any $y \in U$. □

Denote by $\mathcal{B}_\bullet^*(X, \mathrm{GL}^*(A)) \subset \mathcal{B}^*(X, \mathrm{GL}^*(A))$ the set of functions with bounded (pointwise) inverse.

COROLLARY 5.6.9. *The homomorphism J defined above gives an isometric isomorphism of the groups*

$$\mathrm{GL}(A_0(X)) \cong \mathcal{B}_\bullet(X, \mathrm{GL}(A)), \qquad \mathrm{GL}^*(A_0(X)) \cong \mathcal{B}_\bullet^*(X, \mathrm{GL}^*(A)).$$

PROOF. Since J is a homomorphism of algebras, the statement immediately follows from its unitality. □

Note that one has to be cautious when identifying different classes of operators in the standard Hilbert module over a commutative C^*-algebra with continuous sets of operators of the same class on a Hilbert space. For example, although an operator of finite rank on the Hilbert module $H_{C(X)}$ defines a continuous set of operators of finite rank on a compact space X, the inverse statement is not true, as can be seen from the following example obtained by D. Kucerovsky [**67**]. It is interesting that this example is of topological origin.

Denote by ξ_n the standard tautological vector bundle over the complex projective space $\mathbf{C}P(n)$. For information about vector bundles and their characteristic classes we refer the reader, for example, to the books [**49, 60**]. Let $\Gamma(\xi_n)$ be the Hilbert $C(\mathbf{C}P(n))$-module of sections of the bundle ξ_n.

LEMMA 5.6.10 ([**67**]). *Let K be a compact operator with an algebraically n-generated image on the Hilbert $C(\mathbf{C}P(n))$-module $\Gamma(\xi_n)$. Then there exists a point $x_0 \in \mathbf{C}P(n)$ such that $K(x_0) = 0$, where $K(x) \in C(\mathbf{C}P(n), \mathcal{K})$ is the $\mathcal{K}$-valued function defined by the operator K.*

PROOF. The operator K has the form $\sum_{k=1}^n s_k \langle r_k, \cdot\rangle$, where s_k, r_k are continuous sections of the bundle ξ_n. Let $E = \xi_n \oplus \cdots \oplus \xi_n$ be the vector bundle equal to the direct sum of n copies of the bundle ξ_n. Then $s_1 \oplus \cdots \oplus s_n$ is a section of the bundle E. Let us calculate the higher order Chern class of the bundle E. We have

$$c_n(E) = c_n(\xi_n \oplus \cdots \oplus \xi_n) = c_1(\xi_n)^n \neq 0,$$

because $c_1(\xi_n) \neq 0$. The nonvanishing of $c_n(E)$ means that any section of the bundle E vanishes at some point. In particular, the section $s_1 \oplus \cdots \oplus s_n$ vanishes at some point $x_0 \in \mathbf{C}P(n)$, hence all the sections s_i, $i = 1, \ldots, n$, vanish at the point x_0 too. □

EXAMPLE 5.6.11 ([**67**]). Let

$$X = \coprod_{n=1}^{\infty} \mathbf{C}P(n)$$

be the disjoint union of complex projective spaces and let X^+ be the one-point compactification of X. Define a Hilbert $C(X^+)$-module $\mathcal{H}$ as a direct sum of spaces of sections of the bundles ξ_n, $\mathcal{H} = \bigoplus_{n=1}^{\infty} \Gamma(\xi_n)$. The module $\mathcal{H}$ is countably generated, so it can be realized as an orthogonally complemented submodule of the standard Hilbert module $H_{C(X^+)}$. Define the compact operator K on the module $\mathcal{H}$ by the formula

$$K\left(\bigoplus_{n=1}^{\infty} s_n\right) = \bigoplus_{n=1}^{\infty} \frac{1}{n} s_n,$$

where $s = \bigoplus_{n=1}^{\infty} s_n \in \mathcal{H}$. The operator K defines a continuous family of rank one operators over the space X^+. However, since the set $K(x)$ does not vanish at any point of X, by Lemma 5.6.10 the operator K is not an operator of a finite rank on the module $\mathcal{H}$. At the same time, since the set $K(x)$ is continuous, $K \in \mathcal{K}(\mathcal{H})$. Extending the operator K by zero, one can obtain a compact operator on the module $H_{C(X^+)}$ possessing the same property.

Now we consider nonadjointable operators in the commutative case [**54**]. Let $A = C(X)$, where X is a compact Hausdorff space. We will use the shortcut $\mathcal{K}_A$ for the space $\mathcal{K}(H_A)$ of A-compact operators on the Hilbert module $H_A = l_2(A)$. We also will use the characterization of compact operators from Proposition 2.2.1. As before, let L_n be the module generated by the first n vectors of the standard basis and let p_n be the orthogonal projection onto that module (for $A = \mathbf{C}$ we use the notation p_n^C). The space of all operators $F : l_2(A) \to l_2(A)$ satisfying the conditions of Definition 2.7.4, except, possibly, the condition to be adjointable, will be denoted by $\mathcal{F}_A$.

THEOREM 5.6.12. *An operator $K \in \operatorname{End} H_A$ is compact if and only if, for the map S introduced in the proof of Theorem* 5.6.8, *$S(K)$ maps X to $\mathcal{K}$ and $S(K) : X \to \mathcal{K}$ is norm continuous.*

PROOF. Let us prove that $S(K)$ is norm continuous if $K \in \mathcal{K}_A$. Choose an arbitrary point $x \in X$ and a number $\varepsilon > 0$. There exists a number $n \in \mathbf{N}$ such that $\|K(1-p_n)\| < \varepsilon/4$. Note that $S(K)$ is strongly continuous. Hence, $S(Kp_n)(x) = S(K)(x)p_n^C$ is norm continuous. Thus, we can choose a neighborhood $U_x \ni x$ such that $\|S(Kp_n)(y) - S(Kp_n)(x)\| < \varepsilon/2$ for each $y \in U_x$. Then, for these y, the following inequality holds:

$$\begin{aligned}\|S(K)(y) - S(K)(x)\| &\le \|S(K(1-p_n))(y)\| \\ &\quad + \|S(K(1-p_n))(x)\| + \|S(Kp_n)(y) - S(Kp_n)(x)\| \\ &< \frac{\varepsilon}{4} + \frac{\varepsilon}{4} + \frac{\varepsilon}{2} = \varepsilon.\end{aligned}$$

Now let us show that $S(K)(x) \in \mathcal{K}$ for each $x \in X$. For an arbitrary $\varepsilon > 0$ choose a number n such that $\|K(1-p_n)\| < \varepsilon$. Then, for any $x \in X$, one has

$$\|S(K)(x)(1-p_n)\| = \|S(K(1-p_n))(x)\| \le \|K(1-p_n)\| < \varepsilon.$$

Hence, $S(K)(x)$ is a compact operator.

Now let a map $D : X \to \mathcal{K}$ be norm continuous. Let us prove that $J(D) \in \mathcal{K}_A$, i.e., $\|J(D)(1-p_n)\| \to 0$ as $n \to \infty$. Choose $\varepsilon > 0$ and for each $x \in X$ find a number n_x such that

$$\|D(x)(1-p_n^C)\| < \varepsilon/2 \quad \text{for} \quad n > n_x$$

and an open neighborhood $U_x \ni x$ such that the inequality

$$\|D(y) - D(x)\| < \varepsilon/2$$

holds for any $y \in U_x$. Let $U_{x_1}, \dots, U_{x_s}$ be a finite sub-covering of $\{U_x\}$ and $m := \max\{n_{x_1}, \dots, n_{x_s}\}$. Then, for each $y \in X$, we can choose x_i such that

$$\begin{aligned}\|J(D)(1-p_m)\| &= \sup_{y\in X} \|D(y)(1-p_m^C)\| \\ &\le \sup_{y\in X} [\|D(x_i)(1-p_m^C)\| + \|(D(y)-D(x_i))(1-p_m^C)\|] \\ &\le \sup_{y\in X} (\varepsilon/2 + \varepsilon/2) = \varepsilon.\end{aligned}$$

□

THEOREM 5.6.13 ([**54**]). *For each $F \in \mathcal{F}_A$, there exists an invertible operator $G \in \mathrm{GL}$ such that the map $S(GF) : X \to \mathcal{F}_{\mathbf{C}}$ is norm continuous.*

PROOF. One can assume that F has the following form:

$$F = \begin{pmatrix} F_1 & 0 \\ 0 & F_4 \end{pmatrix} : M_1 \widetilde{\oplus} N_1 = l_2(A) \to l_2(A) = M_2 \widetilde{\oplus} N_2,$$

where $F_1 : M_1 \cong M_2$ and N_1, $N_2 = L_n$ are finitely generated projective modules. The proof of this fact is similar to the case of adjointable operators (Section 2.7). Since N_1 is a finitely generated projective module, $N_1 \oplus N_1^\perp = l_2(A)$ (by Lemma 1.4.3). We can consider N_1 as a locally trivial vector bundle over X and $N_1^\perp$ as a locally trivial bundle over X with a Hilbert space as a fiber. Indeed, let $P : l_2(A) \to N_1$ be an orthogonal projection. By Lemma 5.6.14 (see below), the map $S(P) : X \to \mathrm{Proj}(l_2)$ is norm continuous. Hence, for an arbitrary point $x_0 \in X$ there exists an open neighborhood U_{x_0} such that the inequality $\|S(P)(x) - S(P)(x_0)\| < 1$ holds for each $x \in U_{x_0}$. Denote the fiber of the bundle N_1 by $(N_1)_x$. For $x \in U_{x_0}$, the restriction of the map $1 - S(P)(x)$ to $(N_1)^\perp_{x_0}$ is an isomorphism onto $(N_1)^\perp_{x_0}$ and the map $(1 - S(P)) : X \to \mathrm{Proj}(l_2)$ is norm continuous.

By Kuiper's theorem [**68**], the bundle $N_1^\perp$ is trivial. Hence, there exists an isomorphism of A-modules $T : N_1^\perp \to L_n^\perp$ such that the map $S(T(1-P)) : X \to \mathrm{End}_{\mathbf{C}}\, l_2$ is norm continuous.

The operator F has the matrix form

$$F = \begin{pmatrix} \bar F_1 & 0 \\ \bar F_3 & \bar F_4 \end{pmatrix}, \quad \text{where } \bar F_1 : N_1^\perp \to M_2 \text{ is an isomorphism}$$

with respect to the decomposition $l_2(A) = N_1^\perp \oplus N_1$, $l_2(A) = M_2 \widetilde{\oplus} L_n$. Then $\psi F = \begin{pmatrix} \bar F_1 & 0 \\ 0 & \bar F_4 \end{pmatrix}$, where

$$\psi := \begin{pmatrix} 1 & 0 \\ -\bar F_3 \bar F_1^{-1} & 1 \end{pmatrix} : M_2 \widetilde{\oplus} L_n \to M_2 \widetilde{\oplus} L_n.$$

Hence, F has a diagonal form,

$$F = \begin{pmatrix} \bar F_1 & 0 \\ 0 & \bar F_4 \end{pmatrix} : N_1^\perp \oplus N_1 \to \bar M_2 \widetilde{\oplus} \bar N_2,$$

where

$$\bar M_2 := \psi^{-1}(M_2), \qquad \bar N_2 := \psi^{-1}(L_n) = L_n.$$

Define the operator $G \in \mathrm{GL}$ by the formula

$$G := \begin{pmatrix} T\bar F_1^{-1} & 0 \\ 0 & 1 \end{pmatrix} : \bar M_2 \widetilde{\oplus} L_n \to L_n^\perp \oplus L_n.$$

Then

$$GF = \begin{pmatrix} T & 0 \\ 0 & \bar{F}_4 \end{pmatrix} : N_1^\perp \oplus N_1 \to L_n^\perp \oplus L_n.$$

Since $\bar{F}_4 = GFP$ is a compact operator, by Theorem 5.6.12 the operator $S(\bar{F}_4)$ is norm continuous. Hence, the map $S(GF) = S(T(1-P)) + S(\bar{F}_4)$ is norm continuous too. □

LEMMA 5.6.14 ([**33**, Lemma 5, p. 253]). *Let H be a Hilbert space, $n > 0$ an integer and $\mathcal{P}_n$ the space of rank n projections in H. The strong topology and the norm topology coincide on $\mathcal{P}_n$.*

PROOF. Take $p_0 \in \mathcal{P}_n$ and $\varepsilon > 0$. Let $(\xi_1, \dots, \xi_n)$ be an orthonormal basis of $p_0(H)$. If a projection $p \in \mathcal{P}_n$ is sufficiently close to p_0 with respect to the strong topology, the norms $\|p\xi_i - \xi_i\|$ are arbitrarily small and $p\xi_i$, $i = 1, \dots, n$, is a basis of pH. Hence, there exists a neighborhood (with respect to the strong topology) V of p_0 in $\mathcal{P}_n$ such that the inequalities

$$\|\xi_i - \eta_i\| \le \frac{\varepsilon}{2n}, \qquad i = 1, \dots, n,$$

hold for the orthogonalization $(\eta_1, \dots, \eta_n)$ of the basis $(p\xi_1, \dots, p\xi_n)$. Let $\xi \in H$ be an arbitrary element with $\|\xi\| \le 1$. Then, for $p \in V$, one has

$$\begin{aligned} \|p(\xi) - p_0(\xi)\| &= \left\| \sum_{i=1}^n \langle \xi, \eta_i \rangle \eta_i - \langle \xi, \xi_i \rangle \xi_i \right\| \\ &\le \sum_{i=1}^n |\langle \xi, \eta_i \rangle| \cdot \|\eta_i - \xi_i\| + \sum_{i=1}^n |\langle \xi, \eta_i - \xi_i \rangle| \cdot \|\xi_i\| \le 2n \frac{\varepsilon}{2n} = \varepsilon. \end{aligned}$$

□

EXAMPLE 5.6.15 ([**54**]). The following example shows that there exists a strongly continuous map $F : X \to \Phi_{\mathbf{C}}$ such that $J(F) \notin \Phi_A$.

Let us identify l_2 and $L^2[0,1]$. Let H_t be the set of all functions $f \in L^2[0,1]$ such that $f(x) = 0$ for $x \ge t$. Let us denote by P_t the orthogonal projection $L^2 \to H_t$. Define the following family of isometries [**33**]:

$$U_t : H_t \to L^2[0,1], \quad (U_t f)(x) := \sqrt{t} f(tx), \quad t \in (0,1], \quad f \in H_t.$$

Let $R \in \Phi_{\mathbf{C}}$ be a Fredholm operator with $\operatorname{index} R = 1$. Suppose also that $RR^* = 1$ and $R^*R = 1 - p'$, where p' is a one-dimensional projection. Define a strongly continuous map $F : [0,1] \to \Phi_{\mathbf{C}}$ by the formula

$$F(t) := \begin{cases} (1 - P_t) + U_t^{-1} R U_t P_t & \text{for } t > 0, \\ 1 & \text{for } t = 0. \end{cases}$$

Since $F^*(t) = (1 - P_t) + P_t U_t^{-1} R^* U_t$, one has

$$F^*(t)F(t) = 1 - P_t + P_t U_t^{-1} R^* R U_t P_t = 1 + K_1,$$

$$F(t)F^*(t) = 1 - P(t) + P(t) = 1,$$

where K_1 is some rank one operator. Hence, $F(t) \in \Phi_{\mathbf{C}}$. We have

$$\operatorname{index} F(0) = 0 \ne 1 = \operatorname{index} F(1)$$

and, by Theorem 5.6.13, we see that $J(F) \notin \Phi_A$, $A = C[0,1]$.

Thus, an arbitrary strongly continuous map $F : X \to \Phi_{\mathbf{C}}$ does not define an operator in $\Phi_{C(X)}$. Nevertheless, one can define a new topology on $\Phi_{\mathbf{C}}$ (independent of X!) such that the continuity of $F : X \to \Phi_{\mathbf{C}}$ with respect to this new topology implies $J(F) \in \Phi_{C(X)}$ and vice versa.

DEFINITION 5.6.16 ([**54**]). An *F-topology* is the topology on $\Phi_{\mathbf{C}}$ generated by the following pre-base:

$$U_{\varepsilon,a_1,\dots,a_n,D} = \{R \in F_{\mathbf{C}} \mid \|(R-D)(a_i)\| < \varepsilon,\ i = 1,\dots,n\};$$

$$U_{\varepsilon,V,D} = \{R \in F_{\mathbf{C}} | \text{ there exists an operator } T \in \mathrm{GL}(l_2), \text{ such that } T(V) \subset V \text{ and } \|TR - D\| < \varepsilon\},$$

where $a_1, \dots, a_n$ are some vectors from l_2 and V is a finite-dimensional subspace of l_2.

We complete this section with the following two theorems. The proofs can be found in [**54**].

THEOREM 5.6.17 ([**54**]). *The set of all F-continuous maps $f : X \to \Phi_{\mathbf{C}}$ coincides with $S(\Phi_{C(X)})$.*

THEOREM 5.6.18 ([**54**]). *The spaces ($\Phi_{\mathbf{C}}$, norm) and ($\Phi_{\mathbf{C}}, F$-topology) are weakly homotopy equivalent.*

5.7. Inner products on Hilbert C^*-modules

We have already encountered the case where several inner products on the same Hilbert module were used. In this section we will study the relations between different inner products more thoroughly.

Let $\mathcal{M}$ be a module over a C^*-algebra A, on which two sesquilinear maps $\langle\cdot,\cdot\rangle_1$ and $\langle\cdot,\cdot\rangle_2$ are defined in such a way that $\mathcal{M}$ is a pre-Hilbert module with respect to each of these maps.

DEFINITION 5.7.1. Two inner products, $\langle\cdot,\cdot\rangle_1$ and $\langle\cdot,\cdot\rangle_2$, are called *equivalent* if the norms defined by these inner products are equivalent.

Note that if the inner products are equivalent to each other and if $\{\mathcal{M}, , \langle\cdot,\cdot\rangle_1\}$ is a Hilbert C^*-module, then $\{\mathcal{M}, , \langle\cdot,\cdot\rangle_2\}$ is a Hilbert C^*-module too.

Let us consider first the case of different inner products on self-dual Hilbert modules.

PROPOSITION 5.7.2 ([**42**]). *Let $\mathcal{M}$ be a self-dual Hilbert C^*-module over a C^*-algebra A with the inner product $\langle\cdot,\cdot\rangle_1$. If $\langle\cdot,\cdot\rangle_2$ is another inner product equivalent to the given one, then there exists a unique invertible positive operator $S \in \mathrm{End}^*_A(\mathcal{M})$ such that $\langle x, y\rangle_1 = \langle Sx, Sy\rangle_2$ for all $x, y \in \mathcal{M}$.*

PROOF. For $x \in \mathcal{M}$ consider the functional on $\mathcal{M}$ defined by the formula $y \longmapsto \langle x, y\rangle_2$. Since the module $\mathcal{M}$ is self-dual, there exists an element $Bx \in \mathcal{M}$ such that

$$\langle x, y\rangle_2 = \langle Bx, y\rangle_1$$

for all $y \in \mathcal{M}$. The map $x \longmapsto Bx$ is an A-homomorphism. Denote by $\|\cdot\|_i$ the norm defined by the inner product $\langle\cdot,\cdot\rangle_i$, $i = 1, 2$. By our assumption, there exist constants $k, l > 0$ such that

$$\|x\|_1 \le k\,\|x\|_2 \le l\,\|x\|_1$$

for any $x \in \mathcal{M}$. Then

$$\|Bx\|_1^2 = \|\langle Bx, Bx\rangle_1\| = \|\langle x, Bx\rangle_2\| \le \|x\|_2 \|Bx\|_2 \le \frac{l^2}{k^2} \|x\|_1 \|Bx\|_1 .$$

Therefore $\|Bx\|_1 \le \frac{l^2}{k^2} \|x\|_1$, i.e., the map B is bounded. The equality $\langle Bx, x\rangle_1 = \langle x, x\rangle_2 \ge 0$ implies that the operator B is positive with respect to the initial inner product. The inequality

$$\|x\|_1^2 \le k^2 \|x\|_2^2 = k^2 \|\langle x, x\rangle_2\| = k^2 \|\langle Bx, x\rangle_1\| \le k^2 \|Bx\|_1 \|x\|_1$$

shows that the estimate $\|Bx\|_1 \ge \frac{1}{k^2} \|x\|_1$ holds, from which we obtain, by [**92**] (cf. the proof of Theorem 2.3.3), invertibility of the operator B. To complete the proof it remains to put $S = B^{-1/2}$. □

PROPOSITION 5.7.3 ([**44**]). *Let $\mathcal{M}$ be a Hilbert module with an inner product $\langle\cdot,\cdot\rangle_1$. Let $\langle\cdot,\cdot\rangle_2$ be another inner product equivalent to the initial one. Then the map $\mathcal{M} \longrightarrow \mathcal{M}'$ given by the formula $x \longmapsto \langle x,\cdot\rangle_2$, $x \in \mathcal{M}$, defines an invertible positive element in $\mathbf{QM}(\mathcal{K}(\mathcal{M})) \subset \mathcal{K}(\mathcal{M})^{**}$. Conversely, any invertible positive element in $\mathbf{QM}(\mathcal{K}(\mathcal{M})) \subset \mathcal{K}(\mathcal{M})^{**}$ (which can be identified with an element T of the set $\mathrm{End}_A(\mathcal{M}, \mathcal{M})$) defines an inner product $\langle x, y\rangle = T(x)(y)$, $x, y \in \mathcal{M}$.*

PROOF. Let $\phi : \mathrm{End}_A(\mathcal{M}, \mathcal{M}') \longrightarrow \mathbf{QM}(\mathcal{K}(\mathcal{M}))$ be the isometric isomorphism defined in Theorem 5.4.1, $\theta_{x,y}\phi(T)\theta_{z,t} = \theta_{x,t\cdot T(z)(y)}$, $x, y, z, t \in \mathcal{M}$. The map $x \longmapsto \langle x,\cdot\rangle_2$ is bounded, hence it defines a map $T : \mathcal{M} \longrightarrow \mathcal{M}'$ and the element $\phi(T) \in \mathbf{QM}(\mathcal{K}(\mathcal{M}))$ is defined by the formula

$$\theta_{x,y}\phi(T)\theta_{z,t} = \theta_{x,t\cdot\langle z,y\rangle_2}$$

(note that the elementary operators of the form $\theta_{x,y}$ are considered with respect to the initial inner product $\langle\cdot,\cdot\rangle_1$). Then for $s \in \mathcal{M}$ one has

$$\begin{aligned}\langle \theta_{x,x\cdot\langle y,y\rangle_2}(s), s\rangle_1 &= \langle x \cdot \langle x \cdot \langle y, y\rangle_2, s\rangle_1, s\rangle_1 \\ &= \langle x \cdot \langle y, y\rangle_2, s\rangle_1^* \langle x, s\rangle_1 = \langle x, s\rangle_1^* \langle y, y\rangle_2 \langle x, s\rangle_1 \ge 0.\end{aligned}$$

Since linear combinations of elementary operators are dense in the algebra $\mathcal{K}(\mathcal{M})$, the operator $\phi(T)$ is positive. Let us show that it is invertible. Let us pass to the Hilbert module $\mathcal{M}^\#$ over the enveloping W^*-algebra A^{**}. Both inner products can be extended to the module $\mathcal{M}^\#$ and to the self-dual module $(\mathcal{M}^\#)'$. By Proposition 5.7.2, there exists an invertible operator $S \in \mathrm{End}^*_{A^{**}}((\mathcal{M}^\#)')$ such that these extensions of the inner product are related by $\langle x, y\rangle_1 = \langle Sx, Sy\rangle_2$ for any $x, y \in (\mathcal{M}^\#)'$. But the image of the operator $\phi(T)$ under the inclusion $\mathbf{QM}(\mathcal{K}(\mathcal{M})) \subset \mathrm{End}^*_{A^{**}}((\mathcal{M}^\#)')$ obviously coincides with the product S^*S. Since the operator S is invertible, the spectrum of the operator $\phi(T)$ is separated from zero, hence $\phi(T)$ is invertible. In the opposite direction the statement can be proved similarly. □

COROLLARY 5.7.4 ([**44**]). *Let $\mathcal{M}$ be a Hilbert C^*-module with an inner product $\langle\cdot,\cdot\rangle_1$. The following conditions are equivalent:*

(i) *any other inner product $\langle\cdot,\cdot\rangle_2$ equivalent to the initial one is defined by an invertible operator $S \in \mathrm{End}_A(\mathcal{M})$ and is given by the formula $\langle x, y\rangle_2 = \langle Sx, Sy\rangle_1$, $x, y \in \mathcal{M}$;*

(ii) *each positive invertible quasi-multiplier* $T \in \mathbf{QM}(\mathcal{K}(\mathcal{M}))$ *can be decomposed into a product* $T = S^*S$ *for some invertible left multiplier* $S \in \mathbf{LM}(\mathcal{K}(\mathcal{M}))$.

THEOREM 5.7.5 ([**44**], see also [**17**]). *Let* $\mathcal{M}$ *be a countably generated Hilbert* C^**-module with an inner product* $\langle\cdot,\cdot\rangle_1$. *Then for any inner product* $\langle\cdot,\cdot\rangle_2$ *equivalent to the initial one, there exists an invertible operator* $S \in \mathrm{End}_A(\mathcal{M})$ *such that* $\langle x,y\rangle_2 = \langle Sx,Sy\rangle_1$.

PROOF. By assumption, the C^*-algebra $\mathcal{K}(\mathcal{M})$ is σ-unital, hence it contains a strictly positive element $H \in \mathcal{K}(\mathcal{M})$. It is sufficient to show that each positive invertible quasi-multiplier admits a decomposition $T = S^*S$ with some left multiplier S. Put $K = (HTH)^{1/2} \in \mathcal{K}(\mathcal{M})$, $V_n = K\left(H^2 + \frac{1}{n}\right)^{-1} H \in \mathcal{K}(\mathcal{M})$. Then $\|V_n\| \le \|T\|^{1/2}$ and the sequence (V_nH) converges to K with respect to the norm. Then, for any $K' \in H \cdot \mathcal{K}(\mathcal{M})$, the sequence (V_nK') is norm-convergent to KK'. Since $H \cdot \mathcal{K}(\mathcal{M})$ is dense in $\mathcal{K}(\mathcal{M})$, we conclude that the sequence (V_n) is convergent to some element $S \in \mathbf{LM}(\mathcal{K}(\mathcal{M}))$ with respect to the left strict topology and $SH = K$. Therefore $HS^*SH = K^*K = HTH$ and, finally, $S^*S = T$. □

As we can see from the following example of a nontrivial inner product, the requirement for Hilbert C^*-modules to be countably generated is essential in Theorem 5.7.5.

EXAMPLE 5.7.6 ([**17, 44**]). Let H be a *nonseparable* Hilbert space. Consider the space

$$X = \{x \in \mathcal{B}(H) : 1/2 \le x \le 1\}$$

equipped with the weak topology and the standard Hilbert $C(X)$-module $H_{C(X)}$. Let us show that there exist inner products on $H_{C(X)}$ equivalent to the standard one, which cannot be written as $\langle S\cdot, S\cdot\rangle$ for any operator $S \in \mathrm{End}_{C(X)}(H_{C(X)})$. For this purpose it is sufficient to find a quasi-multiplier $T \in \mathbf{QM}(\mathcal{K} \otimes C(X))$ that cannot be represented as $T = S^*S$, $S \in \mathbf{LM}(\mathcal{K} \otimes C(X))$. Let us use the identification of $\mathbf{LM}(\mathcal{K} \otimes C(X))$ (resp. $\mathbf{QM}(\mathcal{K} \otimes C(X))$) with the set of bounded maps from X to $\mathcal{B}(H)$ continuous in the strong (resp. weak) topology (it is discussed in detail in Section 5.6). Define the new inner product on the module $H_{C(X)}$ by the formula

$$\langle y,z\rangle_o(x) = \langle y, x(x)\rangle(x), \tag{5.9}$$

where $y,z \in H_{C(X)}$, $x \in X$. It is easy to see that $\frac{1}{2}\langle y,y\rangle \le \langle y,y\rangle_o \le \langle y,y\rangle$. This inner product defines a positive invertible quasi-multiplier T. Suppose that $T = S^*S$ for some $S \in \mathbf{LM}(\mathcal{K} \otimes C(X))$. Let us show that one can choose a separable infinite-dimensional Hilbert space $H_T \subset H$ such that $T(x)H_T \subset H_T$ and $T^{-1}(x)H_T \subset H_T$ for all $x \in X$. Let $\{e_1,\ldots,e_k,\ldots\}$ be a basis of some separable subspace $H_0 \subset H$. Since X is compact, the sets $T(X)e_k$ and $T^{-1}(X)e_k$ are compact subsets in H for each number k. Therefore they generate a separable Hilbert subspace $H_1 \subset H$ such that $T(X)H_0 \subset H_1$, $T^{-1}(X)H_0 \subset H_1$. Further, by induction we find separable subspaces $H_n \subset H$ such that $T(X)H_n \subset H_{n+1}$, $T^{-1}(X)H_n \subset H_{n+1}$. Finally, set $H_T := \left(\bigcup_n H_n\right)^{-}$, i.e., the closure of the union of all H_n.

Let us denote by $X_0 \subset X$ the subset

$$X_0 = \left\{ \begin{pmatrix} 3/4 & r \\ r^* & 3/4 \end{pmatrix} : r : H_T \longrightarrow H_T^{\perp},\ \|r\| \leq 1/4,\ r \text{ is linear} \right\}.$$

The restriction of the operator S onto the subspace X_0 has the form

$$S|_{X_0} = \begin{pmatrix} s_1 & s_2 \\ 0 & s_3 \end{pmatrix},\ s_1^* s_1 = 3/4,\ s_2^* s_2 + s_3^* s_3 = 3/4,\ s_1^* s_2 = r,$$

with respect to the decomposition $H = H_T \oplus H_T^{\perp}$. Since the subspace H_T is invariant under the action of T^{-1}, the operator $s_1 \in \mathcal{B}(H_T)$ is invertible and the operator $\frac{2}{\sqrt{3}} s_1$ is unitary. Since the map $u \longmapsto u^*$ is continuous in the strong topology on the group of unitary elements, we conclude that s_1^* is continuous on X_0. Therefore the map $r = s_1^* s_2$ is also strongly continuous as a map from X_0 to $\mathcal{B}(H_T^{\perp}, H_T)$. Thus, the assumption that $T = S^* S$ implies that any weakly continuous bounded (by $1/4$) linear map $r : H_T \longrightarrow H_T^{\perp}$ turns out to be strongly continuous. But, since the strong and the weak topologies do not coincide on the ball of radius $1/4$ in $\mathcal{B}(H_T^{\perp}, H_T)$, the obtained contradiction shows that the inner product (5.9) is not related to the standard inner product on the module $H_{C(X)}$ by any invertible bounded operator $S \in \mathrm{End}_A(H_{C(X)})$.

If one considers various equivalent inner products on a Hilbert C^*-module, the problem of whether an operator is adjointable depends on the concrete inner product. By $\mathrm{End}_A^{*(i)}(\mathcal{M})$ (resp. $\mathcal{K}^{(i)}(\mathcal{M})$) we denote the C^*-algebra of adjointable operators (resp., compact operators) with respect to the inner product $\langle \cdot, \cdot \rangle_i$, $i = 1, 2$. The adjoint operator for the operator T with respect to this inner product is denoted by $T^*_{(i)}$.

PROPOSITION 5.7.7 ([**44**]). *Let $\mathcal{M}$ be a Hilbert A-module over a C^*-algebra A with the inner product $\langle \cdot, \cdot \rangle_1$. Let $S \in \mathrm{End}_A(\mathcal{M})$ be an invertible operator defining the inner product $\langle \cdot, \cdot \rangle_2 = \langle S \cdot, S \cdot \rangle_1$. Then the operator S is adjointable with respect to the first inner product if and only if it is adjointable with respect to the second one.*

If S is adjointable, then the sets $\mathrm{End}_A^{*(1)}$ and $\mathrm{End}_A^{*(2)}$ coincide and the same is true for $\mathcal{K}^{(1)}(\mathcal{M})$ and $\mathcal{K}^{(2)}(\mathcal{M})$.

PROOF. Let the operator S be adjointable with respect to the inner product $\langle \cdot, \cdot \rangle_1$. Then one has

$$\langle Sx, y \rangle_2 = \langle S^2 x, Sy \rangle_1 = \langle Sx, S(S^{-1} S_1^* S) y \rangle_1 = \langle x, (S^{-1} S^*_{(1)} S) y \rangle_2$$

for any $x, y \in \mathcal{M}$, and the operator $S^*_{(2)} = S^{-1} S^*_{(1)} S$ is an adjoint for S with respect to the second inner product. The converse statement can be proved similarly.

Let us assume now that $S \in \mathrm{End}_A^*(\mathcal{M})$. Let $B \in \mathrm{End}_A^{*(1)}(\mathcal{M})$. Then

$$\begin{aligned} \langle Bx, y \rangle_2 = \langle SBx, Sy \rangle_1 &= \langle Sx, S(S^{-1} (S^{-1})^*_{(1)} B^*_{(1)} S^*_{(1)} S) y \rangle_1 \\ &= \langle x, (S^{-1} (S^{-1})^*_{(1)} B^*_{(1)} S^*_{(1)} S) y \rangle_2 \end{aligned}$$

for any $x, y \in \mathcal{M}$, hence $B^*_{(2)} = S^{-1} (S^{-1})^*_{(1)} B^*_{(1)} S^*_{(1)} S$, i.e., $B \in \mathrm{End}_A^{*(2)}(\mathcal{M})$. The statement about compact operators can be proved in the same way. □

However, if the inner product is defined with the help of a nonadjointable operator S, then operators adjointable with respect to one of the equivalent inner products need not be adjointable with respect to another. Thus a problem of whether a functional on $\mathcal{M}$ can be represented as an inner product by elements from $\mathcal{M}$ arises. More precisely, we define the set $F \subset \mathcal{M}'$ by

$$F = \bigcup_{\beta \in \mathbf{B};\, y \in \mathcal{M}} \langle y, \cdot \rangle_\beta,$$

where $\mathbf{B}$ is the set of all inner products $\langle \cdot, \cdot \rangle_\beta$ equivalent to the initial one. The functional $f \in \mathcal{M}'$ is called *representable* if $f \in F$. We study the set F for the standard Hilbert C^*-module H_A. Denote the extension of the initial inner product from the module H_A to $H_A^{\#} = H_{A^{**}}$ and to its adjoint module $H'_{A^{**}}$ still by $\langle \cdot, \cdot \rangle$. It is easy to see that $H'_A \subset H'_{A^{**}}$.

PROPOSITION 5.7.8. *If $f \in H'_A$ is representable, then there exists an element z in the module H_A, for which the operator inequality*

$$\alpha\langle z, z\rangle \le \beta\langle f, f\rangle \le \langle f, z\rangle \le \gamma\langle f, f\rangle \le \delta\langle z, z\rangle \tag{5.10}$$

holds with some positive constants α, β, γ, δ.

PROOF. By Theorem 5.7.5, any inner product equivalent to the given one has the form $\langle x, y\rangle_\beta = \langle Sx, Sy\rangle$, where $S \in \mathrm{End}_A(H_A)$ is an invertible bounded operator. If f is representable, then $f = S^*Sz$ for some S and some $z \in H_A$, the operator $\langle f, z\rangle \in A$ is positive, and we have

$$\langle f, z\rangle = \langle S^*Sz, z\rangle = \langle Sz, Sz\rangle = \langle z, z\rangle_\beta.$$

Since S is invertible, there exist positive numbers a and b such that

$$a\langle z, z\rangle \le \langle z, z\rangle_\beta \le b\langle z, z\rangle,$$

hence

$$a\langle z, z\rangle \le \langle f, z\rangle \le b\langle z, z\rangle. \tag{5.11}$$

Let us now estimate $\langle f, f\rangle = \langle S^*Sz, S^*Sz\rangle$. Since $a^2 \le (S^*S)^2 \le b^2$, we have

$$a^2\langle z, z\rangle \le \langle f, f\rangle \le b^2\langle z, z\rangle. \tag{5.12}$$

Combining (5.11) and (5.12), we obtain the estimate (5.10). □

A functional $f \in H'_A$ is called *nonsingular* if there exists an element z in H_A such that the spectrum of the element $\langle f, z\rangle \in A$ is separated from zero (then one can assume that the operator inequality $\langle f, z\rangle \ge c > 0$ holds for some number c). The following example shows that there exist singular functionals with the property $\langle f, f\rangle = 1$.

EXAMPLE 5.7.9. Let $A = L^\infty([0;1])$. Let us define $f \in H'_A$ as a sequence of functions $f = (f_k(t))$,

$$f_k(t) = \begin{cases} 1 & \text{for } t \in \left[1/2^k; 1/2^{k-1}\right], \\ 0 & \text{for remaining } t. \end{cases}$$

The property $\langle f, f\rangle = 1$ is obvious. Let us show that the spectrum of the operator $\langle f, z\rangle$ is not separated from zero for any $z = (z_k) \in H_A$. Since the series $\sum_{k=1}^\infty z_k^* z_k$ is norm-convergent, for any $\varepsilon > 0$ there exists a number n such that $\|\sum_{k=n+1}^\infty z_k^* z_k\| < \varepsilon$. But then the estimation $|f_k(t)z_k(t)| < \varepsilon$ holds for $t < 1/2^k$. Hence f is singular. Condition (5.10) for f is false, thus f is not representable.

PROPOSITION 5.7.10. *Let $\mathcal{M}$ be a Hilbert C^*-module and let $f \in \mathcal{M}'$ be a nonsingular functional. Then it is representable.*

PROOF. Nonsingularity and the Cauchy–Bunyakovskii inequality for Hilbert modules give us the estimate $0 < c \leq \langle f, z\rangle \leq \|f\| \langle z, z\rangle^{1/2}$, from which it follows that the module $\operatorname{Span}_A z$ is isomorphic to A. Let us show that the decomposition $\mathcal{M} = \operatorname{Span}_A z \widetilde{\oplus} \operatorname{Ker} f$ into a (nonorthogonal) direct sum holds. If $x \in \mathcal{M}$, then put $a = \langle f, z\rangle^{-1} \cdot \langle f, x\rangle$; $y = x - za$. Then $x = za + y$ and $y \in \operatorname{Ker} f$. Uniqueness of this decomposition is clear. Set $\mathcal{M}_1 = \operatorname{Span}_A z$ and $\mathcal{M}_2 = \operatorname{Ker} f$ and let us choose, using Corollary 2.8.18, a new inner product in such a way that the submodules $\mathcal{M}_1$ and $\mathcal{M}_2$ become orthogonal. Then $z \perp \operatorname{Ker} f$. Put $z' = z \cdot \langle z, z\rangle_\beta^{-1} \cdot \langle f, z\rangle$. Then $\langle f, x\rangle = \langle z', x\rangle_\beta$, i.e., f is representable. □

Recall that a C^*-algebra A is of *topological stable rank one* [**111**] (notation $tsr(A) = 1$) if invertible elements of A are dense in A.

PROPOSITION 5.7.11. *Let a C^*-algebra A be of topological stable rank one. Then the representable functionals are dense in H_A' with respect to the initial norm.*

PROOF. It is sufficient to verify that nonsingular functionals are dense in H_A'. If $f = (f_i) \in H_A'$, then one can find an invertible element g_1 in A such that $\|g_1 - f_1\| < \varepsilon$. Putting $g = (g_1, f_2, f_3, \dots) \in H_A'$ and taking $z = e_1 = (1, 0, 0, \dots) \in H_A'$, we obtain that $\langle g, z\rangle$ is invertible and $\|g - f\| < \varepsilon$. □

The situation with representability of functionals in general is more complicated. Consider the following example.

EXAMPLE 5.7.12. Let A be the C^*-algebra of bounded operators on an infinite-dimensional Hilbert space. As is shown in 2.5.6, there exists an isomorphism of Hilbert modules $S : A \longrightarrow H_A'$. Let $a \in A$, $f = S(a)$. Then the condition $\langle f, x\rangle = 0$ can be written as

$$\langle S(a), x\rangle = \langle S(a), S(S^{-1}x)\rangle = \langle a, S^{-1}x\rangle = a^* \cdot S^{-1}x = 0. \tag{5.13}$$

If $a \in A$ is invertible, then it follows from (5.10) that $S^{-1}x = 0$, i.e., $\operatorname{Ker} f = 0$. But if f is representable, $\langle f, \cdot\rangle = \langle z, \cdot\rangle_\beta$ with $z \in H_A$, then the kernel of f cannot vanish. Therefore the functional $f = S(1_A)$ is not representable, moreover, it has an open neighborhood consisting of nonrepresentable functionals.

To complete this section we show how the averaging Theorem 2.8.14 can be generalized from compact groups to amenable ones in the case of Hilbert W^*-modules.

THEOREM 5.7.13 ([**82**]). *Let $\mathcal{M}$ be a Hilbert module over a W^*-algebra $\mathcal{A}$, and let $\rho : G \longrightarrow \operatorname{End}_{\mathcal{A}} \mathcal{M}$ be a uniformly bounded representation of an amenable group G, $\|\rho(g)\| \leq C$, $g \in G$. Then there exists an inner product $\langle \cdot, \cdot\rangle_\beta$ equivalent to the initial one and such that the representation ρ is unitary with respect to it.*

PROOF. For any normal linear functional $\phi \in \mathcal{A}_*$, where $\mathcal{A}_*$ denotes the predual Banach space for $\mathcal{A}$, let us define the function $f_{x,y}$ on the group G by the formula

$$f_{x,y}(g) = \phi(\langle \rho(g)x, \rho(g)y\rangle),$$

$x, y \in \mathcal{M}$, $g \in G$. By our assumption, this function is bounded. Put

$$\phi_{x,y} = M(f_{x,y}),$$

where M is an invariant mean on the group G. By fixing x and y, we obtain a linear bounded map

$$a_{x,y} : \mathcal{A}_* \longrightarrow \mathbf{C}; \qquad \phi \longmapsto \phi_{x,y}.$$

This map is an element of $(\mathcal{A}_*)^* = \mathcal{A}$. Define the new inner product on the module $\mathcal{M}$ by $\langle x, y\rangle_\beta = a_{x,y} \in \mathcal{A}$. Let us verify that it is well defined. Its sesquilinearity is obvious. If $\phi \in \mathcal{A}_*$ is a state, then $f_{x,x}(g) \geq 0$ for any $g \in G$, hence $\phi(\langle x, x\rangle_\beta) = \phi_{x,x} \geq 0$. Suppose that $\langle x, x\rangle_\beta = 0$ for some $x \in \mathcal{M}$. Then $\phi_{x,x} = 0$. But, since

$$\langle x, x\rangle = \langle \rho(g^{-1})(\rho(g)x), \rho(g^{-1})(\rho(g)x)\rangle \leq C^2 \langle \rho(g)x, \rho(g)x\rangle$$

for any $g \in G$, the estimate $\dfrac{1}{C^2} f_{x,x}(e) \leq f_{x,x}(g)$ holds, where e denotes the neutral element of G, and

$$M(f_{x,x}) \geq \frac{1}{C^2} f_{x,x}(e).$$

Hence, $\phi_{x,x} \geq \frac{1}{C^2} f_{x,x}(e)$ and $f_{x,x}(e) = 0$, i.e., $\phi(\langle x, x\rangle) = 0$ for an arbitrary state ϕ. But then $\langle x, x\rangle = 0$, hence $x = 0$. Therefore $\langle \cdot, \cdot\rangle_\beta$ is an inner product. The property $\langle \rho(g)x, \rho(g)y\rangle_\beta = \langle x, y\rangle_\beta$ obviously holds, hence the representation ρ is unitary. Equivalence of $\langle \cdot, \cdot\rangle$ and $\langle \cdot, \cdot\rangle_\beta$ follows immediately from the estimate

$$\frac{1}{C^2} \langle x, x\rangle \leq \langle \rho(g)x, \rho(g)x\rangle \leq C^2 \langle x, x\rangle,$$

which is valid for any $g \in G$. □

CHAPTER 6

Diagonalization of Operators over C^*-Algebras

6.1. Problem of diagonalizing operators in Hilbert C^*-modules

It is well known that in the commutative case, i.e., for $A = C(X)$ a commutative C^*-algebra of continuous functions on a compact topological space X, compact selfadjoint operators acting on the Hilbert C^*-module H_A (which are the same as norm continuous families of classical compact selfadjoint operators $K(x)$, $x \in X$ [**1**]) can be diagonalized pointwise and that diagonal elements determine continuous functions $\lambda(x) \in A$ by the theorem of the continuity of spectra. We consider here a noncommutative analog of this well-known fact. We prove in particular that a compact positive operator acting on a Hilbert C^*-module over a finite W^*-algebra $\mathbf{A}$ can be reduced to diagonal form with diagonal entries from the algebra $\mathbf{A}$.

If $A \subset \mathbf{A}$ is a weakly dense C^*-subalgebra in a finite W^*-algebra, then operators acting on the Hilbert C^*-module $H_A \subset H_{\mathbf{A}}$ can be diagonalized in a larger $\mathbf{A}$-module. When we pass to noncommutative C^*-algebras, "eigenvalues" are not defined uniquely (only up to the unitary equivalence), but in some cases one can choose them from the initial C^*-algebra A instead of the larger W^*-algebra $\mathbf{A}$. We describe some cases, when such possibility can be realized. We also show how these results can be applied to describe spectra of operators commuting with an action of some C^*-algebras.

The following theorem due to Kadison was the first result about diagonalization of operators in the noncommutative case, which initiated a series of further results in this direction [**50, 95, 137, 46, 79, 81, 83**].

THEOREM 6.1.1 ([**57**]). *Let $\mathcal{A}$ be a σ-finite W^*-algebra and let K be a selfadjoint $(n\times n)$-matrix with the entries from $\mathcal{A}$. Then there is a unitary $(n\times n)$-matrix U with the entries from $\mathcal{A}$ such that a matrix U^*KU is diagonal with "eigenvalues" from $\mathcal{A}$.*

Consider a simple example showing what kind of difficulties can occur while diagonalizing matrices over C^*-algebras. Let $A = C[-1,1]$. Put

$$F(t) = \begin{cases} T & \text{if } t > 0, \\ 0 & \text{if } t \le 0. \end{cases}$$

Define the operator $K : A \oplus A \longrightarrow A \oplus A$ by

$$K = \begin{pmatrix} 1+t & f(t) \\ f(t) & 1+|t| \end{pmatrix}.$$

For $t < 0$, the eigenvectors of an operator K are $(1,0)$ and $(0,1)$, and for $t > 0$, they are $(\frac{\sqrt{2}}{2}, \frac{\sqrt{2}}{2})$ and $(\frac{\sqrt{2}}{2}, -\frac{\sqrt{2}}{2})$. One can glue these eigenvectors together in such a way that their coordinates are continuous functions on $[-1,1]$, hence one can diagonalize the operator K over the C^*-algebra A.

Let A be a C^*-algebra. A compact operator K acting on a Hilbert C^*-module $\mathcal{M}$ is called *strictly positive* if the element $\langle Kx, x\rangle \in A$ is positive for any $x \in \mathcal{M}$ and if the spectral projection of the operator K corresponding to the origin is equal to zero. This condition means that the operator K is a strictly positive element of the C^*-algebra $\mathrm{Hom}^*_A(\mathcal{M})$.

By Theorem 3.2.1 due to Paschke [**100**], in the case where A is a W^*-algebra, the inner product can be extended in a natural way to the dual module $\mathcal{M}'$.

For modules $L_n(A)$ and H_A we use common notation $\mathcal{M}_A$.

DEFINITION 6.1.2. Let $\mathbf{A}$ be a W^*-algebra. A selfadjoint operator K in a Hilbert $\mathbf{A}$-module $\mathcal{M}_{\mathbf{A}}$ is called *diagonalizable* if there exists a set $\{x_i\}$ of elements in $\mathcal{M}'_{\mathbf{A}}$ and a set of elements $\lambda_i \in \mathbf{A}$ such that

(i) the set $\{x_i\}$ is orthonormal, i.e., $\langle x_i, x_j\rangle = \delta_{ij}$,
(ii) the module $\mathcal{M}'_{\mathbf{A}}$ coincides with the $\mathbf{A}$-module $\mathcal{N}'$ dual to the module $\mathcal{N}$ generated by the set $\{x_i\}$,
(iii) $Kx_i = x_i\lambda_i$,
(iv) for any unitary elements u_i, $u_{i+1} \in \mathbf{A}$ one has the operator inequality

$$u_i^*\lambda_i u_i \geq u_{i+1}^*\lambda_{i+1}u_{i+1}. \tag{6.1}$$

Elements x_i are called "*eigenvectors*" and the operators λ_i "*eigenvalues*" of the operator K. We should keep in mind that the "eigenvectors" and the "eigenvalues" are not defined uniquely. Condition (6.1) means that our notion of diagonalizability includes the natural ordering of "eigenvalues". For example, if $A = L^\infty(X)$ is a commutative W^*-algebra and K is an (already diagonal) operator of the form

$$\begin{pmatrix} f(x) & 0 \\ 0 & g(x) \end{pmatrix}$$

on the module $A \oplus A$, then, after diagonalizing this operator, we get the operator

$$\begin{pmatrix} \max(f(x), g(x)) & 0 \\ 0 & \min(f(x), g(x)) \end{pmatrix}.$$

Before we proceed, we give a few additional examples explaining Definition 6.1.2.

EXAMPLE 6.1.3. Consider the C^*-algebra A of all complex (2×2)-matrices, the module $\mathcal{M} = A \oplus A$ with the usual A-valued inner product and a selfadjoint operator K on $\mathcal{M}$ acting by the formula $K = \theta_{x,x} + \theta_{y,y}$, where

$$x = \left(\begin{pmatrix} 1 & 0 \\ 0 & 3 \end{pmatrix}, \begin{pmatrix} 0 & 0 \\ 0 & 0 \end{pmatrix}\right), \qquad y = \left(\begin{pmatrix} 0 & 0 \\ 0 & 0 \end{pmatrix}, \begin{pmatrix} 2 & 0 \\ 0 & 2 \end{pmatrix}\right).$$

If we take $x, y \in A \oplus A$ as "eigenvectors" of the operator K, then the corresponding "eigenvalues" would be

$$\lambda_x = \begin{pmatrix} 1 & 0 \\ 0 & 9 \end{pmatrix}, \qquad \lambda_y = \begin{pmatrix} 4 & 0 \\ 0 & 4 \end{pmatrix},$$

but they are not comparable as elements of the positive cone of A. By choosing the "eigenvectors" in another way,

$$x_1 = \left(\begin{pmatrix} 0 & 0 \\ 0 & 1 \end{pmatrix}, \begin{pmatrix} 1 & 0 \\ 0 & 0 \end{pmatrix}\right), \qquad x_2 = \left(\begin{pmatrix} 1 & 0 \\ 0 & 0 \end{pmatrix}, \begin{pmatrix} 0 & 0 \\ 0 & 1 \end{pmatrix}\right),$$

we are in the situation of Definition 6.1.2. In this case the "eigenvalues" are

$$\lambda_1 = \begin{pmatrix} 4 & 0 \\ 0 & 9 \end{pmatrix}, \qquad \lambda_2 = \begin{pmatrix} 1 & 0 \\ 0 & 4 \end{pmatrix}.$$

These "eigenvalues" are ordered and the "eigenvectors" x_1, x_2 satisfy the condition $\langle x_i, x_j \rangle = \delta_{ij}$. Throwing out requirement (i) of Definition 6.1.2, one can choose K-invariant submodules in $\mathcal{M}$ and generating "eigenvectors"

$$x_1 = \left(\begin{pmatrix} 0 & 0 \\ 1 & 1 \end{pmatrix}, \begin{pmatrix} 1 & 1 \\ 0 & 0 \end{pmatrix} \right), \qquad x_2 = \left(\begin{pmatrix} 1 & 1 \\ 0 & 0 \end{pmatrix}, \begin{pmatrix} 0 & 0 \\ 1 & 1 \end{pmatrix} \right).$$

Then the corresponding "eigenvalues" are

$$\lambda_1 = \begin{pmatrix} 4 & 0 \\ 0 & 9 \end{pmatrix}, \qquad \lambda_2 = \begin{pmatrix} 1 & 0 \\ 0 & 4 \end{pmatrix}.$$

They are ordered in a positive cone of algebra A. But the "eigenvectors" generating the K-invariant submodules in $\mathcal{M}$ do not satisfy the condition $\langle x_i, x_j \rangle = \delta_{ij}$.

The following example shows the necessity (even in the commutative case) of using dual Hilbert modules for diagonalizing compact operators.

EXAMPLE 6.1.4. Let $A = L^\infty([0;1])$ and let b_k be a monotone sequence of positive numbers converging to zero. Put

$$A_k = \begin{cases} 1, & t \in (\frac{1}{2^k}; \frac{1}{2^{k-1}}], \\ 0, & \text{for other } \ t, \end{cases}$$

and let $f_k(t) = b_k \cdot a_k(t)$. Let K be a compact operator given by the matrix

$$K = \begin{pmatrix} f_1 & f_2 & \cdots & f_n & \cdots \\ f_2 & 0 & \cdots & 0 & \cdots \\ \vdots & \vdots & & \vdots & \\ f_n & 0 & \cdots & 0 & \cdots \\ \vdots & \vdots & & \vdots & \end{pmatrix}$$

in the standard basis of the module H_A. This operator can be easily diagonalized pointwise. The eigenvector corresponding to the maximal eigenvalue has the form $x = (x_n(t))$, where

$$X_1(t) = a_1(t) + \frac{\sqrt{2}}{2} \sum_{k>1} a_k(t) \quad \text{and} \quad x_n = \frac{\sqrt{2}}{2} a_n(t), \ n > 1.$$

Then $\langle x, x \rangle = \sum_k a_k(t) = 1$. This series is not norm-convergent in A, so $x \in H'_A \setminus H_A$, hence the operator K cannot be diagonalized in the module H_A.

This example leads to the following statement.

PROPOSITION 6.1.5 ([**46**]). *Let* $\mathbf{A}$ *be a finite* W^**-algebra with infinite center. Consider a Hilbert* $\mathbf{A}$*-module* $\mathcal{M}$ *such that* $H_{\mathbf{A}} \subset \mathcal{M} \subset H'_{\mathbf{A}}$. *If each strictly positive compact operator acting on* $\mathcal{M}$ *can be diagonalized in* $\mathcal{M}$, *then the module* $\mathcal{M}$ *is self-dual, i.e.,* $\mathcal{M} = H'_{\mathbf{A}}$.

6.2. Quadratic forms on H'_A related to selfadjoint operators

If A is a C^*-algebra with a faithful finite trace τ and if D is a selfadjoint operator acting on a Hilbert A-module $\mathcal{M}$, then a quadratic form on $\mathcal{M}$ can be defined by the formula $Q(x) = \tau(\langle Dx, x\rangle)$ for $x \in \mathcal{M}$. Below we will see that such C^*-module quadratic forms have properties similar to those of the usual Hilbert space quadratic forms. By $B_1(\mathcal{M})$ we denote the unit ball of the module $\mathcal{M}$.

PROPOSITION 6.2.1. *Let D be a positive operator on $\mathcal{M}$ such that* $\operatorname{Ker} D = 0$. *If the quadratic form $Q(x)$ reaches its supremum over $B_1(\mathcal{M})$ at some element x, then the operator $\langle x, x\rangle$ is a projection.*

PROOF. Denote $\langle x, x\rangle$ by $h \in A$. Then we have $\|h\| \leq 1$, $h \geq 0$ and $h^* = h$. Suppose that the spectrum of the operator h contains some number c different from zero and one. Define the function $f(t)$ on the segment $[0;1] \supset \operatorname{Sp} h$ by the formula

$$f(t) = \begin{cases} \frac{1}{\sqrt{\varepsilon}}, & 0 \leq t \leq \varepsilon, \\ \frac{1}{\sqrt{t}}, & \varepsilon \leq t \leq 1, \end{cases}$$

where $0 < \varepsilon < c$. Put $a = f(h)$, $x' = xa$. Then $\langle x', x'\rangle = aha = ha^2$. This operator is equal to the value of the function $t \cdot f^2(t)$ applied to the operator h. Since $t \cdot f^2(t) \leq 1$ for $0 \leq t \leq 1$, we obtain the estimate $\|ha^2\| \leq 1$ and the element x' lies in $B_1(\mathcal{M})$. By assumption, $\langle Dx, x\rangle$ is a positive operator. Denote it by k^2, where $k \geq 0$. Then

$$\begin{aligned} Q(x') - Q(x) &= \tau\Big(\langle Dxa, xa\rangle - \langle Dx, x\rangle\Big) = \tau(ak^2a - k^2) \\ &= \tau(a^2k^2 - k^2) = \tau(ka^2k - k^2) \\ &= \tau(k(a^2 - 1)k). \end{aligned}$$

By definition, the operator $a^2 - 1$ is also positive. Denote it by b^2, where $b \geq 0$. Then

$$\tau(k(a^2-1)k) = \tau(kb^2k) = \tau\Big(b\langle Dx, x\rangle b\Big) = \tau\Big(\langle Dxb, xb\rangle\Big),$$

hence

$$Q(x') - Q(x) = \tau\Big(\langle Dxb, xb\rangle\Big). \tag{6.2}$$

But $\langle xb, xb\rangle = b\langle x, x\rangle b = bhb$ and the operator bhb corresponds to the function $t(f^2(t) - 1)$. This function differs from zero at $t = c$, so the operator bhb differs from zero, hence xb differs from zero as well. Note that the operator $D^{1/2}$ is well defined and that $\operatorname{Ker} D^{1/2} = 0$. Therefore

$$\tau\Big(\langle Dxb, xb\rangle\Big) = \tau\Big(\langle D^{1/2}xb, D^{1/2}xb\rangle\Big) > 0,$$

so we obtain from (6.2) the inequality $Q(x') - Q(x) > 0$, which contradicts the assumption that $Q(x)$ is a supremum of the quadratic form Q over $B_1(\mathcal{M})$. Thus, we have proved that $\operatorname{Sp} h$ does not contain values other than zero and one, so h is a projection. □

PROPOSITION 6.2.2. *Let $x \in B_1(\mathcal{M})$ be a vector at which the quadratic form $Q = \langle D\cdot, \cdot\rangle$ reaches its supremum over B_1, and let $\mathcal{L} \subset \mathcal{M}$ denote the submodule generated by the vector x. Then $\mathcal{M} = \mathcal{L} \oplus \mathcal{L}^\perp$ and the submodules $\mathcal{L}$ and $\mathcal{L}^\perp$ are D-invariant, i.e., $D\mathcal{L} \subset \mathcal{L}$, $D\mathcal{L}^\perp \subset \mathcal{L}^\perp$.*

PROOF. By the previous proposition, the operator $\langle x, x\rangle$ is a projection. Therefore $\mathcal{L}$ is a projective module and, by the Dupré – Fillmore Theorem 1.4.6, $\mathcal{M} = \mathcal{L} \oplus \mathcal{L}^\perp$. Choose an element $y \in \mathcal{L}^\perp$ such that $\|y\| = 1$. Put $x_t = x\cos t + y \sin t$. Since Q reaches its maximum at the point $x_0 = x$, one should have $\frac{d}{dt}Q(x_t) = 0$ at $t = 0$. It is easy to see that

$$\frac{d}{dt}Q(x_t)|_{t=0} = \tau\Big(\langle Dx, y\rangle + \langle Dx, y\rangle^*\Big).$$

If $\langle Dx, y\rangle \neq 0$, then put $a = \frac{1}{\|\langle Dx,y\rangle\|}\langle Dx, y\rangle$. Obviously, $\|a\| = 1$. Further, put $z = ya^*$ and $\bar{x}_t = x\cos t + z\sin t$. Then

$$\begin{aligned} 0 = \frac{d}{dt}Q(\bar{x}_t)|_{t=0} &= \tau\Big(\langle Dx, z\rangle + \langle Dx, z\rangle^*\Big) \\ = \tau\Big(\langle Dx, y\rangle a^* + a\langle Dx, y\rangle^*\Big) &= \tau\Big(2aa^* \cdot \|\langle Dx, y\rangle\|\Big) \\ = 2\,\|\langle Dx, y\rangle\| \cdot \tau(aa^*). \end{aligned}$$

Since the trace τ is faithful, it follows that $a = 0$. Therefore the vector Dx is orthogonal to any $y \in \mathcal{L}^\perp$, hence $Dx \in \mathcal{L}$. Since the operator D is selfadjoint, we conclude that $\langle Dx, y\rangle = 0$ for any $y \in \mathcal{L}^\perp$. Therefore $D\mathcal{L}^\perp \subset \mathcal{L}^\perp$. □

PROPOSITION 6.2.3. *Let $x \in B_1(H_A)$ be a vector, at which the quadratic form Q reaches its supremum over the set $B_1(H_A)$. Then $\langle x, x\rangle = 1$.*

PROOF. If the operator $\langle x, x\rangle$ is less than one, then there exists a vector $y \in x^\perp$ such that $\|y\| \leq 1$, $y \neq 0$ and $yq = y$, where $q = 1 - \langle x, x\rangle$ is a projection. Then $\langle x + y, x + y\rangle = \langle x, x\rangle + \langle y, y\rangle \leq 1$, so $x + y \in B_1(H_A)$. But by the previous proposition, one has $Q(x + y) = Q(x) + Q(y)$ and since $y \neq 0$ and $\operatorname{Ker} D = 0$, the inequality $Q(y) > 0$ should be valid, hence $Q(x + y) > Q(x)$ and the obtained contradiction proves the proposition. □

PROPOSITION 6.2.4. *Let the operator D on the module $\mathcal{M}$ be positive and let its "eigenvectors" form a basis in $\mathcal{M}$. If these "eigenvalues" satisfy the inequality $\operatorname{Sp}\lambda_i \geq \operatorname{Sp}\lambda_{i+1}$, then the supremum of the quadratic form Q over $B_1(\mathcal{M})$ is reached at the first "eigenvector" x_1 and is equal to $\tau(\lambda_1)$.*

PROOF. Any vector $x \in B_1(\mathcal{M})$ can be written as $x = \sum_i x_i a_i$, where $a_i \in A$. Then

$$\begin{aligned} Q(x) = \tau\Big(\langle Dx, x\rangle\Big) &= \tau\left(\sum_i a_i^*\langle Dx_i, x_i\rangle a_i\right) \\ = \tau\left(\sum_i a_i^*\lambda_i a_i\right) &\leq \tau(a_1^*\lambda_1 a_1) + \sum_{i>1}\tau(a_i^*\lambda_i a_i). \end{aligned}$$

Let $\operatorname{Sp}\lambda_1 \geq d \geq \operatorname{Sp}\lambda_2$. Then

$$Q(x) = \tau(a_1^*\lambda_1 a_1) + \sum_{i>1}\tau(a_i^* d a_i) \leq \tau(a_1^*\lambda_1 a_1) + d\tau(1 - a_1^* a_1),$$

since the inequality $\sum_i a_i^* \leq 1$ follows from $\|X\| \leq 1$. Further, we have

$$\begin{aligned} Q(x) &\leq \tau(a_1^*\lambda_1 a_1) + d(1-\tau(a_1^*a_1)) = \tau(a_1^*\lambda_1 a_1 - a_1^* d a_1) + d \\ &= \tau(a_1^*(\lambda_1 - d)a_1) + d = \tau\Big((\lambda_1-d)^{1/2} a_1 a_1^* (\lambda_1 - d)^{1/2}\Big) + d \\ &\leq \|a_1\|^2 \cdot \tau(\lambda_1 - d) + d \leq \tau(\lambda_1 - d) + d = \tau(\lambda_1). \end{aligned}$$

Thus, $\tau(\lambda_1)$ is the supremum of the form $Q(x)$ over $B_1(\mathcal{M})$ and it is reached at the vector x_1. □

6.3. Diagonalizing operators in the W^*-case

THEOREM 6.3.1 ([**81**]). *Let* $\mathbf{A}$ *be a* W^**-algebra that admits a decomposition into a direct integral of finite factors. Then a compact strictly positive operator* K *in the module* $H_{\mathbf{A}}$ *can be diagonalized in* $H'_{\mathbf{A}}$ *and its "eigenvalues" are defined uniquely up to the unitary equivalence.*

PROOF. Note that the W^*-algebra $\mathrm{End}^*_{\mathbf{A}}(H'_{\mathbf{A}})$ is semifinite and its center is isomorphic to the center Z of the algebra $\mathbf{A}$. Therefore (see [**123**]) both these algebras can be decomposed into a direct integral of factors over a Borel space Γ with a measure $d\gamma$ such that $L^\infty(\Gamma) = Z$. The operator K then also admits the decomposition

$$K = \int_\Gamma^\oplus K(\gamma)\, d\gamma.$$

Put $\bar{T} = T \otimes \mathrm{tr}$, where T is the standard Z-valued finite trace on $\mathbf{A}$ and tr is the standard trace in the Hilbert space l_2. This is a semifinite center-valued trace on the W^*-algebra $\mathrm{End}^*_{\mathbf{A}}(H'_{\mathbf{A}})$. Let us show first that if we separate the spectrum of K from zero, then the resulting operator would lie in the ideal of operators with a finite trace in the algebra $\mathrm{End}^*_{\mathbf{A}}(H'_{\mathbf{A}})$. By χ_E we denote, as usual, the characteristic function of a set $E \subset \mathbf{R}$.

LEMMA 6.3.2. *For each* $\varepsilon > 0$*, the inequality*

$$\bar{T}(\chi_{(\varepsilon;+\infty)}(K)) < \infty$$

holds almost everywhere on Γ.

PROOF. Denote the spectral projection $\chi_{(\varepsilon;+\infty)}(K)$ by P. Then the operator inequality

$$K|_{\mathrm{Im}\, P} \geq \varepsilon \tag{6.3}$$

holds on the $\mathbf{A}$-submodule $\mathrm{Im}\, P \subset H'_{\mathbf{A}}$ by the spectral theorem. The compactness of the operator K implies that the module $H_{\mathbf{A}}$ can be decomposed into the direct sum $H_{\mathbf{A}} = L_n(\mathbf{A}) \oplus \mathcal{R}$ in such a way that $\|K|_{\mathcal{R}}\| < \varepsilon$ for sufficiently large n. Passing to dual modules, we obtain the estimate

$$\|K|_{\mathcal{R}'}\| < \varepsilon, \tag{6.4}$$

where $H'_{\mathbf{A}} = L_n(\mathbf{A}) \oplus \mathcal{R}'$. Denote by Q the projection onto $\mathcal{R}'$ in $H'_{\mathbf{A}}$. Then the projection on $\mathrm{Im}\, P \cap \mathcal{R}'$ equals $P \wedge Q$ and the projection onto $(\mathrm{Ker}\, P \cap L_n(\mathbf{A}))^\perp$ is equal to $P \vee Q$ (here we use standard notation for projections onto intersections and sums of subspaces; see, e.g., [**123**]). It follows from Theorem V.1.37 of [**123**] that

$$\bar{T}(P \vee Q - Q) = \bar{T}(P - P \wedge Q). \tag{6.5}$$

Since $P \vee Q \leq 1$, where 1 denotes the unit operator in $\mathrm{End}^*_{\mathbf{A}}(H'_{\mathbf{A}})$, we get the inequality $\bar{T}(P \vee Q - Q) \leq \bar{T}(1 - Q)$. But $1 - Q$ is the projection onto $L_n(\mathbf{A})$, so its trace is equal to n. Therefore it follows from (6.5) that $\bar{T}(P - P \wedge Q) \leq n$. Comparing (6.3) with (6.4), we see that $\mathrm{Im}\, P \cap \mathcal{R}' = 0$, hence $P \wedge Q = 0$ and, finally, we can conclude that $\bar{T}(P) \leq n$. □

Now we need one simple fact about measurable functions.

LEMMA 6.3.3. *Let Γ be a Borel space and let $\psi : \Gamma \times \mathbf{R} \longrightarrow \mathbf{R}$ be a function such that*

(i) *for each $\lambda \in \mathbf{R}$ the function $\psi(\gamma; \lambda)$ is measurable on Γ;*
(ii) *the function $\psi(\gamma; \lambda)$ is right-semicontinuous and monotone nonincreasing in the second argument for almost all γ.*

For each real number α put

$$c_\alpha(\gamma) = \inf\{\lambda : \psi(\gamma; \lambda) \leq \alpha\}. \tag{6.6}$$

Then the function $c_\alpha(\gamma)$ is measurable.

PROOF. We should show that the set $V = \{\gamma : c_\alpha(\gamma) \leq \beta\}$ is measurable for each $\beta \in \mathbf{R}$. But the definition of the function $c_\alpha(\gamma)$ and (ii) imply that $V = \{\gamma : \inf\{\lambda : \psi(\gamma; \lambda) \leq \alpha\} \leq \beta\} = \{\gamma : \psi(\gamma; \beta) \leq \alpha\}$. The measurability follows from (i). □

Recall that the operator K is decomposable over Γ. Let

$$P_1(\gamma; \lambda) = \chi_{(\lambda; +\infty)}(K(\gamma)), \qquad P_2(\gamma; \lambda) = \chi_{[\lambda; +\infty)}(K(\gamma))$$

be the spectral projections of the operator $K(\gamma)$ corresponding to the sets $(\lambda; +\infty)$ and $[\lambda; +\infty)$ respectively. Put

$$P_1(\lambda) = \chi_{(\lambda; +\infty)}(K), \qquad P_2(\lambda) = \chi_{[\lambda; +\infty)}(K)$$

and $\phi(\gamma; \lambda) = \bar{T}(P_1(\lambda))$. Then this function satisfies the conditions of Lemma 6.3.3, hence the function

$$\lambda(\gamma) = \inf\{\lambda : \phi(\gamma; \lambda) \leq 1\} \tag{6.7}$$

is measurable.

Now let us define two new projections in $H'_{\mathbf{A}}$:

$$\begin{aligned} P_1 &= \int_\Gamma^\oplus \chi_{(\lambda(\gamma); +\infty)}(K(\gamma))\, d\gamma &= \int_\Gamma^\oplus P_1(\gamma; \lambda(\gamma))\, d\gamma, \\ P_2 &= \int_\Gamma^\oplus \chi_{[\lambda(\gamma); +\infty)}(K(\gamma))\, d\gamma &= \int_\Gamma^\oplus P_2(\gamma; \lambda(\gamma))\, d\gamma \end{aligned} \tag{6.8}$$

and let us check that they are well defined.

LEMMA 6.3.4. *The operator-valued functions $P_1(\gamma; \lambda(\gamma))$ and $P_2(\gamma; \lambda(\gamma))$ are measurable.*

PROOF. We can assume that the W^*-algebra $\mathbf{A}$ is faithfully represented in the direct integral of Hilbert spaces $H = \int_\Gamma^\oplus H(\gamma)\, d\gamma$ equipped with the scalar product $(\cdot, \cdot)$. We have to show that the function

$$\gamma \longmapsto \Big(P_1(\gamma; \lambda(\gamma))\, \xi(\gamma), \xi(\gamma)\Big) \tag{6.9}$$

is measurable for all $\xi = \int_\Gamma^\oplus \xi(\gamma)\, d\gamma \in H$. By Theorem XIII.85 in [**107**] the function

$$\psi(\gamma;\lambda) = \Big(P_1(\gamma;\lambda(\gamma))\,\xi(\gamma),\xi(\gamma)\Big)$$

satisfies the conditions of Lemma 6.3.3. The measurability of (6.9) follows from the measurability of the set $U = \{\gamma : \psi(\gamma;\lambda(\gamma)) \le \alpha\}$ for each α. But the definition of the function $c_\alpha(\gamma)$ (6.6) implies that

$$U = \{\gamma : \lambda(\gamma) \ge c_\alpha(\gamma)\} = \{\gamma : \lambda(\gamma) - c_\alpha(\gamma) \ge 0\}.$$

This set is measurable because of the measurability of the function $\lambda(\gamma) - c_\alpha(\gamma)$. The case of the projection P_2 can be dealt with in the same way. □

COROLLARY 6.3.5. *The projections P_1 and P_2 given by formula (6.8) are well defined and*

$$\bar{T}(P_1) \le 1,\ \bar{T}(P_2) \ge 1,\ P_1 \le P_2.$$

These two projections define the decomposition of the module $H'_{\mathbf{A}}$ into the direct sum of three submodules,

$$H'_{\mathbf{A}} = H_- \oplus H_0 \oplus H_+, \tag{6.10}$$

where $H_+ = \operatorname{Im} P_1$, $H_0 = \operatorname{Im}(P_2 - P_1)$, $H_- = \operatorname{Ker} P_2$. The operator K commutes with these projections, since the operators $K(\gamma)$ commute with the projections $P_1(\gamma;\lambda(\gamma))$ and $P_2(\gamma;\lambda(\gamma))$ for almost all γ. Therefore the operator K can be written, with respect to the decomposition (6.10), as

$$K = \begin{pmatrix} K_+ & 0 & 0 \\ 0 & K_0 & 0 \\ 0 & 0 & K_- \end{pmatrix},$$

and the operator K_0 is an operator of multiplication by the scalar $\lambda(\gamma)$ for almost all γ. Therefore any submodule of the module H_0 is invariant for the operator K. We can conclude from Corollary 6.3.5 that there exists a projection P such that $P_1 \le P \le P_2$ and $\bar{T}(P) = 1$. Then the operator K is diagonal with respect to the decomposition $H'_{\mathbf{A}} = \operatorname{Im} P \oplus \operatorname{Ker} P$,

$$K = \begin{pmatrix} K_1 & 0 \\ 0 & K' \end{pmatrix}.$$

Note that the module $\operatorname{Im} P$ is isomorphic to $\mathbf{A}$, since the projections onto these two submodules in $H'_{\mathbf{A}}$ both have the same center-valued traces equal to 1, hence these projections are equivalent [**123**]. Let $x_1 \in H'_{\mathbf{A}}$ be a generator of the module $\operatorname{Im} P$, $\langle x_1, x_1\rangle = 1$. If this generator is fixed, then one can represent the operator $K_1 : \operatorname{Im} P \longrightarrow \operatorname{Im} P$ as the operator of multiplication by some element $\lambda_1 \in \mathbf{A}$, i.e., $K_1 x_1 a = x_1 \lambda_1 a$ for $a \in \mathbf{A}$, $x_1 a \in \operatorname{Im} P$. By Theorem 3.7.1, the module $\operatorname{Ker} P$ is isomorphic to the module $H'_{\mathbf{A}}$ and the operator K' is obviously compact in the module $\operatorname{Ker} P$, so we can apply Lemma 6.3.2 to K'. One also has the operator inequality $K_1 = \lambda_1 \ge K'$.

Proceeding by induction, we can find elements $x_i \in H'_{\mathbf{A}}$ and operators $\lambda_i \in \mathbf{A}$ such that $\langle x_i, x_j\rangle = \delta_{ij}$, $K x_i = x_i \lambda_i$ and $\lambda_{i+1} \le \lambda_i$. Denote by $\mathcal{N}$ the Hilbert $\mathbf{A}$-module generated by all these elements x_i. It is clear that $\mathcal{N} \cong H_{\mathbf{A}}$. Note that the operator $K|_{\mathcal{N}}$ is not necessarily compact. We need to show that $\mathcal{N}' = H'_{\mathbf{A}}$.

LEMMA 6.3.6. *The norms of the operators λ_i tend to zero as $i \to \infty$.*

PROOF. Since the sequence $\|\lambda_i\|$ is monotone nonincreasing, it converges to some number $b \geq 0$. Suppose that $b \neq 0$. The operators λ_i can be decomposed into a direct integral over Γ. From the construction of the operators λ_i we can conclude that there exist numbers $d_i(\gamma)$ such that

$$\lambda_i(\gamma) \geq d_i(\gamma) \geq \lambda_{i+1}(\gamma). \tag{6.11}$$

If we decompose x_i coordinatewise into the direct integral, $x_i = \int_\Gamma^\oplus x_i(\gamma)\, d\gamma$, then, for almost all γ, the vectors $x_i(\gamma)$ are orthonormal in $H'_{\mathbf{A}}$ and $K(\gamma)x_i(\gamma) = x_i(\gamma)\lambda_i(\gamma)$. Define the function $b(\gamma)$ as the limit of the norms $\|\lambda_i(\gamma)\|$ in the algebras $\mathbf{A}(\gamma)$. Then

$$\|\lambda_i(\gamma)\| = \|\langle K(\gamma)x_i(\gamma), x_i(\gamma)\rangle\| \geq b(\gamma), \tag{6.12}$$

where the inner product is taken also in the $\mathbf{A}(\gamma)$-modules $H'_{\mathbf{A}(\gamma)}$. Let x be an element of the module $\mathcal{N}$. Then it can be written as

$$x(\gamma) = \sum_i x_i(\gamma)a_i(\gamma) \quad \text{for some} \quad a_i = \int_\Gamma^\oplus a_i(\gamma)\, d\gamma \in \mathbf{A}.$$

If $\langle x, x\rangle = 1$ for almost all γ, then

$$\sum_i a_i^*(\gamma)a_i(\gamma) = 1. \tag{6.13}$$

From (6.11) and (6.12) we can conclude that the operator inequality $\lambda_i(\gamma) \geq b(\gamma)$ is valid for all i. Therefore, since $b(\gamma)$ is a scalar, due to (6.13) one has

$$\begin{aligned} \langle K(\gamma)x(\gamma), x(\gamma)\rangle &= \sum_i a_i^*(\gamma)\lambda_i(\gamma)a_i(\gamma) \\ &\geq \sum_i a_i^*(\gamma)a_i(\gamma)b(\gamma) \;=\; b(\gamma). \end{aligned}$$

Further, we obtain that

$$\|\langle Kx, x\rangle\| = \operatorname{ess\,sup} \|\langle K(\gamma)x(\gamma), x(\gamma)\rangle\| \geq \operatorname{ess\,sup} b(\gamma) = b,$$

and, since $b > 0$ by assumption, we have

$$\|\langle Kx, x\rangle\| \geq b \tag{6.14}$$

for all $x \in \mathcal{N}$ satisfying $\langle x, x\rangle = 1$. Consider now the projection $P_n : \mathcal{N} \longrightarrow L_n(\mathbf{A})$. If the spectrum of the operator K is separated from zero, then P_n defines an inclusion of the module $\mathcal{N}$ into the module L_n, which is impossible in the case of finite W^*-algebras. Therefore for any $\varepsilon > 0$, we can find a vector $x \in \mathcal{N}$ such that $\langle x, x\rangle = 1$ and $\|P_n x\| < \varepsilon$. Put $x' = P_n x$, $x'' = x - x'$. Then $\|x'\| < \varepsilon$, $\|x''\| \leq 1$. Estimate the norm of $\langle Kx, x\rangle$:

$$\begin{aligned} \|\langle Kx, x\rangle\| &\leq \|\langle Kx', x'\rangle\| \;+\; 2\,\|\langle Kx', x''\rangle\| \;+\; \|\langle Kx'', x''\rangle\| \\ &\leq \|K\|\,\|x'\|^2 \;+\; 2\,\|K\|\,\|x'\|\,\|x''\| \;+\; \|\langle Kx'', x''\rangle\| \\ &\leq \|K\|\varepsilon^2 \;+\; 2\,\|K\|\varepsilon \;+\; \|\langle Kx'', x''\rangle\|. \end{aligned}$$

Since $x'' \perp L_n$, due to the compactness of the operator K, one should have the estimate $\|\langle Kx'', x''\rangle\| < \varepsilon$ for sufficiently large n. Therefore $\|\langle Kx, x\rangle\| < \varepsilon'$, where $\varepsilon' = \|K\|\varepsilon^2 + 2\|K\|\varepsilon + \varepsilon$. If we choose ε sufficiently small, then this estimate contradicts (6.14), hence our assumption $b > 0$ is false. □

We have shown that the norm of the restriction of the operator K onto the orthogonal complement to the vectors $x_1, \dots, x_n$ tends to zero. Hence $x \in \mathcal{N}^\perp$ implies that $\|Kx\| = 0$. But since the spectral projection of K corresponding to the origin is zero, $\operatorname{Ker} K = 0$ and hence $\mathcal{N}^\perp = 0$. Therefore it follows from Proposition 3.4.3 that $\mathcal{N}' = H'_{\mathbf{A}}$. □

The "eigenvalues" λ_i of the operator K are obviously not defined uniquely and the same is true for the "eigenvectors" x_i. If, for example, we take $x'_i = x_i u_i$, where $u_i \in \mathbf{A}$ is a unitary element, then the elements $\lambda'_i = u_i^* \lambda_i u_i$ are also the "eigenvalues" of the operator K. Nevertheless the following statement shows that if the "eigenvalues" are ordered, then they are unique up to the unitary equivalence.

PROPOSITION 6.3.7. *Let λ_i and x_i be the "eigenvalues" and "eigenvectors" constructed in the proof of Theorem* 6.3.1 *and let μ_i be "eigenvalues" of the operator K corresponding to another "basis" $\{y_i\}$ of the module $H'_{\mathbf{A}}$ (i.e., the orthogonal complement to the submodule generated by $\{y_i\}$ is zero). If the operator inequality $v_i^* \mu_i v_i \geq v_{i+1}^* \mu_{i+1} v_{i+1}$ holds for all unitaries $v_i \in \mathbf{A}$ and for all numbers i, then λ_i and μ_i coincide up to the unitary equivalence.*

PROOF. It is easy to verify that, by assumption, the inequality $\inf \operatorname{Sp} \mu_i(\gamma) \geq \sup \operatorname{Sp} \mu_{i+1}(\gamma)$ is valid in the factors $\mathbf{A}(\gamma)$ for almost all $\gamma \in \Gamma$. Therefore each projection onto a submodule generated by each vector y_i in the module $H'_{\mathbf{A}}$ is a spectral projection for the operator K. Denote the projection onto $\operatorname{Span}_{\mathbf{A}}(y_1)$ by Q. Then obviously $P_1 \leq Q \leq P_2$, where P_1 and P_2 were defined by formulas (6.8). We can decompose Q into the sum $Q = P_1 \oplus R$ and the projection P onto $\operatorname{Span}_{\mathbf{A}}(x_1)$ can be written as the sum $P = P_1 \oplus S$, where R and S are some projections. Since $\bar{T}(P) = \bar{T}(Q) = 1$, R and S are equivalent and $\operatorname{Im} R \cong \operatorname{Im} S$. This module isomorphism commutes with the action of the operator K, since its restriction onto these modules is scalar and coincides with multiplication by $d_1(\gamma)$ for almost all $\gamma \in \Gamma$. Therefore there exists a unitary element $u_1 \in \mathbf{A}$ such that it induces this isomorphism between the modules $\operatorname{Im} P$ and $\operatorname{Im} Q$ and $\lambda_1 = u_1^* \mu_1 u_1$. Proceeding by induction, we obtain the unitary equivalence of the two sets of "eigenvalues". □

To finish this section let us discuss the theorem about diagonalizing operators in the case when the requirement of the strict positiveness of the operator K is discarded. If K is an arbitrary selfadjoint compact operator in $H_{\mathbf{A}}$ or $H'_{\mathbf{A}}$, then the module $H'_{\mathbf{A}}$ can be decomposed into the direct sum $H'_{\mathbf{A}} = H_- \oplus \operatorname{Ker} K \oplus H_+$ in such a way that the restriction of K onto H_+ (resp. H_-) is positive (resp. negative). We can find two sets of the "eigenvectors" separately in H_+ and in H_-, but we should have to dismiss the requirement for the "eigenvectors" to be normalized, i.e., the inner products of such vectors by themselves are now equal not to unit but to some projections. Let us consider the following example.

EXAMPLE 6.3.8. Let $p \in \mathbf{A}$ be a nontrivial projection, $p \neq 0$, $p \neq 1$. Consider the module $\mathcal{M} = H_{\mathbf{A}} \oplus p\mathbf{A}$. By the Kasparov stabilization theorem (Theorem 1.4.2), we have $\mathcal{M} \cong H_{\mathbf{A}}$. Let $\{e_i\}$ be a basis in the module $H_{\mathbf{A}}$, $e = (0, p) \in H_{\mathbf{A}} \oplus p\mathbf{A}$. Define the operator K on $\mathcal{M}$ by the formulas $Ke_i = \lambda_i$, $Ke = 0$, where $\{\lambda_i\}$ is a sequence of positive numbers converging to zero. Then the operator K cannot be diagonalized in the sense of Definition 6.1.2, since it is impossible to simultaneously satisfy conditions (i) and (iv) of this definition.

Note also that in [**46**] it is shown that any selfadjoint operator acting on a self-dual Hilbert module over a W^*-algebra can be diagonalized, but without the requirement for "eigenvectors" to satisfy the condition $\langle x, x\rangle = 1$, and "eigenvalues" are defined not uniquely (even as classes of the unitary equivalence).

6.4. Continuity of "eigenvalues"

In this section we show that "eigenvalues" of compact operators on Hilbert C^*-modules over W^*-algebras enjoy the property of continuity. Throughout this section $\mathbf{A}$ denotes a W^*-algebra that admits a decomposition into a direct integral of finite factors.

LEMMA 6.4.1. *Let $K_1 = \sum \alpha_l^{(1)} P_l^{(1)}$, $K_2 = \sum \alpha_l^{(2)} P_l^{(2)}$ be strictly positive operators on $L_n(\mathbf{A})$ with a finite spectrum and let $\|K_1 - K_2\| < \varepsilon$. Then*

(i) *there exists a unitary operator U in $L_n(\mathbf{A})$ mapping the "eigenvectors" of K_2 to the "eigenvectors" of K_1 such that $\|U^* K_1 U - K_2\| < \varepsilon$,*
(ii) *the "eigenvalues" $\{\lambda_i^{(r)}\}$ of the operators K_r $(r = 1, 2)$ can be chosen to satisfy the inequality $\left\|\lambda_i^{(1)} - \lambda_i^{(2)}\right\| < \varepsilon$.*

PROOF. Since the algebra $\mathbf{A}$ can be decomposed into a direct integral of finite factors, it is sufficient to prove the lemma for the case where $\mathbf{A}$ is a type II_1 factor (the proof for type I_n factors is obvious). Denote by $E_K(\lambda)$ the spectral projection of the operator K corresponding to the set $(-\infty, \lambda)$. The faithful finite trace τ on $\mathbf{A}$ can be extended to the (infinite) trace $\bar{\tau} = \mathrm{tr} \otimes \tau$ on the algebra $\mathrm{End}_{\mathbf{A}}(H'_{\mathbf{A}})$ and then restricted to a finite trace on the smaller algebra $\mathrm{End}_{\mathbf{A}}(L_n(\mathbf{A}))$, where $\bar{\tau}(1) = n$. Put

$$\varepsilon_K(\alpha) = \inf_{\bar{\tau}(E_K(\lambda)) \geq \alpha} \lambda, \quad 0 \leq \alpha \leq n.$$

Using the continuous minimax principle [**96**], we obtain

$$(6.15) \qquad \varepsilon_K(\alpha) = \inf_{P \in \mathcal{P},\, \bar{\tau}(P) \geq \alpha} \left\{ \sup_{\xi \in \mathrm{Im}\, P,\, \|\xi\| = 1} (K\xi, \xi) \right\},$$

where $\mathcal{P}$ is the set of projections in $\mathrm{End}_{\mathbf{A}}(L_n(\mathbf{A}))$ and where $(\cdot, \cdot)$ denotes the scalar product in the Hilbert space where the algebra $\mathrm{End}_{\mathbf{A}}(L_n(\mathbf{A}))$ is represented. It follows from (6.15) that $\|K_1 - K_2\| < \varepsilon$ implies the estimate

$$(6.16) \qquad |\varepsilon_{K_1}(\alpha) - \varepsilon_{K_2}(\alpha)| < \varepsilon.$$

Let $Q_i^{(r)}$ be the projections onto the "eigenvectors" $x_i^{(r)}$ of the operators K_r corresponding to the greatest "eigenvalues" $\lambda_i^{(r)}$, $\bar{\tau}(Q_i^{(r)}) = 1$. Out of the two partitions of unity $\{P_l^{(1)}, Q_i^{(1)}\}$ and $\{P_l^{(2)}, Q_i^{(2)}\}$ given by decompositions of K_1 and K_2, one can construct a finer partition of unity. By [**123**], there exist two sets of mutually orthogonal projections $R_m^{(r)} \in \mathrm{End}_{\mathbf{A}}(L_n(\mathbf{A}))$ such that

(i) $\bigoplus_m R_m^{(r)} = 1$,
(ii) $\bar{\tau}(R_m^{(1)}) = \bar{\tau}(R_m^{(2)})$,
(iii) for each m one has $R_m^{(r)} \leq Q_i^{(r)}$ or $R_m^{(r)} \leq P_j^{(r)}$ for some i or j.

Then (after possible renumbering) the operators K_r can be written in the form $K_r = \sum \alpha_m^{(r)} R_m^{(r)}$, where $\alpha_1^{(r)} \le \alpha_2^{(r)} \le \dots \le \alpha_m^{(r)} \in \mathbf{R}$. This enables us to define a unitary operator $U : L_n(\mathbf{A}) \longrightarrow L_n(\mathbf{A})$ such that

$$U(\operatorname{Im} R_m^{(2)}) = \operatorname{Im} R_m^{(1)}. \tag{6.17}$$

Therefore $U(\operatorname{Im} Q_i^{(2)}) = \operatorname{Im} Q_i^{(1)}$ and thus U maps the $\mathbf{A}$-modules generated by the "eigenvectors" $x_i^{(2)}$ into the modules generated by $x_i^{(1)}$, i.e., $Ux_i^{(2)} = x_i^{(1)} \cdot U_i = \bar{x}_i^{(1)}$ for some unitary elements $u_i \in \mathbf{A}$. Put

$$n(\alpha) = \min\{n | \bar{\tau}(\bigoplus_{m \ge n} R_m^{(r)}) \ge \alpha\}.$$

Then $\varepsilon_{K_r(\alpha)} = \alpha_{n(\alpha)}^{(r)}$ and it follows from (6.16) that $|\alpha_{n(\alpha)}^{(1)} - \alpha_{n(\alpha)}^{(2)}| < \varepsilon$. But, changing α, we obtain that

$$|\alpha_m^{(1)} - \alpha_m^{(2)}| < \varepsilon \tag{6.18}$$

for all m. Taking $\alpha = 1$ (then $i = 1$), we obtain

$$K_r|_{\operatorname{Im} Q_1^{(r)}} = \Lambda_1^{(r)} = \sum_{m \ge n(1)} \alpha_m^{(r)} P_m^{(r)}.$$

It follows from (6.17) and (6.18) that

$$\left\| U^* \Lambda_1^{(1)} U - \Lambda_1^{(2)} \right\| = \left\| \sum_{m \ge n(1)} (\alpha_m^{(1)} - \alpha_m^{(2)}) P_m^{(2)} \right\| \le \varepsilon \left\| \bigoplus_{m \ge n(1)} P_m^{(2)} \right\| = \varepsilon. \tag{6.19}$$

Choosing $\lambda_1^{(r)}$ to satisfy the conditions $\Lambda_1^{(1)} \bar{x}_1^{(1)} = \bar{x}_1^{(1)} \lambda_1^{(1)}$ and $\Lambda_1^{(2)} x_1^{(2)} = x_1^{(2)} \lambda_1^{(2)}$, we get the estimate

$$\left\| \lambda_1^{(1)} - \lambda_1^{(2)} \right\| < \varepsilon. \tag{6.20}$$

In the same way we can obtain the estimates (6.19), (6.20) for other i. □

THEOREM 6.4.2. *Let $K_r : H_{\mathbf{A}} \longrightarrow H_{\mathbf{A}}$, $r = 1, 2$, be compact strictly positive operators and let $\|K_1 - K_2\| < \varepsilon$. Then*

(i) *there exists a unitary operator U in $H'_{\mathbf{A}}$ such that it maps the "eigenvectors" of K_2 into the "eigenvectors" of K_1 and $\|U^* K_1 U - K_2\| < \varepsilon$;*

(ii) *the "eigenvalues" $\{\lambda_i^{(r)}\}$ of the operators K_r $(r = 1, 2)$ can be chosen to satisfy the inequality $\left\| \lambda_i^{(1)} - \lambda_i^{(2)} \right\| < \varepsilon$.*

PROOF. Let $L_n^{(r)}(\mathbf{A}) \in H'_{\mathbf{A}}$ denote the Hilbert submodule generated by the first n "eigenvectors" of the operator K_r, $L_n^{(r)}(\mathbf{A}) \cong L_n(\mathbf{A})$. By Theorem 3.7.1, the orthogonal complement to this submodule is isomorphic to $H'_{\mathbf{A}}$ and the norm of the restriction of the compact operator K_r onto the orthogonal complement to $L_n^{(r)}(\mathbf{A})$ in $H'_{\mathbf{A}}$ tends to zero. Therefore it is sufficient to consider only the case of operators acting on the module $L_n(\mathbf{A})$ and there one can approximate these operators by operators with finite spectrum. □

6.5. Case of infinite W^*-algebras

Let us consider now the case of Hilbert C^*-modules over infinite W^*-algebras. Results obtained jointly with M. Frank [**46**] show that diagonalizing operators in self-dual Hilbert modules over infinite W^*-algebras is significantly different both from the case of finite W^*-algebras and from the classical case of Hilbert spaces. This explains why the case of infinite W^*-algebras is of lesser interest in applications.

PROPOSITION 6.5.1 ([**46**]). *Let* $\mathbf{A}$ *be a* W^**-algebra with an infinite number of pairwise orthogonal nontrivial projections* $\{p_i : i \in \mathbf{N}\}$ *equivalent to* $1_{\mathbf{A}}$*, the sum of which is equal to* $1_{\mathbf{A}}$ *(the infinite sum is understood in the sense of the* w^**-convergence). Then the Hilbert* $\mathbf{A}$*-module* $H'_{\mathbf{A}}$ *equipped with the standard* $\mathbf{A}$*-valued inner product is isomorphic to the Hilbert* $\mathbf{A}$*-module* $\{\mathbf{A}, \langle .,. \rangle_{\mathbf{A}}\}$*, where* $\langle a, b\rangle_{\mathbf{A}} = a^* b$*.*

PROOF. Suppose that the equivalence of projections $\{p_i : i \in \mathbf{N}\}$ and $1_{\mathbf{A}}$ is realized by partial isometries $\{u_i : i \in \mathbf{N}\} \in \mathbf{A}$, $1_{\mathbf{A}} = u_i u_i^*$, $p_i = u_i^* u_i$. Then the map

$$S : H'_{\mathbf{A}} \longrightarrow \mathbf{A}, \quad \{a_i\} \longmapsto w^*\text{-}\lim_n \sum_{i=1}^n u_i^* a_i$$

with the inverse map

$$S^{-1} : \mathbf{A} \longrightarrow H'_{\mathbf{A}}, \quad a \longmapsto \{u_i a\}$$

gives a Hilbert $\mathbf{A}$-module isomorphism of the modules $H'_{\mathbf{A}}$ and $\mathbf{A}$. □

COROLLARY 6.5.2. *Let* $\mathbf{A}$ *be an infinite* W^**-algebra. Then each bounded operator* T *on the module* $H'_{\mathbf{A}}$ *can be diagonalized and the formula*

$$T(\{a_i\}) = \{u_i\}\lambda_T \langle \{u_i\}, \{a_i\}\rangle$$

holds for all $\{a_i\} \in H'_{\mathbf{A}}$*, for some* $\lambda_T \in \mathbf{A}$ *and for partial isometries* $\{u_i\} \in \mathbf{A}$ *described in the proof of Proposition* 6.5.1.

PROOF. A W^*-algebra of one of the types I_∞, II_∞ or III possesses a set of partial isometries with properties described in the proof of Proposition 6.5.1. It follows from this proposition that the operator T acting on the module $H'_{\mathbf{A}}$ is unitarily equivalent to the operator STS^{-1} acting on the module $\mathbf{A}$. But each bounded operator in $\mathbf{A}$ is an operator of multiplication (from the left) by some element of the algebra $\mathbf{A}$. □

COROLLARY 6.5.3. *Let* $\mathbf{A}$ *be a* W^**-algebra not containing type* I_n*,* $n < \infty$*, and* II_1 *factors in its decomposition into the direct integral. Let* $\mathcal{M}$ *be a self-dual Hilbert* $\mathbf{A}$*-module possessing a countably generated* $\mathbf{A}$*-pre-dual Hilbert* $\mathbf{A}$*-module. Then each bounded operator* T *on* $\mathcal{M}$ *can be diagonalized and the formula*

$$T(x) = u\lambda_T \langle u, x\rangle \tag{6.21}$$

holds for all $x \in \mathcal{M}$*, for some* $\lambda_T \in \mathbf{A}$ *and for an "eigenvector"* $u \in \mathcal{M}$*, common for all* T*.*

PROOF. Since the module $\mathcal{M}$ possesses a countably generated pre-dual Hilbert $\mathbf{A}$-module, $\mathcal{M}$ is a direct summand of the Hilbert $\mathbf{A}$-module $H'_{\mathbf{A}}$ by the Kasparov stabilization theorem [**63**] (see Theorem 1.4.2). Therefore it is sufficient to prove the statement only for the self-dual Hilbert $\mathbf{A}$-module $H'_{\mathbf{A}}$. For convenience we denote the projection onto the module $\mathcal{M}$ in the module $H'_{\mathbf{A}}$ by P. Consider the central decomposition of the algebra $\mathbf{A}$ into a direct integral. Each fiber in it is a factor of one of the types I_∞, II_∞, or III by assumption. It is easy to see that the modules $H'_{\mathbf{A}}$ and $\mathbf{A}$ are isomorphic: we can either apply Corollary 6.5.2 fiberwise or directly find an appropriate set of partial isometries $\{u_i\} \in \mathbf{A}$. Then any bounded operator T in the module $H'_{\mathbf{A}}$ can be diagonalized. To get formula (6.21), it remains to put $u = P(\{u_i\})$. □

REMARK 6.5.4. Let $\mathbf{A} = \mathcal{B}(H)$ be a I_∞-factor. Then there are selfadjoint operators $T \in \mathbf{A}$ that cannot be diagonalized in a stronger sense. More precisely, it is impossible to write such operators as the sum $\sum \lambda_i P_i$, where $\lambda_i \in \mathbf{C} = \mathrm{Z}(\mathbf{A})$ and $P_i = P_i^* = P_i^2 \in \mathbf{A}$ (compare with the Weyl theorem). Therefore, in the case of infinite W^*-algebras, Corollaries 6.5.2 and 6.5.3 are the best results one could hope for.

6.6. Case of C^*-algebras of real rank zero

It is known that in the commutative case, i.e., when $A = C(X)$, the compact selfadjoint operators cannot be diagonalized in the module H_A. However this becomes possible if we pass to a larger module over the larger W^*-algebra $L^\infty(X) = \mathbf{A} \supset A$, which is obtained from the algebra A by completion (on bounded sets) with respect to the weak topology in the natural representation of the algebra A in the Hilbert space $H = L^2(X)$ with a new (integral) norm assigned to a finite faithful trace (measure) on A. Unlike the "eigenvectors", the "eigenvalues" are continuous, i.e., they lie in the initial C^*-algebra A. This leads to the following definition.

Let A be a separable C^*-algebra with a finite faithful trace τ. Let $(a,b)_\tau = \tau(a^*b)$, $a,b \in A$, be a nondegenerate inner product on A and let $\|\cdot\|_\tau = (\cdot,\cdot)_\tau^{1/2}$ be the norm defined by this trace. Completing A with respect to this norm we obtain the Hilbert space $L^2(A)$ and the faithful representation of the algebra A on this space. Let $\mathbf{A} = L^\infty(A)$ be the corresponding trace finite W^*-algebra containing A as a weakly closed subalgebra. Let K be a compact strictly positive operator in the module H_A. This operator admits a natural extension to the larger module $H_{\mathbf{A}}$ over the larger algebra $\mathbf{A}$, where it is still compact and strictly positive, and where, by Theorem 6.3.1, it can be diagonalized in this module.

DEFINITION 6.6.1. Let $\mathbf{A}$ be a W^*-algebra that admits a decomposition into a direct integral of finite factors and let a separable C^*-algebra A be a weakly dense subalgebra in $\mathbf{A}$. Let K be a compact strictly positive operator on H_A. We call A an algebra admitting *weak diagonalization* if, while diagonalizing K in $H'_{\mathbf{A}}$, its diagonal entries can be chosen within the C^*-algebra $A \in \mathbf{A}$.

Note that if a C^*-algebra admits weak diagonalization, then each selfadjoint operator of a finite rank $K \in M_n \otimes A$ can also be diagonalized over the W^*-algebra $\mathbf{A}$ in such a way that its diagonal entries can be chosen within the C^*-algebra A.

Denote by $\mathcal{Z} = C(Z)$ the center of the W^*-algebra $\mathbf{A}$ and let $T : \mathbf{A} \longrightarrow C(Z)$ be the standard faithful $\mathcal{Z}$-valued trace on $\mathbf{A}$ with $T(\mathbf{1}) = 1$. Assume that a C^*-subalgebra $A \in \mathbf{A}$ possesses the following property:

(∗) for any two projections $p, q \in A$ one can find projections $r_p \sim R_q$, $r_p \leq p$, $r_q \leq q$ in the C^*-algebra A such that they are equivalent in A and $T(r_p) = T(r_q) = \min\{T(p)(z), T(q)(z)\}$, $z \in Z$.

Below we show that the class of C^*-algebras admitting weak diagonalization contains, besides commutative algebras, also weakly dense separable C^*-subalgebras of real rank zero with property $(*)$ in finite W^*-algebras.

Recall that real rank zero ($RR(A) = 0$) means [**19**] that each selfadjoint operator in A can be approximated by operators with finite spectrum, i.e., of the form $\sum_i \alpha_i p_i$, where $p_i \in A$ are selfadjoint mutually orthogonal projections and $\alpha_i \in \mathbf{R}$. By [**19**], in this case one also has $RR(\mathrm{End}_A(L_n(A))) = 0$.

THEOREM 6.6.2. *Let A be a weakly dense separable C^*-subalgebra with property $(*)$ and with $RR(A) = 0$ in a W^*-algebra $\mathbf{A}$ that admits a decomposition into a direct integral of finite factors. If K is a compact strictly positive operator in the A-module H_A, then the "eigenvalues" $\{\lambda_i\}$ of the natural extension of the operator K to the $\mathbf{A}$-module $H'_{\mathbf{A}}$ can be chosen in such a way that $\lambda_i \in A$.*

PROOF. By results of S. Zhang [**137, 19**], an operator K can be approximated by operators $K_n \in \mathrm{End}_A(L_n(A))$ with a finite spectrum. By [**137**, Corollary 3.5], there exist unitary operators $U_n \in \mathrm{End}_A(L_n(A))$ such that the operators

$$U_n^* K_n U_n = \begin{pmatrix} \lambda_1^{(n)} & & 0 \\ & \ddots & \\ 0 & & \lambda_n^{(n)} \end{pmatrix}$$

are diagonal and $\lambda_i^{(n)} \in A$ have a finite spectrum. Let us show that, using property $(*)$ and choosing such U_n in an appropriate way, one can satisfy condition (6.1) for the "eigenvalues" $\{\lambda_i^{(n)}\}$. Let $\lambda_a = \sum \alpha_k q_k$, $\lambda_b = \sum \beta_l r_l$, where q_k, $r_l \in A$ are projections. Suppose that $a < b$, but for some m and n one has $\beta_m > \alpha_n$. Diagonalizing projections [**137**], we can find projections $s_l \in A$ such that they are equivalent to r_l, and $s_l = \bigoplus_k s_k^{(l)}$ and $s_k^{(l)} \leq q_k$. Put

$$\lambda'_a = \sum_{k \neq n} \alpha_k q_k \oplus \sum_{l \neq m} \alpha_n s_n^{(l)} \oplus \beta_n s_n^{(m)},$$

$$\lambda'_b = \sum_{l \neq m} \beta_l s_l \oplus \sum_{k \neq n} \beta_k s_k^{(m)} \oplus \alpha_n s_n^{(m)}$$

and note that the operators

$$\begin{pmatrix} \lambda_1 & & 0 \\ & \ddots & \\ 0 & & \lambda_n \end{pmatrix} \quad \text{and} \quad \begin{pmatrix} \lambda'_1 & & 0 \\ & \ddots & \\ 0 & & \lambda'_n \end{pmatrix}$$

are unitarily equivalent. Repeating this procedure each time when $\beta_l > \alpha_k$, one can satisfy (6.1) for λ'_a and λ'_b. In the same way one can order all "eigenvalues" of K_n within A. By property $(*)$, if $\|K_n - K_{n-1}\| < \varepsilon_n$, then one can find unitaries $u_{i,n}$ in A such that

$$\left\| u_{i,n}^* \lambda_i^{(n)} u_{i,n} - \lambda_i^{(n-1)} \right\| < \varepsilon_n. \tag{6.22}$$

Therefore $u_{i,n}^* \lambda_i^{(n)} u_{i,n} \in A$. Taking (if necessary) a subsubsequence $\{K_n\}$, we can assume that $\varepsilon_n = \frac{1}{2^n}$ in (6.22). Then the sequence

$$\bar{\lambda}_i^{(1)} = \lambda_i^{(1)}, \ \bar{\lambda}_i^{(2)} = u_{i,2}^* \lambda_i^{(2)} u_{i,2}, \ \bar{\lambda}_i^{(3)} = u_{i,3}^* u_{i,2}^* \lambda_i^{(3)} u_{i,2} u_{i,3}, \ \ldots$$

is a Cauchy sequence in A. Denote its limit by $\bar{\lambda}_i \in A$. By Theorem 6.4.2, for any K_n one can find a unitary operator U_n mapping first n "eigenvectors" of K into the "eigenvectors" of K_n. Put $K'_n = U_n^* K_n U_n \in \operatorname{End}_{\mathbf{A}}(H'_{\mathbf{A}})$. Then

$$K'_n x_i = x_i \bar{\lambda}_i^{(n)} \tag{6.23}$$

and $\|K'_n - K\| \to 0$. Passing to the limit in (6.23) we obtain $Kx_i = x_i\bar{\lambda}_i$. Therefore $\bar{\lambda}_i$ are the "eigenvalues" of the operator K. □

Note that condition $(*)$ is necessary for a C^*-algebra to be weakly diagonalizable. Indeed, if the operator K is the direct sum of two projections,

$$K = \begin{pmatrix} p & 0 \\ 0 & q \end{pmatrix},$$

then its "eigenvalues" can be ordered only if the "common part" of projections $1-p$ and q is a projection from the algebra A.

6.7. Case of continuous fields of trace C^*-algebras

Let X be a compact Hausdorff space and let $\{A(x), x \in X\}$ be a family of unital C^*-algebras equipped with faithful finite traces τ_x, $\tau_x(1_x) = 1$. By $\prod_{x\in X} A(x)$ we denote the set of functions $a = a(x)$ defined on X and such that $a(x) \in A(x)$ for any $x \in X$.

DEFINITION 6.7.1. Let $\mathcal{A} \subset \prod_{x\in X} A(x)$ be a subset with the following properties:

(i) $\mathcal{A}$ is a $*$-subalgebra in $\prod_{x\in X} A(x)$;
(ii) the set $\{a(x), a \in \mathcal{A}\}$ is dense in the C^*-algebra $A(x)$ for any $x \in X$;
(iii) the function $x \longmapsto \|a(x)\|$ is continuous for each $a \in \mathcal{A}$;
(iv) let $a \in \prod_{x\in X} A(x)$; if for each $x \in X$ and for all $\varepsilon > 0$ there exists an element $a' \in \mathcal{A}$ such that $\|a(x) - a'(x)\| < \varepsilon$ in some neighborhood of the point x, then $a \in \mathcal{A}$;
(v) the function $x \longmapsto \tau_x(a(x))$ is continuous for each $a \in \mathcal{A}$.

Then the triple $(A(x), X, \mathcal{A})$ is called a continuous field of trace C^*-algebras.

Note that the first four properties of this definition make the standard definition of a continuous field of C^*-algebras [**31**]. It is known that $\mathcal{A}$ is a C^*-algebra. For the sake of brevity we denote by $\|a(x)\|$ the norm of an element $a = a(x) \in \mathcal{A}$, i.e., $\sup_x \|a(x)\|$. If $V \subset X$ is a closed subset, then by $\mathcal{A}|_V$ we denote the restriction of a continuous field of C^*-algebras onto V.

It is clear that $\tau : A \longmapsto \tau_x(a(x))$ is a trace on $\mathcal{A}$ with values in $C(X) \subset \mathcal{A}$.

This trace defines an inner product on $A(x)$ by

$$(b_1, b_2)_{\tau_x} = \tau_x(b_1^* b_2), \qquad \|b\|^2_{\tau_x} = (b, b)_{\tau_x},$$

and a $C(X)$-valued inner product on $\mathcal{A}$,

$$(b_1(x), b_2(x))_\tau = \tau_x(b_1^*(x) b_2(x)) \in C(X) \tag{6.24}$$

with the norm

$$\|b(x)\|_\tau = \sup_x (b(x), b(x))_\tau^{1/2} = \sup_x \|b(x)\|_{\tau_x}.$$

Let $H(x) = L^2(A(x))$ be the completion of $A(x)$ with respect to the norm $\|\cdot\|_{\tau_x}$. Then the algebra $A(x)$ is faithfully represented in the Hilbert space $H(x)$ and we can pass to the appropriate W^*-algebra $B(x) = L^\infty(A(x))$, i.e., the weak closure of $A(x)$ (the notation L^∞ is used by analogy with the commutative case).

Now let dx be a σ-finite Borel measure on X. Note that the function (6.24) is continuous. Therefore one can put

$$(b_1, b_2)_\tau = \int_X (b_1(x), b_2(x))_{\tau_x}\, dx, \qquad \|b\|_\tau^2 = (b, b)_\tau.$$

Denote the completion of the algebra $\mathcal{A}$ with respect to this norm by $H = L^2(\mathcal{A})$. The algebra $\mathcal{A}$ is faithfully represented on the Hilbert space H and we denote the corresponding W^*-algebra by $\mathbf{A} = L^\infty(\mathcal{A}) \supset \mathcal{A}$. It is clear that

$$\mathbf{A} = \int_X^\oplus B(x)\, dx.$$

Since the $L^\infty(X)$-valued trace $\overline{\tau} = \int_X \tau_x\, dx$ defined by the family of traces τ_x on $\mathbf{A}$ is faithful and finite, $\mathbf{A}$ is a finite W^*-algebra.

We are interested here only in the case, when X is either a segment or a circle. Then, if X is divided by points $\{x_k\}$ into segments $D_k = [x_k; x_{k+1}]$ and if $a_k(x) \in \mathcal{A}|_{D_k}$ is a continuous field for each k, we call a set $a(x) = \{a_k(x)\}$ a *piecewise continuous* field on X. Such piecewise continuous fields clearly belong to the W^*-algebra $\mathbf{A}$. The distance from such a field to the C^*-algebra $\mathcal{A}$ is given by the formula

$$\operatorname{dist}(a(x), \mathcal{A}) = \sup_k \{\|a_k(x_k) - a_{k+1}(x_k)\|\}.$$

For convenience of notation we write $K_1(x) \sim K_2(x)$ for two piecewise continuous fields $K_1(x)$ and $K_2(x)$ if there exists a piecewise continuous unitary field $u(x)$ such that $u^*(x)K_1(x)u(x) = K_2(x)$. Recall that $tsr(A) = 1$ means that invertible elements are dense in the algebra A.

Further we are going to use the following result on almost commuting operators.

PROPOSITION 6.7.2 ([**84**]). *Let A be a C^*-algebra with the properties*

(i) *$RR(A) = 0$ and $tsr(A) = 1$;*
(ii) *the unitary group of the C^*-algebra pAp is connected for every projection $p \in A$.*

Let $h \in A$ be selfadjoint, let $u \in A$ be unitary and let the estimate

$$\|u^* h u - h\| < \delta \tag{6.25}$$

hold. Then there exist a constant C depending only on $\|h\|$ and a path $u(t)$ connecting u with 1 such that for sufficiently small δ, one has

$$\|u^*(t) h u(t) - h\| < C\sqrt[4]{\delta}$$

for all t.

THEOREM 6.7.3 ([**83**]). *Let $(A(x), X, \mathcal{A})$ be a continuous field of trace C^*-algebras over a segment or a circle. Let $(B(x), X, \mathcal{B})$ be the corresponding continuous field of W^*-algebras and let $\mathbf{A}$ be the corresponding W^*-algebra. Suppose that for each point $x \in X$, the C^*-algebra $A(x)$ has the following properties:*

(i) *$RR(A(x)) = 0$ and $tsr(A(x)) = 1$;*
(ii) *the unitary group of the C^*-algebra $pA(x)p$ is path-connected for any projection $p \in A(x)$;*
(iii) *the trace τ_x on $A(x)$ is finite and faithful for any $x \in X$;*

(iv) *the homomorphism* $K_0(A(x)) \longrightarrow K_0(B(x))$ *induced by the inclusion* $A(x) \subset B(x) = L^\infty(A(x))$ *is injective and* $K_0(A(x))$ *is a sublattice in* $K_0(B(x))$.

Then the C^**-algebra* $\mathcal{A} \in \mathbf{A}$ *has the weak diagonalization property.*

PROOF. We denote the algebra of $(n\times n)$-matrices by M_n. Let us begin the proof with the following lemmas.

LEMMA 6.7.4. *Let* $K_1(x)$, $K_2(x)$ *be two continuous fields of selfadjoint operators with a finite spectrum in the algebra* $M_n \otimes \mathcal{A}$ *such that* K_k *is unitarily equivalent to* $\operatorname{diag}\{\lambda_i^{(k)}\}$, $k = 1, 2$. *If* $\|K_1(x) - K_2(x)\| < \varepsilon$, *then for any* $x_0 \in X$ *there exists a closed neighborhood* W *of the point* x_0 *and continuous unitary fields* $u_i(x)$ *in* W *such that*

$$\left\|\lambda_i^{(1)}(x) - u_i^*(x)\lambda_i^{(2)}(x)u_i(x)\right\| < \varepsilon. \tag{6.26}$$

PROOF. By assumption, we can write the operators $K_k(x)$ in the form

$$K_1(x) = \sum \alpha_m p_m(x), \quad K_2(x) = \sum \beta_j q_j(x),$$

where $p_m(x)$, $q_j(x)$ are spectral projections and the eigenvalues are ordered. Since $K_0(A(x_0)) \subset K_0(B(x_0))$ is a sublattice by assumption, we can find projections $r_l(x_0) \in A(x_0)$ such that $\sum_l r_l(x_0) = 1 \in M_n \otimes A(x_0)$ and each of the projections $p_m(x_0)$, $q_j(x_0)$ and also each of the projections from $M_n \otimes 1$ is unitarily equivalent to a sum of certain projections $r_l(x_0)$. By renumbering the eigenvalues of $K_k(x_0)$ (counting them with multiplicities), we can write

$$K_1(x_0) = \sum \alpha_l r_l(x_0); \quad K_2(x_0) = \sum \beta_l r_l'(x_0); \quad r_l(x_0) \sim r_l'(x_0).$$

Since in the W^*-algebra $M_n \otimes B(x_0) = M_n \otimes L^\infty(A(x_0))$ one has

$$\|K_1(x_0) - K_2(x_0)\| < \varepsilon$$

(cf. [**122**]), we obtain

$$|\alpha_l - \beta_l| < \varepsilon \tag{6.27}$$

for all l. Decompose the set of the projections $\{r_l(x_0)\}$ into s subsets,

$$\{r_1(x_0), \dots, r_{l_1}(x_0)\}; \dots ; \{r_{l_{s-1}+1}(x_0), \dots, r_{l_s}(x_0)\}$$

in such a way that the sum of all projections in each subset is unitarily equivalent to a one-dimensional projection in $M_n \otimes 1$. Then each subset corresponds to the "eigenvalue" $\lambda_i^{(k)}(x_0)$:

$$\lambda_i^{(1)}(x_0) = \alpha_{l_{i-1}+1} r_{l_{i-1}+1}(x_0) + \ldots + \alpha_{l_i} r_{l_i}(x_0);$$
$$\lambda_i^{(2)}(x_0) = \beta_{l_{i-1}+1} r'_{l_{i-1}+1}(x_0) + \ldots + \beta_{l_i} r'_{l_i}(x_0).$$

Choose unitaries $u_i(x_0) \in A(x_0)$ such that $r_l'(x_0) = u_i^*(x_0) r_l(x_0) u_i(x_0)$ for $l_{i-1}+1 \leq l \leq l_i$. Then the estimate (6.26) follows from (6.27). □

LEMMA 6.7.5. *Let* $D = [x_k; x_{k+1}]$ *be a segment in* X, *and let* $\lambda(x)$, $\mu'(x) \in \mathcal{A}|_D$ *be two continuous fields of selfadjoint operators with a finite spectrum on* D *such that* $\|\lambda(x) - \mu'(x)\| < \varepsilon$. *Let* $\mu \in A(x_k)$ *be a selfadjoint element with a finite spectrum and let* $u \in A(x_k)$ *be unitary and satisfy*

$$\|\mu - u^*\mu'(x_k)u\| < \delta \quad \textit{and} \quad \|\lambda(x_k) - \mu\| < \varepsilon.$$

Then for sufficiently small δ and ε there exists a piecewise continuous unitary field $u(x)$ such that

(i) $u(x_k) = u$, $u(x_{k+1}) = 1$;

(ii) *the estimate*

$$\|u^*(x)\mu'(x)u(x) - \lambda(x)\| < C\sqrt[4]{2\varepsilon + \delta} \tag{6.28}$$

holds on D, where C is a constant depending only on the norm of $\lambda(x)$.

PROOF. By assumption, we have $\|\mu - \mu'(x_k)\| < 2\varepsilon$. Therefore

$$\|\mu'(x_k) - u^*\mu'(x_k)u\| < 2\varepsilon + \delta.$$

By Proposition 6.7.2, there exist a constant C depending only on the norm of $\lambda(x_k)$ (or of $\mu'(x_k)$ since these elements are close to each other) and a path u_t, $0 \leq t \leq 1$, connecting u with the unit in the unitary group of the algebra $A(x_k)$ such that

$$\|u_t^*\mu'(x_k)u_t - \lambda(x_k)\| < C\sqrt[4]{2\varepsilon + \delta}$$

for any $t \in [0, 1]$. Continuity implies that there exists a point $x'_{k+1} \in (X_k; x_{k+1})$ such that there exists a unitary continuous field $u_t(x)$ over the segment $[x_k; x'_{k+1}]$ such that it extends u_t and the estimate

$$\|u_t^*(x)\mu'(x)u_t(x) - \lambda(x)\| < C\sqrt[4]{2\varepsilon + \delta}$$

still holds for all $x \in [x_k; x'_{k+1}]$. Define the unitary field $u(x)$ by the formula

$$u(x) = \begin{cases} u_t(x) & \text{for} \quad t = \frac{x'_{k+1} - x}{x'_{k+1} - x_k}, \quad \text{where} \quad x \in [x_k; x'_{k+1}], \\ 1 & \text{for} \quad x \in [x'_{k+1}; x_{k+1}]. \end{cases}$$

It is a unitary continuous field and the estimate (6.28) holds on $[x_k; x'_{k+1}]$. On the other hand, if $x \geq x'_{k+1}$, then $u(x) = 1$, so this estimate is also valid there. $\square$

LEMMA 6.7.6. *Let $K(x)$ be a selfadjoint operator on the module $L_n(\mathcal{A})$, $K(x) \in M_n \otimes \mathcal{A}$. Then for any $\varepsilon > 0$ there exists a piecewise continuous field $K'(x)$ of operators with a discrete spectrum such that it locally lies in $M_n \otimes \mathcal{A}$, $\|K'(x) - K(x)\| < \varepsilon$ and the operator $K'(x)$ is diagonalizable on $L_n \otimes \mathbf{A}$ with piecewise continuous "eigenvalues" $\lambda_i(x)$ with* $\operatorname{dist}(\lambda_i(x), \mathcal{A}) < 2\varepsilon$.

PROOF. Fix $\varepsilon > 0$ and choose a point $x_0 \in X$. Since $RR(A(x_0)) = 0$, we can find an operator $K'(x_0) \in M_n \otimes A(x_0)$ with a finite spectrum, $K'(x_0) = \sum_m \alpha_m p_m(x_0)$ (where $p_m(x_0)$ are spectral projections of $K'(x_0)$) such that one has $\|K'(x_0) - K(x_0)\| < \frac{\varepsilon}{2}$. There exists a neighborhood of the point x_0 such that the projections $p_m(x_0)$ can be continuously extended in this neighborhood up to projections $p_m(x)$. Put $K'(x) = \sum_m \alpha_m p_m(x)$. Then one can find a smaller neighborhood of the point x_0, where

$$\|K'(x) - K(x)\| < \varepsilon. \tag{6.29}$$

So, for each point in X there exist a neighborhood W and a continuous field on W such that the inequality (6.29) is valid. By taking a finite covering of X by such neighborhoods, one can obtain a partition of X by segments Δ_k and a piecewise continuous field (continuous on each segment) $K'(x) = \{K'_k(x)\}$, $K'_k(x) \in M_n \otimes \mathcal{A}|_{D_k}$, such that $\operatorname{dist}(K'(x), M_n \otimes \mathcal{A}) < 2\varepsilon$. Let us diagonalize the operator field $K'(x)$ on each segment D_k, $K'(x) = \operatorname{diag}\{\lambda_i(x)\}$. Then the fields $\lambda_i(x) = \{\lambda_i^{(k)}(x)\}$, $\lambda_i^{(k)}(x) \in \mathcal{A}|_{D_k}$, are piecewise continuous and, since $\left\|K'_{k-1}(x_k) - K'_k(x_k)\right\| < 2\varepsilon$, using Lemmas 6.7.4 and 6.7.5, we can replace the diagonal entries $\lambda_i(x)$ on each interval by unitarily equivalent ones in such a way that the estimate

$$\operatorname{dist}(\lambda_i(x), \mathcal{A}) = \sup_k \left\| \lambda_i^{(k-1)}(x_k) - \lambda_i^{(k)}(x_k) \right\| < 2\varepsilon$$

is valid. □

LEMMA 6.7.7. *Let $K_1(x), K_2(x) \in M_n \otimes \mathbf{A}$ be piecewise continuous fields of operators with a finite spectrum and let $\|K_1(x) - K_2(x)\| < \varepsilon$, $K_1(x) \sim \operatorname{diag}\{\lambda_i(x)\}$, $K_2(x) \sim \operatorname{diag}\{\mu_i(x)\}$. Suppose that their diagonal elements $\lambda_i(x)$ and $\mu_i(x)$ are piecewise continuous and $\operatorname{dist}(\mu_i(x), \mathcal{A}) < \delta$. Then there exist piecewise continuous unitary fields $u_i(x)$, piecewise continuous fields $\mu_i'(x)$ and a piecewise continuous field of operators $K_2'(x) \sim \operatorname{diag}\{\mu_i'(x)\}$ such that*

(i) $\|K_2'(x) - K_2(x)\| < \delta$;
(ii) $\|u_i^*(x)\mu_i'(x)u_i(x) - \mu_i(x)\| < \delta$;
(iii) $\|\lambda_i(x) - \mu_i'(x)\| < C\sqrt[4]{2\varepsilon + \delta}$;
(iv) $\operatorname{dist}(\mu_i'(x), \mathcal{A}) < \delta$.

PROOF. Let $x_0 \in X$ be a point of continuity for $K_2(x)$ and let $D \supset x_0$ be an interval in X. By Lemma 6.7.4, we can diagonalize the operator $K_2(x_0)$ in such a way that its "eigenvalues" $\widetilde{\mu}_i(x_0)$ satisfy the inequality

$$\|\widetilde{\mu}_i(x_0) - \lambda(x_0)\| < \varepsilon.$$

Let us extend these "eigenvalues" arbitrarily up to continuous fields $\widetilde{\mu}_i(x)$. Then the inequality

$$\|\widetilde{\mu}_i(x) - \lambda_i(x)\| < \varepsilon$$

is still valid in some neighborhood of the point x_0. By assumption, there exist unitary elements $w_i \in A(x_0)$ such that $w_i^*\widetilde{\mu}(x_0)w_i = \mu_i(x_0)$. Take unitary extensions $w_i(x)$ of the elements w_i, $w_i(x_0) = w_i$, and choose a neighborhood of the point x_0 in such a way that the estimate

$$\|w_i^*(x)\widetilde{\mu}_i(x)w_i(x) - \mu_i(x)\| < \delta$$

holds in this neighborhood. Thus we obtain a partition of X into segments $D_k = [x_k; x_{k+1}]$, piecewise continuous fields $\widetilde{\mu}_i(x)$, $\widetilde{\mu}_i(x)|_{D_k} = \widetilde{\mu}_i^{(k)}(x)$, $\widetilde{\mu}_i^{(k)}(x) \in \mathcal{A}|_{D_k}$ and piecewise continuous unitary fields $w_i(x) = \{w_{i,k}(x)\}$, $w_{i,k}(x) \in \mathcal{A}|_{D_k}$, such that the estimates

$$\left\| \widetilde{\mu}_i^{(k)}(x) - \lambda_i(x) \right\| < \varepsilon \tag{6.30}$$

and

$$\left\| w_{i,k}^*(x)\widetilde{\mu}_i^{(k)}(x)w_{i,k}(x) - \mu_i(x) \right\| < \delta/2 \tag{6.31}$$

hold on D_k. There also exists a piecewise continuous field of operators $\widetilde{K}_2(x)$ unitarily equivalent to the operator $\operatorname{diag}\{\widetilde{\mu}_i(x)\}$ and such that

$$\left\| \widetilde{K}_2(x) - K_2(x) \right\| < \delta/2.$$

It follows from (6.31) that at the point x_k one has

$$\left\| w_{i,k-1}^*(x_k)\widetilde{\mu}_i^{(k-1)}(x_k)w_{i,k-1}(x_k) - w_{i,k}^*(x_k)\widetilde{\mu}_i^{(k)}(x_k)w_{i,k}(x_k) \right\| < \delta, \tag{6.32}$$

and (6.32) implies existence of unitary elements $v_{i,k}(x_k) \in A(x_k)$ such that

$$\left\| v_{i,k}^*(x_k)\widetilde{\mu}_i^{(k-1)}(x_k)v_{i,k}(x_k) - \widetilde{\mu}_i^{(k)}(x_k)\right\| < \delta.$$

By Lemma 6.7.5, we can find such extensions of the unitary elements $v_{i,k}(x_k)$ to some neighborhood of the point x_k, for which, by (6.30), one has the estimate

$$\left\| v_{i,k}^*(x)\widetilde{\mu}_i^{(k)}(x)v_{i,k}(x) - \lambda_i(x)\right\| < C\sqrt[4]{2\varepsilon} + \delta.$$

Put

$$\overline{\mu}_i^{(k)}(x) = v_{i,k}(x)\widetilde{\mu}_i^{(k)}(x)v_{i,k}^*(x).$$

Then

$$\left\| \overline{\mu}_i^{(k)}(x) - \widetilde{\mu}_i^{(k-1)}(x)\right\| < \delta \tag{6.33}$$

and

$$\left\| \overline{\mu}_i^{(k)}(x) - \lambda_i(x)\right\| < C\sqrt[4]{2\varepsilon} + \delta.$$

Proceeding by induction and passing from D_{k-1} to D_k, we substitute the fields $\widetilde{\mu}_i^{(k)}(x)$ by $\overline{\mu}_i^{(k)}(x)$ on each D_k and we obtain new piecewise continuous fields $\mu_i'(x) = \{\overline{\mu}_i^{(k)}(x)\}$ such that there exist piecewise continuous fields of unitary elements $u_i(x)$ for which

$$\|u_i^*(x)\mu_i'(x)u_i(x) - \mu_i(x)\| < \delta.$$

By setting $K_2'(x) = \operatorname{diag}\{u_i(x)\mu_i'(x)u_i^*(x)\}$, we get statement (i) and the estimate $\operatorname{dist}(\mu_i'(x), \mathcal{A}) < \delta$ follows from (6.33). □

Now let $K(x)$ be a strictly positive compact operator on the module $H_{\mathcal{A}}$. Take a sequence $\varepsilon_m > 0$ vanishing as $m \to \infty$. By Lemma 6.7.6, we can find a sequence of piecewise continuous fields of diagonalizable operators $K_m(x) \in M_{n_m} \otimes \mathbf{A}$ of finite rank with "eigenvalues" $\lambda_{i,m}(x) \in \mathcal{A}$ such that

(i) $\|K_m(x) - K(x)\| < \varepsilon_m$;
(ii) $K_m(x) \sim \operatorname{diag}\{\lambda_{i,m}(x)\}$, where $\operatorname{dist}(\lambda_{i,m}(x), \mathcal{A}) < \varepsilon_m$.

Since

$$\|K_1(x) - K_2(x)\| \le \|K_1(x) - K(x)\| + \|K(x) - K_2(x)\| < \varepsilon_1 + \varepsilon_2,$$

one can find, by Lemma 6.7.7, a piecewise continuous operator $K_2'(x) \sim \operatorname{diag}\{\lambda_{i,2}'(x)\}$ such that

$$\|K_2'(x) - K_2(x)\| < \varepsilon_2, \quad \left\|\lambda_{i,2}'(x) - \lambda_{i,2}(x)\right\| < \varepsilon_2, \quad \operatorname{dist}(\lambda_{i,2}'(x), \mathcal{A}) < \varepsilon_2,$$

and

$$\left\|\lambda_{i,2}'(x) - \lambda_{i,1}(x)\right\| < C\sqrt[4]{2(\varepsilon_1+\varepsilon_2)} + \varepsilon_2 < 2C\sqrt[4]{\varepsilon_1+\varepsilon_2}.$$

Then, since

$$\begin{aligned}\left\|K_{m-1}'(x) - K_m(x)\right\| &\le \left\|K_{m-1}'(x) - K_{m-1}(x)\right\| + \|K_{m-1}(x) - K(x)\| \\ &\quad + \|K(x) - K_m(x)\| < 2\varepsilon_{m-1} + \varepsilon_m,\end{aligned}$$

we can find, by induction, a sequence $K_m'(x) \sim \operatorname{diag}\{\lambda_{i,m}'(x)\}$ such that

$$\|K_m'(x) - K_m(x)\| < \varepsilon_m, \quad \left\|\lambda_{i,m}'(x) - \lambda_{i,m}(x)\right\| < \varepsilon_m, \tag{6.34}$$

$$\operatorname{dist}(\lambda_{i,m}'(x), \mathcal{A}) < \varepsilon_m, \tag{6.35}$$

$$\left\|\lambda_{i,m}'(x) - \lambda_{i,m-1}(x)\right\| < C\sqrt[4]{2(2\varepsilon_{m-1}+\varepsilon_m)} + \varepsilon_m < 2C\sqrt[4]{\varepsilon_{m-1}+\varepsilon_m}. \tag{6.36}$$

Since, by (6.34), we have

$$\|K'_m(x) - K(x)\| \le \|K'_m(x) - K_m(x)\| + \|K_m(x) - K(x)\| < 2\varepsilon_m,$$

the sequence of operators $K'_m(x)$ is norm-convergent to the operator $K(x)$ and (6.36) implies that for any i the sequences $\lambda'_{i,m}(x)$ are Cauchy sequences if the numbers ε_m tend to zero sufficiently fast, namely if the series $\sum_m \sqrt[4]{\varepsilon_{m-1} + \varepsilon_m}$ is convergent. Therefore the limits

$$\overline{\lambda}_i(x) = \lim_{m\to\infty} \lambda'_{i,m}(x)$$

exist and, since, by (6.35), $\operatorname{dist}(\lambda'_{i,m}(x), \mathcal{A})$ tends to zero, we conclude that $\overline{\lambda}_i(x) \in \mathcal{A}$. Let us show that $\overline{\lambda}_i(x)$ are the "eigenvalues" of the operator $K(x)$. By Theorem 6.4.2, it follows from the estimate $\|K'_m(x) - K(x)\| < 2\varepsilon_m$ that there exist unitary operators $U_m \in M_{n_m} \otimes \mathbf{A}$ such that they map the "eigenvectors" $x_i \in H'_{\mathbf{A}}$ of $K(x)$ into the "eigenvectors" of $K'_m(x)$ and $\|U^*_m K'_m U_m - K\| < 2\varepsilon_m$. Put $\widetilde{K}_m = U^*_m K'_m U_m \in M_{n_m} \otimes \mathbf{A}$; $\widetilde{K}_m \to K$. Then

$$\widetilde{K}_m x_i = x_i \lambda'_{i,m}. \tag{6.37}$$

Passing to the limit in (6.37), we obtain $Kx_i = x_i\overline{\lambda}_i$. □

6.8. Schrödinger operator as an operator acting on a Hilbert C^*-module

Let D be an operator (possibly unbounded) acting on a Hilbert space H. Suppose that D commutes with operators $U_1, \dots, U_n$. Then we can consider D as an operator acting on a Hilbert C^*-module over the C^*-algebra A generated by $U_1, \dots, U_n$. Let us assume that the operator D is positive and the operator $(1+D)-1$ is A-compact. These assumptions are usually satisfied in the case where the operator D is elliptic in some sense. If the operators $U_1, \dots, U_n$ are normal and commute with each other, then the algebra A is the C^*-algebra of continuous functions on the space $X = \operatorname{Sp} U_1 \times \dots \times \operatorname{Sp} U_n$, so the operator D defines the family of operators D_x, $x \in X$, such that the operators $(1+D_x)^{-1}$ are compact (in the usual sense). Then we can diagonalize pointwise the family of operators D_x and obtain a family of eigenvalues $\lambda_i(x) \in C(X)$. The spectrum of the operator D is determined by the spectra of the operators of multiplication by the functions $\lambda_i(x)$. In the case, where the operators $U_1, \dots, U_n$ generate a noncommutative C^*-algebra, the situation is similar. As an example, we consider in this section an operator commuting with two unitary operators that generate a C^*-algebra A_θ of irrational rotation, namely a perturbed Schrödinger operator with irrational magnetic flow

$$\left(i\frac{\partial}{\partial x} + 2\pi\theta y\right)^2 - \frac{\partial^2}{\partial y^2} + W(x,y) \tag{6.38}$$

with a double-periodic perturbation $W(x,y) = W(x+1,y) = W(x,y+1)$. This operator has been studied in a number of papers (see [**76, 99**]). Applying to the operator (6.38) the Fourier transform in the variable x $(x \longrightarrow \xi)$ and the change of variables $t = -\frac{\xi}{2\pi} + \theta y$, $s = \frac{\xi}{2\pi}$, we obtain the operator

$$D = \Delta + W \tag{6.39}$$

with

$$\Delta = \theta^2\left(\left(\frac{2\pi t}{\theta}\right)^2 - \frac{\partial^2}{\partial t^2}\right) \tag{6.40}$$

and

$$W = \sum_{k,l} w_{kl} T_t^k T_s^{-k} e^{2\pi ilt/\theta} e^{2\pi ils/\theta},$$

where T_t (resp. T_s) denotes the unit translation in variable t (resp. s), $T_t \phi(t,s) = \phi(t+1,s)$, and w_{kl} denote the Fourier series coefficients of the function $W(x,y)$. We suppose that the function $W(x,y)$ is sufficiently smooth to satisfy the condition $\sum_{k,l} |w_{kl}| < \infty$.

Let A_θ be the C^*-algebra generated by two noncommuting unitaries U and V such that $UV = e^{2\pi i\theta} VU$ [**14, 22**] and let $A_\theta^\infty \subset A_\theta$ be its "infinitely smooth" subalgebra of elements of the form $\sum_{k,l} a_{kl} U^k V^l$, where the coefficients a_{kl} are of rapid decay. The Schwartz space $S(\mathbf{R})$ of rapidly decreasing functions on $\mathbf{R}$ can be endowed with a structure of a projective right A_θ^∞-module with one generator [**22**]. We denote this module by $\mathcal{M}^\infty$. The action of A_θ^∞ on $\mathcal{M}^\infty$ is given by formulas

$$(\phi U)(t) = \phi(t+\theta), \quad (\phi V)(t) = e^{2\pi it}\phi(t),$$

where $\phi(t) \in \mathcal{M}^\infty$. The module $\mathcal{M}^\infty$ is generated by a projection $p \in A_\theta^\infty$, $\mathcal{M}^\infty \cong pA_\theta^\infty$ with $\tau(p) = \theta$ and, since $\mathcal{M}^\infty \subset A_\theta$, $\mathcal{M}^\infty$ inherits the norm from A_θ. Its closure $\mathcal{M} = \mathcal{M}^\infty \otimes_{A_\theta^\infty} A_\theta$ with respect to this norm is a Hilbert A_θ-module. Note that the eigenfunctions of the operator Δ (6.40) form an orthonormal basis $\{\phi_i(t)\}$ in $S(\mathbf{R}) \subset L^2(\mathbf{R})$ and the functions from $S(\mathbf{R}^2) = \mathcal{M}^\infty \hat{\otimes} \mathcal{M}^\infty$ can be represented as the series $\sum_i \phi_i(t) m_i(s)$ with $m_i(s) \in M^\infty$. Define the A_θ-valued inner product on $S(\mathbf{R}^2)$ by the formula

$$\Big\langle \sum_i \phi_i(t) m_i(s), \sum_j \phi_j(t) n_j(s) \Big\rangle = \sum_i \langle m_i(s), n_i(s) \rangle,$$

where $n_j(s) \in \mathcal{M}^\infty$. By $S(\mathbf{R};\mathcal{M})$ (resp. $L^2(\mathbf{R};\mathcal{M})$) we denote the Schwartz space of functions (resp. the space of square-integrable functions) taking values in the Banach space $\mathcal{M}$. The inclusion

$$S(\mathbf{R}\times\mathbf{R}) \hookrightarrow S(\mathbf{R};\mathcal{M}) \hookrightarrow L^2(\mathbf{R};\mathcal{M}) \cong \mathcal{N}$$

allows us to consider $S(\mathbf{R}^2)$ as a dense subspace in the Hilbert A_θ-module

$$\mathcal{N} = \{(m_i) : \sum_i \langle m_i, m_i \rangle \text{ is convergent in } A_\theta\}.$$

One can see that, since the C^*-algebra A_θ is simple and $\langle \mathcal{M}, \mathcal{M} \rangle$ should be its ideal, the module $\mathcal{M}$ is full, i.e., $\langle \mathcal{M}, \mathcal{M} \rangle = A_\theta$. By Theorem 2.4.1, one has $\mathcal{N} \cong H_{A_\theta}$.

THEOREM 6.8.1 ([**81**]). *The operator D defined by* (6.39) *is a selfadjoint unbounded operator in $\mathcal{N}$ with a dense domain.*

PROOF. The proof consists of the following five steps.

1. Denote by $\mathcal{N}_1 \subset \mathcal{N}$ the subspace of all sequences (m_i) such that the series $\sum_i i^2 \langle m_i, m_i \rangle$ is norm-convergent to an element of A_θ. If $\xi \in \mathcal{N}_1$, $\xi = \sum_i \phi_i(t) m_i$, then $\Delta\xi = \sum_i (2i-1)\theta\phi_i(t) m_i$ and the series $\sum_i (2i-1)^2\theta^2 \langle m_i, m_i \rangle$ is convergent in A_θ. Therefore Δ is an unbounded operator in $\mathcal{N}$ with the dense domain $\mathcal{N}_1$.

2. Here we show that the action of the operator $C_{kl} = T_s^{-k} e^{2\pi ils/\theta}$ can be extended from $\mathcal{M}^\infty$ to $\mathcal{M}$. Since this operator commutes with the action of the algebra A_θ^∞

on the A_θ^∞-module $\mathcal{M}^\infty$, we have $C_{kl} \in \operatorname{End}_{A_\theta^\infty} \mathcal{M}^\infty$. The image of the generator p of $\mathcal{M}^\infty$ can be written as $C_{kl}(p) = pa_{kl} \in \mathcal{M}^\infty$ for some $a_{kl} \in A_\theta^\infty$ and we have

$$C_{kl}(p) = C_{kl}(p^2) = C_{kl}(p)p = pa_{kl}p.$$

Obviously the map $m \longmapsto pa_{kl}pm = C_{kl}(m)$ can be continuously extended from $\mathcal{M}^\infty$ to $\mathcal{M}$. Besides that, since C_{kl} is a unitary operator, we have $\|C_{kl}\| = 1$ and $\|a_{kl}\| = 1$.

3. Consider now the operator $B_{kl} = T_t^k e^{2\pi i l t/\theta} \cdot C_{kl}$. It is obviously continuous in $S(\mathbf{R}^2)$. Let α_{ij} be the matrix coefficients of the operator $T_t^k e^{2\pi i l t/\theta}$ with respect to the basis $\{\phi_j\}$,

$$T_t^k e^{2\pi i l t/\theta} \phi_i(t) = \sum_j \alpha_{ij}\phi_j(t).$$

Since this operator is unitary, we get $\sum_j \bar{\alpha}_{ij}\alpha_{nj} = \delta_{in}$. Let $\xi = \sum_i \phi_i(t) m_i \in N$. Then $B_{kl}(\xi) = \sum_{i,j} \alpha_{ij}\phi_j(t) C_{kl}(m_i)$. Estimate its norm:

$$\begin{aligned}
\langle B_{kl}(\xi), B_{kl}(\xi)\rangle &= \sum_{i,j} \left\langle \sum_i \alpha_{ij} C_{kl}(m_i), \sum_n \alpha_{nj} C_{kl}(m_n) \right\rangle \\
&= \sum_{i,n,j} \bar{\alpha}_{ij}\alpha_{nj} \langle C_{kl}(m_i), C_{kl}(m_n)\rangle \\
&= \sum_{i,n} \left(\sum_j \bar{\alpha}_{ij}\alpha_{nj} \right) \langle C_{kl}(m_i), C_{kl}(m_n)\rangle \\
&= \sum_{i,n} \delta_{in} \langle C_{kl}(m_i), C_{kl}(m_n)\rangle = \sum_i \langle C_{kl}(m_i), C_{kl}(m_i)\rangle \\
&= \sum_i (a_{kl}m_i)^* a_{kl} m_i = \sum_i m_i^* a_{kl}^* a_{kl} m_i \\
&\le \|a_{kl}\|^2 \sum_i m_i^* m_i = \sum_i m_i^* m_i = \langle \xi, \xi \rangle.
\end{aligned}$$

Hence $\|B_{kl}\| \le 1$ and it is a continuous operator on $\mathcal{N}$.

4. We have

$$\|W\| \le \sum_{k,l} \|w_{kl} B_{kl}\| \le \sum_{k,l} |w_{kl}| \cdot \|B_{kl}\| \le \sum_{k,l} |w_{kl}|.$$

By our assumption, the last sum is finite, hence W is continuous on $\mathcal{N}$.

5. It remains to show that D commutes with the action of the C^*-algebra A_θ on $\mathcal{N}$. This is clear for the operators Δ and B_{kl}. Since the series $W = \sum_{k,l} w_{kl} B_{kl}$ is convergent, W also commutes with the action of A_θ. D is selfadjoint if the function $W(x, y)$ is real-valued. □

Now let $\mathbf{A}$ be a type II_1 factor containing A_θ as a weakly dense subalgebra (cf. [**14**]). This inclusion induces the inclusion of H_{A_θ} into $H_{\mathbf{A}}$ and operators acting in H_{A_θ} can be extended to operators acting in $H_{\mathbf{A}}$. Note that if $\|W\| < c$, then the operator $D + c$ is invertible and its inverse $(\Delta + W + c)^{-1} = (1 + \Delta^{-1}(W + c))^{-1}\Delta^{-1}$ is compact due to the compactness of the operator Δ^{-1}. So, by Theorem 6.3.1, it is diagonalizable in $H'_{\mathbf{A}}$, hence the same is true for the operator D. Slightly changing the proof of that theorem (namely, taking θ instead of 1 in (6.7)) we can obtain the set of "eigenvectors" $\{x_i\}$ for D with $\langle x_i, x_i \rangle = p$. In that case the corresponding "eigenvalues" λ_i can be viewed as elements of $\mathrm{End}^*_{\mathbf{A}}(\overline{\mathcal{N}})$, where $\overline{\mathcal{N}} = p\mathbf{A} = \mathcal{N} \otimes_{A_\theta} \mathbf{A}$.

Since the C^*-algebra A_θ is a C^*-algebra of real rank zero [**20**], the "eigenvalues" λ_i can be chosen from the smaller algebra $\mathrm{End}^*_{A_\theta}(\mathcal{M})$ by Theorem 6.6.2. If $\|W\| < \theta$, then the spectrum of the operator D lies in the set $\bigcup_i (2\theta(i-1); 2\theta i)$. Therefore the spectral projections $P_i = P_{(2\theta(i-1);\, 2\theta i)}(D)$ lie in $\mathrm{End}^*_{A_\theta}(\mathcal{N})$ and the standard methods of perturbation theory [**80**] can be applied. However the problem of analyticity of "eigenvalues", which takes place in the commutative case for the integer values of θ [**99, 107**], is far from being understood.

6.9. Example: A continuous field of rotation algebras

We give an example here of a C^*-algebra without the weak diagonalization property. Let $X = [a, b]$ be a segment containing at least one integer point, for example 1, and let S be a circle. Let the group $\mathbf{Z}$ of integers act on the C^*-algebra $C(X \times S)$ of continuous functions on the cylinder by the formula

$$(\alpha(n)f)(x,t) = f(x, t + nx), \tag{6.41}$$

where $f(x,t) \in C(X \times S)$, $x \in X$, $t \in S$, $n \in \mathbf{Z}$. By A_X we denote the cross-product

$$C(X \times S) \rtimes_\alpha \mathbf{Z}.$$

As is shown in [**112, 37**], the algebra A_X is a continuous field of rotation algebras $A_x = A(x)$ over X. Note that the continuous field $(A(x), X, A_X)$ is a continuous field of trace C^*-algebras and that for irrational x one has $RR(A(x)) = 0$ and $tsr(A(x)) = 1$ [**20, 111**]. Moreover, the algebra $A(x)$ has the weak diagonalization property for each $x \in X$. On the other hand, the unitary group of the C^*-algebra $A(x)$ is not path-connected, but we believe that the obstruction to diagonalizability is not related to this fact. Note that $A_1 \cong C(\mathbf{T}^2)$ is the commutative C^*-algebra of continuous functions on a torus, so it is not a real rank zero algebra, unlike the irrational case, and the homomorphism $K_0(A(1)) \longrightarrow K_0(B(1))$ is not injective. There exists a nontrivial projection in $M_2 \otimes A_1$, which is the Bott generator in the group $K^0(\mathbf{T}^2)$. This projection can be extended ([**111, 78**]) up to a continuous field of projections $p(x)$ in some neighborhood of the point $1 \in X$ in such a way that

$$\tau_x(p(x)) = 1 + x \tag{6.42}$$

for the standard trace τ_x on A_X. We can diagonalize this field of projections in the direct integral of type II_1 factors: $p(x) \sim \mathrm{diag}\{\lambda_1(x), \lambda_2(x)\}$, where $\lambda_1(x) \geq \lambda_2(x)$. Then, for $x = 1$, we should have $\lambda_1(1) = 1$, $\lambda_2(1) = 0$. For arbitrary x the "eigenvalues" $\lambda_i(x)$, $i = 1, 2$, should be projections. If these "eigenvalues" are continuous fields in A_X, then we should have $\lambda_1(x) = \lambda_1(1) = 1$, $\lambda_2(x) = \lambda_2(1) = 0$ for all x, hence

$$\tau_x(\mathrm{diag}\{\lambda_1(x), \lambda_2(x)\}) = 1.$$

But since the trace is invariant with respect to the unitary equivalence, this contradicts (6.42). Therefore the C^*-algebra A_X does not have the weak diagonalization property.

Note that on a dense set of irrational points in X, the C^*-algebras $A(x)$ satisfy all the conditions of Theorem 6.7.3 except condition (ii). Slightly changing the proof, for any given continuous field of operators $K(x)$ and for any $\varepsilon > 0$ we can find a subset $X_\varepsilon \subset X$ such that the measure of the set $X \setminus X_\varepsilon$ does not exceed ε and the operator field $K(x)$ is diagonalizable on X_ε with "continuous" "eigenvalues" (i.e., "eigenvalues" lying in the C^*-algebra A_{X_ε}).

CHAPTER 7

Homotopy Triviality of Groups of Invertible Operators

As before, denote by $\mathrm{End}_A l_2(A)$ (resp. by $\mathrm{End}^*_A l_2(A)$) the Banach algebra of all bounded A-homomorphisms of the Hilbert A-module $l_2(A)$ (resp. the C^*-algebra (cf. Section 2.1) of adjointable operators). Let $\mathrm{GL}(A)$ and $\mathrm{GL}^*(A)$ denote the corresponding groups of invertible operators. The question about contractibility of general linear groups is very important for K-theory in order to construct classifying spaces in terms of Fredholm operators. A series of papers is devoted to this problem: [**88, 61, 127, 89**]. These results were used to construct the classifying spaces for $K^{p,q}(X;A)$ in [**126**] (cf. Section 8.5). In [**30**] Cuntz and Higson proved contractibility of $\mathrm{GL}^*(A)$ for any C^*-algebra A with a strictly positive element (or, equivalently, with a countable approximate unit, i.e., σ-unital).

In this chapter, which is based on [**130**], we give another simple proof of the theorem of Cuntz and Higson based on a generalized construction of a homotopy from [**97**]. We also show that a similar argument can be applied to prove contractibility of $\mathrm{GL}(A)$ in some special cases, in particular, for A a subalgebra of the C^*-algebra of compact operators in a separable Hilbert space and for $A = C_0(M)$, where M is a finite-dimensional manifold.

The classical paper of Dixmier and Douady [**33**] gives a proof of the contractibility of the unitary group of a Hilbert space with respect to the strong topology. We prove several versions of the generalization of this theorem to the case of $\mathrm{GL}(A)$ and $\mathrm{GL}^*(A)$ for a σ-unital C^*-algebra A. There strict topology plays the role of strong topology.

We complete this chapter by a detailed exposition of the modified Neubauer homotopy.

7.1. Technical lemmas

It is known that the set of invertible operators in a Banach space is open with respect to the norm topology, while the set of bounded A-homomorphisms is closed in the set of all endomorphisms. Thus, GL is an open set in a Banach space. The same argument is valid for GL^*. According to Milnor's theorem [**86**], such sets have the homotopy type of CW-complexes and, by Whitehead's theorem, strong and weak homotopy triviality are the same for them. Thus, we have proved the following statement.

LEMMA 7.1.1. *To prove contractibility of* GL (*resp.*, GL^*) *it is sufficient to verify that any continuous map* $f : S \to \mathrm{GL}$ *of a sphere* S *of an arbitrary dimension is homotopic to the trivial* (*single point*) *map* $\mathrm{Id} \in \mathrm{GL}$. *A similar statement holds for* GL^*.

Let us make one more reduction. In order to simultaneously consider the case of GL and of GL* we introduce the following common notation: $\mathcal{G} := \mathrm{GL}$ (resp., GL*), $\mathcal{E}(\mathcal{M}) := \mathrm{End}_A(\mathcal{M})$ (resp., $\mathrm{End}^*_A(\mathcal{M})$).

LEMMA 7.1.2 (a version of the Atiyah theorem about small balls). *Let $f : S \to \mathcal{G}$ be a continuous map of a sphere of arbitrary (but finite) dimension. Then f is homotopic to a map $f' : S \to \mathcal{G}$ such that $f'(S)$ is a finite polyhedron.*

PROOF. Let $\varepsilon > 0$ be such that the ε-neighborhood of the compact set $f(S)$ lies in $\mathcal{G}$. Choose a fine simplicial subdivision of the sphere S such that $\mathrm{diam}(f(\sigma)) < \varepsilon/2$ for any simplex σ of this subdivision. This is possible, since S is compact. Let f' be a piecewise linear map extending the restriction of f onto the 0-dimensional skeleton. Thus $\mathrm{diam}(f'(\sigma)) \le \mathrm{diam}(f(\sigma)) < \varepsilon/2$ for any s. For any point $s \in S$ there exists a vertex $s_i \in S$ such that $\|f(s) - f'(s_i)\| = \|f(s) - f(s_i)\| < \varepsilon/2$ and $\|f'(s) - f'(s_i)\| < \varepsilon/2$, hence the segment $[f(s), f'(s)] \subset \mathcal{G}$ for any point $s \in S$. Therefore the linear homotopy $f_t(s) = tf'(s) + (1-t)f(s)$ lies in $\mathcal{G}$. Passing to a subdivision of $f'(S)$, we obtain the structure of a simplicial complex. □

REMARK 7.1.3. Note that this argument is not valid for other topologies to be considered. For example, with respect to the strong topology on the set of linear operators in a Hilbert space, the sequence Id_n is convergent to Id, where Id_n has the matrix $\mathrm{diag}(1,\dots,1,0,0,\dots)$ (unit up to the n-th place). Hence, the general linear group is not an open set with respect to this topology.

One more step taken from the original work of Kuiper [**68**] is universal.

LEMMA 7.1.4. *Let $V \subset \mathcal{G}$ be defined by*

$$V = \{g \in \mathcal{G} \mid g|_{H'} = \mathrm{Id}_{H'},\ g(H_1) = H_1\},$$

where

$$l_2(A) = H' \oplus H_1, \qquad H' \cong H_1 \cong l_2(A).$$

Then V is contractible in $\mathcal{G}$ to $1 \in \mathcal{G}$.

PROOF. Let us decompose H' as

$$H' = H_2 \oplus H_3 \oplus \cdots, \qquad H_i \cong l_2(A),$$

so that $l_2(A) = H_1 \oplus H_2 \oplus H_3 \oplus \cdots$. The matrix of g with respect to this decomposition has the form

$$m(1,1) = u = g|_{H_1}, \qquad m(i,i) = 1 \in \mathcal{E}(H_i),\ i > 1, \qquad m(i,j) = 0,\ i \ne j,$$
$$g = \mathrm{diag}(u,1,1,1,\dots) = \mathrm{diag}(u, u^{-1}u, 1, u^{-1}u, 1, \dots).$$

We would like to define a homotopy $g_t \in \mathcal{G}$, $t \in [0,\pi]$, in such a way that

$$g_0 = g, \qquad g_{\pi/2} = \mathrm{diag}(u, u^{-1}, u, u^{-1}, u, \dots), \qquad g_\pi = \mathrm{diag}(1,1,1,\dots) = \mathrm{Id} \in \mathcal{G}.$$

For this purpose, put

$$m_t(1,1) = u$$

for $t \in [0, \pi/2]$,

$$\begin{pmatrix} m_t(2i,2i) & m_t(2i,2i+1) \\ m_t(2i+1,2i) & m_t(2i+1,2i+1) \end{pmatrix} = \begin{pmatrix} \cos t & -\sin t \\ \sin t & \cos t \end{pmatrix} \begin{pmatrix} u & 0 \\ 0 & 1 \end{pmatrix} \begin{pmatrix} \cos t & \sin t \\ -\sin t & \cos t \end{pmatrix} \begin{pmatrix} u^{-1} & 0 \\ 0 & 1 \end{pmatrix}$$

for $i \geq 1$, and

$$m_t(r,s) = 0 \qquad \text{for the remaining } r,\ s.$$

For $t \in [\pi/2, \pi]$ put

$$\begin{pmatrix} m_t(2i-1,2i-1) & m_t(2i-1,2i) \\ m_t(2i,2i-1) & m_t(2i,2i) \end{pmatrix} = \begin{pmatrix} \cos t & -\sin t \\ \sin t & \cos t \end{pmatrix} \begin{pmatrix} u^{-1} & 0 \\ 0 & 1 \end{pmatrix} \begin{pmatrix} \cos t & \sin t \\ -\sin t & \cos t \end{pmatrix} \begin{pmatrix} u & 0 \\ 0 & 1 \end{pmatrix},$$

and

$$m_t(r,s) = 0 \qquad \text{for remaining } r,\ s.$$

□

LEMMA 7.1.5. *Let $W \subset \mathcal{G}$ be defined by*

$$W = \{g \in \mathcal{G} \mid g|_{H'} = \mathrm{Id}_{H'}\},$$

where

$$l_2(A) = H' \oplus H_1, \qquad H' \cong H_1 \cong l_2(A).$$

Then W is contractible inside $\mathcal{G}$ to

$$V = \{g \in \mathcal{G} \mid g|_{H'} = \mathrm{Id}_{H'},\ g(H_1) = H_1\}.$$

PROOF. With respect to the decomposition $l_2(A) = H' \oplus H_1$ we define the homotopy by

$$f_t(s) = \begin{pmatrix} 1 & \beta(s)(1-t) \\ 0 & \gamma(s) \end{pmatrix}.$$

Let the operator $\begin{pmatrix} \varphi & \psi \\ \chi & \xi \end{pmatrix}$ be the inverse for $\begin{pmatrix} 1 & \beta \\ 0 & \gamma \end{pmatrix}$. Then

$$\begin{pmatrix} \varphi & \psi \\ \chi & \xi \end{pmatrix}\begin{pmatrix} 1 & \beta \\ 0 & \gamma \end{pmatrix} = \begin{pmatrix} \varphi & \varphi\beta + \psi\gamma \\ \chi & \chi\beta + \xi\gamma \end{pmatrix} = \begin{pmatrix} 1 & 0 \\ 0 & 1 \end{pmatrix},$$

$$\begin{pmatrix} 1 & \beta \\ 0 & \gamma \end{pmatrix}\begin{pmatrix} \varphi & \psi \\ \chi & \xi \end{pmatrix} = \begin{pmatrix} \varphi + \beta\chi & \psi + \beta\xi \\ \gamma\chi & \gamma\xi \end{pmatrix} = \begin{pmatrix} 1 & 0 \\ 0 & 1 \end{pmatrix},$$

whence

$$\varphi = 1, \qquad \chi = 0, \qquad \gamma\xi = \xi\gamma = 1,$$
$$\beta + \psi\gamma = 0, \qquad \psi + \beta\xi = 0,$$

and

$$\begin{pmatrix} 1 & \psi(1-t) \\ 0 & \xi \end{pmatrix}\begin{pmatrix} 1 & \beta(1-t) \\ 0 & \gamma \end{pmatrix} = \begin{pmatrix} 1 & \beta(1-t) + (1-t)\psi\gamma \\ 0 & \xi\gamma \end{pmatrix} = \begin{pmatrix} 1 & (1-t)\cdot 0 \\ 0 & 1 \end{pmatrix},$$

$$\begin{pmatrix} 1 & \beta(1-t) \\ 0 & \gamma \end{pmatrix}\begin{pmatrix} 1 & \psi(1-t) \\ 0 & \xi \end{pmatrix} = \begin{pmatrix} 1 & \psi(1-t) + \beta\xi(1-t) \\ 0 & \gamma\xi \end{pmatrix} = \begin{pmatrix} 1 & (1-t)\cdot 0 \\ 0 & 1 \end{pmatrix}.$$

Hence, the homotopy lies in $\mathcal{G}$. □

Let $\mathcal{K}_A$, $\mathbf{LM}(\mathcal{K}_A) \cong \mathrm{End}_A l_2(A)$, $\mathbf{M}(\mathcal{K}_A) \cong \mathrm{End}^*_A l_2(A)$ and

$$\mathbf{QM}(\mathcal{K}_A) \cong \mathrm{End}_A(l_2(A), l_2(A)')$$

denote the C^*-algebra of A-compact operators on $l_2(A)$, the algebra of the left multipliers, the C^*-algebra of multipliers and the space of quasi-multipliers, respectively (see [**17, 73, 44, 63, 92, 100**] and Sections 5.3 and 5.4).

Let α be a strictly positive element (see, e.g., Section 1.1) in a σ-unital C^*-algebra A and let $\alpha_i := \varphi_i(\alpha)$ be a countable approximate unit, where the functions φ_i, $i = 1, 2, \ldots$, are defined by the following graph:

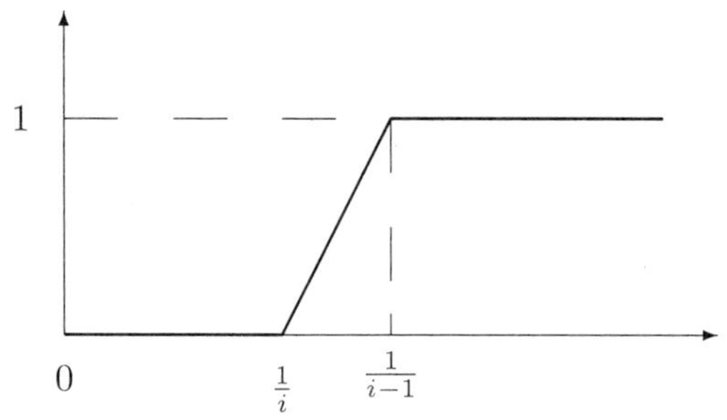

Put $\omega_i := (\alpha_i - \alpha_{i-1})^{1/2}$ for $i \geq 3$ and $\omega_2 = \alpha_2^{1/2}$. Then

$$\begin{aligned} \omega_j \alpha_i &= \alpha_i \omega_j = 0, & j &= i+2, i+3, \ldots, \\ \omega_j \alpha_i &= \alpha_i \omega_j = \omega_j, & j &= 1, \ldots, i-1. \end{aligned} \tag{7.1}$$

If A is not unital, the notion of the "standard basis" $\{e_i\}$ for the module $l_2(A)$ makes no sense. Nevertheless, the elements $e_i\gamma$ are well defined for any $\gamma \in A$, namely,

$$e_i\gamma := (0, \ldots, 0, \gamma, 0, \ldots), \qquad \gamma \text{ at the } i\text{-th place.}$$

Denote by Q_i the orthogonal projections onto these one-dimensional submodules E_i.

LEMMA 7.1.6. *The injection* $i : A \to l_2(A)$ *defined by*

$$x \mapsto \sum_i e_{k(i)}\omega_i x, \qquad k(1) < k(2) < k(3) < \cdots,$$

is adjointable and preserves the inner product. In particular, its image $\mathrm{Im}\, i$ *is defined by the selfadjoint projection of the form*

$$p = ii^*. \tag{7.2}$$

PROOF. First of all,

$$\begin{aligned} \langle ix, iy \rangle &= \left\langle \sum_i e_{k(i)}\omega_i x, \sum_i e_{k(i)}\omega_i y \right\rangle = \sum_i \langle e_{k(i)}\omega_i x, e_{k(i)}\omega_i y \rangle \\ &= \sum_i x^* \omega_i \omega_i y = x^* y = \langle x, y \rangle. \end{aligned}$$

Consider the operator $t : l_2(A) \to A$ of the form

$$t(z) := \sum_i \langle e_{k(i)}\omega_i, z \rangle = \sum_i \omega_i z_{k(i)}.$$

This series satisfies the Cauchy criterion: if the number m is so large that

$$\sum_{i=m+1}^{\infty} z_i^* z_i < \delta,$$

then

$$\left\| \sum_{i=s}^{r} \omega_i z_{k(i)} \right\| \le \left\| \sum_{i=s}^{r} \omega_i^2 \right\|^{1/2} \cdot \left\| \sum_{i=s}^{r} z_{k(i)}^* z_{k(i)} \right\|^{1/2} \le 1 \cdot \delta.$$

The same argument for $s = 1$ implies that $\|t(z)\| \le \|z\|$. Also one has $\langle ix, z \rangle = \langle x, tz \rangle$, i.e., $t = i^*$.

Consider arbitrary elements $x, y \in A$. Then

$$(i^*ix)^* y = \langle i^*ix, y \rangle = \langle ix, iy \rangle = \langle x, y \rangle = x^* y.$$

Since y is arbitrary, $i^*ix = x$ and $i^*i = \mathrm{Id}$. Hence,

$$ii^*ii^* = ii^*,$$

i.e., p is a projection. Since $i^*i = \mathrm{Id}$, i^* is an epimorphism and $\operatorname{Im} i = \operatorname{Im} p$ (see also [**71**, Sect. 3]). □

We need the following stronger version of this lemma.

LEMMA 7.1.7. *The injection* $J : l_2(A) \to l_2(A)$ *given by*

$$(a_1, a_2, \dots) \mapsto \sum_j \sum_i v_{ij} a_j, \qquad \langle v_{ij}, v_{ij} \rangle = \omega_i^2, \qquad v_{ij} \in M_{k(i,j)},$$

$$l_2(A) = M_1 \oplus M_2 \oplus \cdots, \qquad M_r = \{ (0, \dots, 0, a_{s(r)}, \dots, a_{s(r+1)-1}, 0, \dots) \},$$
$$\{k(1,1); k(1,2), k(2,1); k(1,3), k(2,2), k(3,1); \dots\} = \{1, 2, \dots\},$$

is adjointable and preserves the inner product. In particular, its image is defined by the selfadjoint projection of the form JJ^*.

PROOF. Let $x = (a_1, a_2, \dots) \in l_2(A)$, $y = (b_1, b_2, \dots) \in l_2(A)$. Then

$$\langle Jx, Jy \rangle = \left\langle \sum_j \sum_i v_{ij} a_j, \sum_j \sum_i v_{ij} b_j \right\rangle = \sum_j \sum_i a_j^* \omega_i^2 b_j$$
$$= \sum_j a_j^* \left(\sum_i \omega_i^2 \right) b_j = \sum_j a_j^* b_j = \langle x, y \rangle.$$

In particular, J is bounded. Consider the operator $T : l_2(A) \to l_2(A)$ of the form

$$T(z) := (t_1, t_2, \dots), \quad t_j := \sum_i \langle v_{ij}, z \rangle.$$

For this series the Cauchy criterion holds. Indeed, let a number $N = N(z)$ be so large that $\|(1 - p_N)z\| < \delta$ and let m be so large that $s(k(m,j)) > N$ (j is fixed). Then (by Proposition 1.2.4)

$$\left\| \sum_{i=m}^{r} \langle v_{ij}, z \rangle \right\| = \left\| \left\langle \sum_{i=m}^{r} v_{ij}, (1 - p_N) z \right\rangle \right\|$$
$$\le \left\| \left\langle \sum_{i=m}^{r} v_{ij}, \sum_{i=m}^{r} v_{ij} \right\rangle \right\|^{1/2} \cdot \|(1 - p_N) z\| \le 1 \cdot \delta.$$

By 1.2.4, the inequality

$$\sum_{i=1}^{r}\langle v_{ij},z\rangle^*\sum_{i=1}^{r}\langle v_{ij},z\rangle=\left\langle\sum_{i=1}^{r}v_{ij},q_jz\right\rangle^*\left\langle\sum_{i=1}^{r}v_{ij},q_jz\right\rangle\le\langle q_jz,q_jz\rangle$$

holds for any r, where q_j is the orthogonal projection onto $\bigoplus_i M_{k(i,j)}$. Hence

$$t_j^*t_j\le\langle q_jz,q_jz\rangle,\qquad\langle T(z),T(z)\rangle\le\langle z,z\rangle.$$

So, T is bounded and it clearly is the adjoint for J.

The proof of the second statement literally repeats the argument from the previous lemma. □

Consider an operator $F\in\mathrm{GL}$. Then, with respect to the standard decomposition of $l_2(A)$ into the direct sum of $E_i\cong A$, the operator F has a matrix F_j^i with entries from $\mathbf{LM}(A)$. If $F\in\mathrm{GL}^*$, then $F_j^i\in\mathbf{M}(A)$, since $(F^*)_j^i=(F_i^j)^*$.

LEMMA 7.1.8. *For any $F\in\mathrm{GL}^*$, any $\varepsilon>0$ and any $e_k\gamma$ there exists a number $m(k)$ such that for any $m\ge m(k)$ and any $\varphi\in A$ with $\|\varphi\|\le1$, one has*

$$\|\langle e_k\gamma,Fe_m\varphi\rangle\|<\varepsilon.$$

PROOF. Consider the bounded operator F^*. Since $F^*e_k\gamma\in l_2(A)$, there exists a number $m(k)$ such that

$$\|(1-p_{m(k)})F^*e_k\gamma\|<\varepsilon,\qquad\|Q_mF^*e_k\gamma\|<\varepsilon\quad(m>m(k)).$$

Hence,

$$\|\langle e_k\gamma,Fe_m\varphi\rangle\|=\|Q_mF^*e_k\gamma\|\cdot\|\varphi\|<\varepsilon\quad(m>m(k)).$$

□

7.2. Proof of the Cuntz–Higson theorem

LEMMA 7.2.1. *Let $F_r\in\mathrm{GL}^*$, $r=1,\dots,N$, be arbitrary operators and let $\varepsilon>0$ be any number. Then we can choose increasing nonintersecting sequences $i(k)$ and $j(k)$ of natural numbers such that*

$$\|(1-p_{j(s)})F_re_{i(k)}\alpha_k\|<\frac14\cdot\frac{\varepsilon}{2^s\cdot2^k},\qquad s=k,k+1,\dots,\qquad r=1,\dots,N,\tag{7.3}$$

$$\|\langle F_re_{i(k)}\alpha_k,e_{j(s)}\alpha_s\rangle\|<\frac12\cdot\frac{\varepsilon}{2^s\cdot2^k},\qquad s=1,\dots,k-1,\qquad r=1,\dots,N.\tag{7.4}$$

PROOF. Put $i(1):=1$ and choose $j(1)>i(1)$ in such a way that

$$\|(1-p_{j(1)})F_re_{i(1)}\alpha_1\|<\frac14\cdot\frac{\varepsilon}{2^1\cdot2^1},\qquad r=1,\dots,N.$$

Find $i(2)>j(1)$ such that (by Lemma 7.1.8)

$$\|\langle F_re_{i(2)}\alpha_2,e_{j(1)}\alpha_1\rangle\|<\frac12\cdot\frac{\varepsilon}{2^1\cdot2^2},\qquad r=1,\dots,N.$$

Then choose $j(2)>i(2)$ such that

$$\|(1-p_{j(2)})F_re_{i(k)}\alpha_k\|<\frac14\cdot\frac{\varepsilon}{2^2\cdot2^k},\qquad k=1,2,\qquad r=1,\dots,N,$$

and $i(3)>j(2)$ such that

$$\|\langle F_re_{i(3)}\alpha_3,e_{j(s)}\alpha_s\rangle\|<\frac12\cdot\frac{\varepsilon}{2^s\cdot2^k},\qquad s=1,2,\qquad r=1,\dots,N.$$

Let us continue this process by induction. Suppose that $i(1), \dots, i(k-1)$ and $j(1), \dots, j(k-2)$ are already found and satisfy conditions (7.3) and (7.4). Find $j(k-1) > i(k-1)$ such that

$$\|(1 - p_{j(k-1)}) F_r e_{i(m)} \alpha_m\| < \frac{1}{4} \cdot \frac{\varepsilon}{2^{k-1} \cdot 2^m}, \qquad m = 1, \dots, k-1, \quad r = 1, \dots, N,$$

and after that find $i(k) > j(k-1)$ such that

$$\|\langle F_r e_{i(k)} \alpha_k, e_{j(s)} \alpha_s \rangle\| < \frac{1}{2} \cdot \frac{\varepsilon}{2^s \cdot 2^k}, \qquad s = 1, \dots, k-1, \quad r = 1, \dots, N.$$

By induction, we obtain the required statement. $\square$

Let us now define embeddings J and J' similarly to the constructions in Lemma 7.1.7. To define J, we choose the vectors v_{sj} among $e_{i(g)}\alpha_g\omega_s$ that satisfy $g = g(s,j) > s + j$, $g > s$; then $e_{i(g)}\alpha_g\omega_s = e_{i(g)}\omega_s$ and $\langle v_{sj}, v_{sj} \rangle = \omega_s^2$. Let us similarly define v'_{sm} for J', but taking $e_{j(k)}$ instead of $e_{i(k)}$ and denoting by h the corresponding g. From conditions (7.3) and (7.4) we obtain that

$$\begin{aligned} \|\langle F_r v_{st}, v'_{nm} \rangle\| &= \|\langle F_r e_{i(g(s,t))} \alpha_{g(s,t)} \omega_s, e_{j(h(n,m))} \alpha_{h(n,m)} \omega_n \rangle\| \\ &\le \|Q_{j(h(n,m))} F_r e_{i(g(s,t))} \alpha_{g(s,t)}\| \\ &\le \|(1 - p_{j(h(n,m)-1)}) F_r e_{i(g(s,t))} \alpha_{g(s,t)}\| \\ &< \frac{1}{4} \cdot \frac{\varepsilon}{2^{h-1} \cdot 2^g}, \qquad h \ge g,\ r = 1, \dots, N, \end{aligned} \tag{7.5}$$

$$\begin{aligned} \|\langle F_r v_{st}, v'_{nm} \rangle\| &= \|\langle F_r e_{i(g(s,t))} \alpha_{g(s,t)} \omega_s, e_{j(h(n,m))} \alpha_{h(n,m)} \omega_n \rangle\| \\ &< \frac{1}{2} \cdot \frac{\varepsilon}{2^h \cdot 2^g}, \qquad h < g,\ r = 1, \dots, N. \end{aligned} \tag{7.6}$$

Denote by P and P' the corresponding orthogonal projections. Then $PP' = P'P = 0$. Let $x = (a_1, a_2, \dots)$ and $y = (b_1, b_2, \dots)$ be arbitrary norm one vectors from $l_2(A)$. Then, for any $r = 1, \dots, N$, by (7.5), (7.6) one has

$$\begin{aligned} \|\langle F_r J x, J' y \rangle\| &= \left\| \left\langle \sum_t \sum_s F_r v_{st} a_t, \sum_m \sum_n v'_{nm} b_m \right\rangle \right\| \\ &\le \sum_{t,s,n,m} \left(\sum_{h(n,m) \ge g(s,t)} \|\langle F_r v_{st}, v'_{nm} \rangle\| + \sum_{h(n,m) < g(s,t)} \|\langle F_r v_{st}, v'_{nm} \rangle\| \right) \\ &\le \sum_{t,s,n,m} \frac{\varepsilon}{2^{h(n,m)} \cdot 2^{g(t,s)}} < \varepsilon, \end{aligned}$$

since $h(n,m) > n + m$, $g(t,s) > t + s$ by the construction. Therefore one has

$$\|P' F_r P\| < \varepsilon, \qquad r = 1, \dots, N. \tag{7.7}$$

As was shown in Lemma 7.1.2 and Lemma 7.1.5, it suffices to know how to construct a homotopy of any piecewise linear map with the image in a finite polyhedron in GL^* with vertices $F_1, \dots, F_N$ into a map taking values in a compact set $\{D(x)\} \subset \mathrm{GL}^*$ such that

$$PD(x) = D(x)P = P \quad \forall x \in S.$$

For this purpose we can apply a homotopy of Neubauer type (see Section 7.7). By (7.7), we have to verify only that there exists an operator $H_0 : P'(l_2(A)) \to P(l_2(A))$ such that operators $H_0 P'$ and $H_0^{-1} P$ are adjointable. Let us assume $H_0 = JJ'^*$.

Then $H_0 P' = JJ'^* J' J'^* = JJ'^*$, where J'^* is an isomorphism $P'(l_2(A)) \to l_2(A)$ and $J : l_2(A) \cong P(l_2(A))$.

We have proved the following statement.

THEOREM 7.2.2 ([**30**]). *Let A be a σ-unital C^*-algebra. Then* $\mathrm{GL}^*(A)$ *is contractible with respect to the norm topology.*

7.3. The case $A \subset \mathcal{K}$

Let a C^*-algebra A be a subalgebra of the algebra $\mathcal{K}$ of compact operators on a separable Hilbert space H. Under these restrictions we can prove the following statement.

LEMMA 7.3.1. *Let $a, b \in A$, $(f_1, f_2, \dots) \in l_2'(A)$. Then*

$$\|a f_i b\| \to 0 \qquad (i \to \infty).$$

PROOF. Since $a^* \in \mathcal{K}$, for any $\varepsilon > 0$ we can find a number $N = N(\varepsilon)$ and a basis $h_1, h_2, \dots$ in H such that

$$\|p_N' a^*\| < \frac{\varepsilon}{2 \cdot \sup \|f_i\|}, \qquad H_N = \mathrm{span}_{\mathbf{C}} \langle h_1, \dots, h_N \rangle, \qquad H_N' = H_N^\perp,$$

p_N and p_N' are the corresponding projections on H_N and H_N'. Since the partial sums of the series $\sum_i f_i f_i^*$ form a nondecreasing uniformly bounded sequence of positive operators in $\mathcal{B}(H)$ (see [**42**]), $f_i f_i^*$ is strongly convergent to the zero operator. Hence,

$$\|f_i^* h\| = \langle f_i^* h, f_i^* h \rangle = \langle f_i f_i^* h, h \rangle \to 0$$

for any $h \in H$. Thus, f_i^* is strongly convergent to 0. Let i_0 be so large that the estimate

$$\|f_i^* p_N\| < \frac{\varepsilon}{2\|a\|}$$

holds for $i > i_0$. Then

$$\|a f_i\| = \|f_i^* p_N a^*\| + \|f_i^* p_N' a^*\| < \frac{\varepsilon}{2\|a\|} \cdot \|a^*\| + \|f_i^*\| \cdot \frac{\varepsilon}{2 \cdot \sup \|f_i\|} \leq \varepsilon.$$

□

Note that similar properties for matrix elements themselves (which belong to $\mathbf{LM}(\mathcal{K}) = \mathcal{B}(H)$) do not hold even for operators from GL^*. Moreover, the following example shows that all matrix elements can have norm equal to 1.

EXAMPLE 7.3.2 (A. V. Buchina, [**105**]). Consider the standard Hilbert $\mathcal{K}$-module

$$l_2(\mathcal{K}) = \left\{ (k_1, k_2, k_3, \dots) \mid k_i \in \mathcal{K}, \|\sum_i k_i^* k_i\| < \infty \right\}.$$

We are going to construct an invertible operator $F : L_2(\mathcal{K}) \longrightarrow l_2(\mathcal{K})$ with matrix elements (with respect to the standard decomposition of $l_2(\mathcal{K})$) satisfying the inequality $\|F_{i,j}\| \geq 1$ for any i and j. These elements belong to $\mathbf{LM}\,(\mathcal{K}) = \mathcal{B}(H)$, i.e., to the algebra of all bounded operators.

Let $\{e_i\}$ be a basis of H and let p_i denote the rank one projection onto the i-th vector. Define F by the matrix

$$\begin{pmatrix} p_1 & p_2 & p_3 & p_4 & p_5 & p_6 & p_7 & p_8 & \cdots \\ p_2 & p_1 & p_4 & p_3 & p_6 & p_5 & p_8 & p_7 & \cdots \\ p_3 & p_4 & p_1 & p_2 & p_7 & p_8 & p_5 & p_6 & \cdots \\ p_4 & p_3 & p_2 & p_1 & p_8 & p_7 & p_6 & p_5 & \cdots \\ p_5 & p_6 & p_7 & p_8 & p_1 & p_2 & p_3 & p_4 & \cdots \\ p_6 & p_5 & p_8 & p_7 & p_2 & p_1 & p_4 & p_3 & \cdots \\ p_7 & p_8 & p_5 & p_6 & p_3 & p_4 & p_1 & p_2 & \cdots \\ p_8 & p_7 & p_6 & p_5 & p_4 & p_3 & p_2 & p_1 & \cdots \\ \vdots & \vdots & \vdots & \vdots & \vdots & \vdots & \vdots & \vdots & \ddots \end{pmatrix},$$

where the first line is the ordered sequence of projections, the second line is obtained from the first line by permutations in pairs p_{2i-1} and p_{2i} for $i = 1, 2, \dots,$ the third line is obtained from the first line by permutations of adjacent pairs, the fourth line is obtained by permutations of pairs in the third line, the fifth line is obtained by permutations of 4-tuples in the first line, and so on. Let us introduce the notation $F_{ij} := p_{\sigma_i(j)}$, and note that, for $i_1 \neq i_2$, one has $p_{\sigma_{i_1}(j)} \neq p_{\sigma_{i_2}(j)}$ for any j. Let us show that this operator F satisfies all necessary conditions.

1. Let us prove that the image of $(k_1, k_2, k_3, \dots) \in l_2(\mathcal{K})$ is compact for the action of each line of the operator F, i.e., let us verify that the inequality $\|\sum_j p_{\sigma_i(j)} k_j\| < \infty$ holds for each i.

For any $\varepsilon > 0$ we can find $N \in \mathbf{N}$ such that the estimate

$$\begin{aligned} \left\| \sum_{j=n}^{n+p} p_{\sigma_i(j)} k_j \right\|^2 &= \left\| \left(\sum_{j=n}^{n+p} p_{\sigma_i(j)} k_j \right)^* \left(\sum_{r=n}^{n+p} p_{\sigma_i(r)} k_r \right) \right\| \\ &= \left\| \sum_{j=n}^{n+p} k_j^* p_{\sigma_i(j)} k_j \right\| \leq \left\| \sum_{j=n}^{n+p} k_j^* k_j \right\| < \varepsilon \end{aligned}$$

holds for any $n > N$ and any $p \in \mathbf{N}$ and, by the Cauchy criterion, it follows from the norm-convergence of $\sum_j k_j^* k_j$ that $\sum_j p_{\sigma_i(j)} k_j$ is also norm-convergent. Thus, the series is norm-convergent and its entries are compact operators, hence $\sum_j p_{\sigma_i(j)} k_j$ is compact as well.

2. Let us verify that the image of a vector $(k_1, k_2, k_3, \dots) \in l_2(\mathcal{K})$ under the action of F belongs to $l_2(\mathcal{K})$, i.e., let us prove the norm-convergence of the series

$$\sum_{i=1}^{\infty} \left(\sum_j p_{\sigma_i(j)} k_j \right)^* \left(\sum_r p_{\sigma_i(r)} k_r \right).$$

We have

$$\begin{aligned} &\left\| \sum_{i=1}^{\infty} \left(\sum_j p_{\sigma_i(j)} k_j \right)^* \left(\sum_r p_{\sigma_i(r)} k_r \right) \right\| \\ &\qquad = \left\| \sum_{i=1}^{\infty} \sum_j k_j^* p_{\sigma_i(j)} \sum_r p_{\sigma_i(r)} k_r \right\| = \left\| \sum_i \sum_j k_j^* p_{\sigma_i(j)} k_j \right\|. \end{aligned}$$

Since $\|\sum_j k_j^* k_j\| < \infty$, the inequalities

$$\begin{aligned}\infty > \left\|\sum_j k_j^* k_j\right\| &\geq \left\langle \sum_j k_j^* k_j x, x\right\rangle = \sum_j \langle k_j x, k_j x\rangle \\ &= \sum_j \|k_j x\|^2 = \sum_j \sum_i |(k_j x)_{\sigma_i(j)}|^2 = \sum_i \sum_j |(k_j x)_{\sigma_i(j)}|^2\end{aligned}$$

hold for any $x \in H$ of norm 1. Hence, for any $\varepsilon > 0$ and for any x with $\|x\| = 1$, one can find a number $N(x) \in \mathbf{N}$ such that $\sum_{i=n}^{n+p} \sum_j |(k_j x)_{\sigma_i(j)}|^2 < \varepsilon$ for any $n > N(x)$ and any $p \in \mathbf{N}$. Thus,

$$\begin{aligned}\left\langle \sum_{i=n}^{n+p} \sum_j k_j^* p_{\sigma_i(j)} k_j x, x \right\rangle &= \sum_{i=n}^{n+p} \sum_j \langle k_j^* p_{\sigma_i(j)} k_j x, x\rangle \\ &= \sum_{i=n}^{n+p} \sum_j \langle p_{\sigma_i(j)} k_j x, p_{\sigma_i(j)} k_j x\rangle = \sum_{i=n}^{n+p} \sum_j \|(k_j x)_{\sigma_i(j)}\|^2 < \varepsilon.\end{aligned}$$

Consider the operators $B = \sum_j k_j^* k_j$ and $B_n = \sum_{i=1}^{n} \sum_j k_j^* p_{\sigma_i(j)} k_j$ acting on H. The first of them is compact and, consequently, the remaining ones are compact as well, being the sums of norm-convergent series with compact entries. For a fixed vector $x \in H$ of norm 1, the following statements hold:

(i) the inequality

$$\langle B_n x, x\rangle = \sum_j \sum_{i=1}^{n} |k_j x|^2_{\sigma_i(j)} \leq \sum_j \sum_i |(k_j x)_{\sigma_i(j)}|^2 = \langle Bx, x\rangle$$

holds;

(ii) $\lim_{n\to\infty} (B_n x, x) = (Bx, x)$.

Put $E_n := B - B_n$. It is easy to see that

(i) $\langle E_n x, x\rangle \to 0$ as $n \to \infty$;

(ii) $E_n^* = E_n$;

(iii) $E_n \geq 0$;

(iv) $\|E_n\| = \|B - B_n\| \leq \|B\|$.

It follows from (i) and (iii) that $E_n^{1/2} \longrightarrow 0$ with respect to the strong topology. By (iv), we obtain $E_n \longrightarrow 0$ with respect to the strong topology, due to the joint strong continuity of multiplication in both variables on bounded sets. We have proved the strong convergence of the increasing sequence of positive compact operators B_n to the compact operator B. Choose a finite-dimensional projection p such that $\|B(1-p)\| < \varepsilon$ and then choose n such that $\|(B - B_m)p\| < \varepsilon$ for any $m > n$. Then, since the sequence increases,

$$\begin{aligned}\|B - B_m\| &\leq \|(B - B_m)p\| + \|p(B - B_m)\| \\ &\quad + \|p(B - B_m)p\| + \|(1-p)(B - B_m)(1-p)\| \\ &\leq 3\|(B - B_m)p\| + \|(1-p)B(1-p)\| < 4\varepsilon.\end{aligned}$$

Thus, the sequence of operators B_n converges to B, i.e., $\|B - B_n\| \to 0$ as $n \to \infty$. Since $\|B\| < \infty$, we have $\|\sum_i \sum_j k_j^* p_{\sigma_i(j)} k_j\| < \infty$ and we are done.

3. By the construction of the operator F, it is clear that $\|F_{ij}\| = 1$ for any i and j.

4. Note that $F^*F = F^2 = FF^* = \mathrm{Id}$. Therefore F is invertible.

By 1.–4. the operator F constructed above has all the necessary properties.

THEOREM 7.3.3. *The group* $\mathrm{GL}(A)$ *for* $A \subset \mathcal{K}$ *is contractible with respect to the norm topology.*

PROOF. Since Lemma 7.3.1 is the analog of Lemma 7.1.8, the proof can be obtained by literally repeating the arguments from Section 7.2. □

7.4. Some other cases

DEFINITION 7.4.1. Let us say that a C^*-algebra A has *property* (K) if for any functional $f : l_2(A) \to A$, any $\varepsilon > 0$ and any $a \in A$, one can find a vector $x \in l_2(A)$ such that

$$\|f(x)\| < \varepsilon, \qquad \langle x, x\rangle = a^*a.$$

DEFINITION 7.4.2. A C^*-algebra A has *property* (E) if for any functional $f = (f_1, \dots, f_n, \dots) \in l_2'(A)$ and any $\varepsilon > 0$, one can find another functional $g = (g_1, \dots, g_n, \dots) \in l_2'(A)$ and a number $k \in \mathbf{Z}$ such that

$$\|f - g\| < \varepsilon, \qquad f_i = g_i, \quad i = k+1,\ k+2, \dots,$$

and $g|_{L_k} : L_k \to A$ is *epimorphism*, where $L_n = \{\, (a_1, \dots, a_n, 0, 0, \dots)\,\}$.

EXAMPLE 7.4.3. Let A be the algebra of continuous functions on a smooth n-dimensional manifold M. Then A has property (E) (with $k = n+1$).

To establish the relation between properties (E) and (K) in Theorem 7.4.5 below, we require the following lemma.

LEMMA 7.4.4. *Let* $\mathcal{M}$ *be a Hilbert module,* $x \in \mathcal{M}$, $\langle x, x\rangle \geq a \geq 0$, $\|a\| \leq 1$. *Then one can find an element* $y = xb$, $\|b\| \leq 1$, *such that* $\langle y, y\rangle = a^2$.

PROOF. Put

$$\gamma := \langle x, x\rangle, \qquad b := \lim_{n\to\infty} \left(\gamma + \frac{1}{n}\right)^{-1/2} a.$$

This limit (with respect to the norm topology) exists, since

$$\left[\left(\gamma + \frac{1}{n}\right)^{-1/2} - \left(\gamma + \frac{1}{m}\right)^{-1/2}\right] a^2 \left[\left(\gamma + \frac{1}{n}\right)^{-1/2} - \left(\gamma + \frac{1}{m}\right)^{-1/2}\right]$$
$$\leq \left[\left(\gamma + \frac{1}{n}\right)^{-1/2} - \left(\gamma + \frac{1}{m}\right)^{-1/2}\right]^2 \gamma^2 \to 0$$

and

$$\frac{z^2}{z + \frac{1}{n}} - \frac{z^2}{z + \frac{1}{m}} = \frac{\frac{1}{m}z^2 - \frac{1}{n}z^2}{(z + \frac{1}{n})(z + \frac{1}{n})} = \left(\frac{1}{m} - \frac{1}{n}\right) \frac{z^2}{(z + \frac{1}{n})(z + \frac{1}{n})} \leq \frac{1}{m} - \frac{1}{n}$$

for any nonnegative z. The estimate $\|b\| \leq 1$ follows from

$$a\left(\gamma + \frac{1}{n}\right)^{-1} a \leq a^{1/2}\gamma\left(\gamma + \frac{1}{n}\right)^{-1} a^{1/2} \leq a \leq 1.$$

The condition $\langle y, y\rangle = a^2$ is now clear. □

THEOREM 7.4.5. *Property* (E) *implies property* (K).

PROOF. Without loss of generality we can assume that $\|a\| = 1$. Consider an arbitrary functional $f = (f_1, \dots) \in l'_2(A)$ and take some $\varepsilon > 0$. Let g and k be given by property (E) for $\varepsilon/2$. Put $f' := f|_{L_k^\perp}$. Since $L_k^\perp \cong l_2(A)$, there exists, by (E), a functional $g' : L_k^\perp \to A$ such that

$$\|f' - g'\| < \varepsilon/2, \qquad f'_i = g'_i = g_i, \quad i = k'+1,\ k'+2, \dots,$$

and $g'|_{L_k^\perp \cap L_{k'}}$ is an epimorphism. Then the functional

$$h := \begin{cases} g & \text{on } L_k, \\ g' & \text{on } L_k^\perp, \end{cases}$$

satisfies the following conditions: $\|f - h\| < \varepsilon$; h is an epimorphism on L_k and $L_k^\perp \cap L_{k'}$ separately. Without loss of generality we can assume that $\|h\| = 1$. Let $x \in L_k$ and $y \in L_k^\perp \cap L_{k'}$ satisfy $h(x) = h(y) = a$. Then $h(x - y) = 0$ and, by 2.1.4,

$$a^*a = \langle h(x), h(x) \rangle \le \langle x, x \rangle, \qquad a^*a = \langle h(y), h(y) \rangle \le \langle y, y \rangle.$$

By Lemma 7.4.4, one can find an element b such that $\|b\| \le 1$ and $z = (x - y)b$ satisfies $\langle z, z \rangle = a^2$. Thus $h(z) = h((x - y)b) = 0$ and, since $\|z\| = 1$, one has $\|f(z)\| < \varepsilon$. □

REMARK 7.4.6. Let i and i' be adjointable embeddings preserving the inner product and let the corresponding projections $q = ii^*$ and $q' = i'i'^*$ satisfy

$$\|qq'\| < \varepsilon, \qquad \|q'q\| < \varepsilon. \tag{7.8}$$

Note that $qq' = ii^*i'i'^*$, where i is an isometric embedding and i'^* is an epimorphism of norm 1. Therefore the inequalities (7.8) are equivalent to $\|i^*i'\| < \varepsilon$, $\|i'^*i\| < \varepsilon$. Then the map $I := (i, i') : l_2(A) \oplus l_2(A) \to l_2(A)$ is also an adjointable embedding with the adjoint $I^*(x) = (i^*(x), i'^*(x))$. Indeed, the operator I^* given by this formula is continuous and

$$\langle I(x, y), z \rangle = \langle i(x) + i'(y), z \rangle = \langle x, i^*(z) \rangle + \langle y, i'^*(z) \rangle = \langle (x, y), I^*(z) \rangle.$$

One also has

$$I^*I(x, y) = (i^*(ix + i'y), i'^*(ix + i'y)) = (x, y) + (i^*i'y, i'^*ix),$$

so that

$$\| \operatorname{Id} - I^*I \| < 2\varepsilon \tag{7.9}$$

and I^*I is invertible. Therefore I is an embedding. Note that, for invertibility, it is sufficient to have $\varepsilon < 1/2$.

THEOREM 7.4.7. *Let a C^*-algebra A have property* (K). *Then the group* $\mathrm{GL}(A)$ *is norm-contractible.*

PROOF. As above, one has to prove a statement similar to Lemma 7.1.8. In the present situation we argue as follows. Let F_1 be the first line (i.e., a functional) of the matrix of F with respect to the standard decomposition of $l_2(A)$. Note that any vector from $l_2(A)$, up to any $\delta > 0$ given earlier, belongs to L_n for a sufficiently large $n = n(\varepsilon)$. Hence, applying property (K), one can assume from the very beginning that $x \in L_n$. Applying property (K) infinitely many times with constants decreasing as a geometrical progression, we can find a sequence of vectors $x_i \in l_2(A)$ satisfying the following conditions:

(7.10)
$$x_i \in M_i, \quad l_2(A) = M_1 \oplus M_2 \oplus \cdots, \qquad M_i = \{(0,\dots,0,a_{k(i)},\dots,a_{k(i+1)-1},0,\dots)\},$$

$$\langle x_i, x_i\rangle = \alpha_i, \qquad \alpha_i \text{ is an approximate unit for } A, \tag{7.11}$$

$$\|F_1(x_i)\| < \frac{\varepsilon}{2}\cdot\frac{1}{2^i}. \tag{7.12}$$

Note that for $k > k(i)$, one has $\omega_i = \alpha_k^{1/2} d(i,k)$, $\|d(i,k)\| \le 1$. Therefore, similarly to the argument above, the map

$$J_1 : l_2(A) \to l_2(A), \qquad (a_1, a_2, \dots) \mapsto \sum_j \sum_i x_{k(i,j)} d(i, k(i,j)) a_j,$$

where

$$k(1,1); k(1,2), k(2,1); k(1,3), k(2,2), k(3,1); \dots$$

is some increasing sequence, is an adjointable embedding preserving the inner product. If we denote $H_1 := \operatorname{Im} J_1$, then, by (7.12),

$$\|F_1|_{H_1}\| < \frac{\varepsilon}{2}.$$

Let G_1 be the orthogonal complement to the image of the first copy of A under J_1. Let $m(2) > m(1) := 1$ be so large that $\|(1 - p_{m(2)})Fy(1)\| < \varepsilon/2$, where $y_1 := J_1(\alpha_1^{1/2}, 0, \dots)$. Denote by F_2 the restriction of the $m(2)$-th line of the matrix F onto $G_1 \cong l_2(A)$ and find, by the same algorithm, a new embedding J_2 such that its image equals H_2 and

$$\|F_2|_{H_2}\| < \frac{\varepsilon}{2^2}.$$

Let $G_2 \subset H_2$ be the orthogonal complement to the image of the first copy of A under J_2. Let $m(3) > m(2)$ be so large that

$$\|(1 - p_{m(3)})Fy_i\| < \frac{\varepsilon}{2^3 \cdot 2^i}, \qquad i = 1, 2, \qquad y_2 := J_2(\alpha_2^{1/2}, 0, \dots),$$

$$\|(1 - p_{m(3)})y_i\| < \frac{\varepsilon}{2^3 \cdot 2^i}, \qquad i = 1, 2.$$

Proceeding by induction, we obtain sequences $m(j)$ and y_i such that

$$\|(1 - p_{m(j)})Fy_i\| < \frac{\varepsilon}{2^j \cdot 2^i}, \qquad i = 1, \dots, j-1, \tag{7.13}$$

$$\|(1 - p_{m(j)})y_i\| < \frac{\varepsilon}{2^j \cdot 2^i}, \qquad i = 1, \dots, j-1, \tag{7.14}$$

$$\|Q_{m(j)}Fy_i\| < \frac{\varepsilon}{2^j \cdot 2^i}, \qquad j = 1, \dots, i. \tag{7.15}$$

Using ω_j again, we can construct an embedding J of the module $l_2(A)$ on a submodule H of the linear span of y_i and an embedding J' of the module $l_2(A)$ on the submodule $H' := \bigoplus_j E_{m(j)}$. Since these modules are ε-*orthogonal* (i.e., $\|\langle h, h'\rangle\| \le \varepsilon$ for any $h \in H$, $h' \in H'$, $\|h\| \le 1$, $\|h'\| \le 1$), there exist projections p and p' onto them such that $pp' = p'p = 0$. More precisely, note first that the embedding J is adjointable. Indeed, the image of each vector $(a_1, a_2, \dots)$ under J_1 is a sum of the form

$$\sum_j \sum_i v_{ij} a_j, \qquad \langle v_{ij}, v_{ij} \rangle = \omega_i^2, \qquad v_{ij} \in M_{k(i,j)}.$$

For constructing the higher J_s's, the corresponding v_{ij}^s lie again in direct sums of modules M_r and, for v_{i1}^s, these sets of modules are not intersecting. We can apply Lemma 7.1.7. The operator J will be defined by the formula

$$J : (a_1, a_2, \dots) \mapsto \sum_s \sum_i v_{i1}^s a_s, \qquad \sum_i v_{i1}^s a_s = y_s \mu_s a_s. \tag{7.16}$$

Hence, there exist the orthogonal projections q and q' onto H and H', respectively. Note that this argument can be refined as follows. In particular, we have shown that for any J_s and any m there exists no more than one r such that $Q_m J_s Q_r \neq 0$. Therefore, throwing out, if necessary, a finite number of canonical summands in $l_2(A)$ and restricting J_s to the remaining module, we can assume that

$$Q_{m(j)} J_s = 0, \qquad j = 1, \dots, s-1, \tag{7.17}$$

$$Q_{m(j)} y_i = 0, \qquad j = 1, \dots, i. \tag{7.18}$$

Also, $\|qq'\| < \varepsilon$, $\|q'q\| < \varepsilon$. Indeed, consider a vector of the form

$$x = \sum_s \sum_i v_{i1}^s a_s = \sum_s y_s \mu_s a_s, \qquad \|\sum_s a_s^* a_s\| \le 1.$$

It is necessary to show that $\|q'x\| < \varepsilon$. This estimate follows from (7.14), (7.18):

$$\begin{aligned}
\|q'x\| &= \left\| \sum_j Q_{m(j)} \sum_s \sum_i v_{i1}^s a_s \right\| \\
&\le \sum_s \left\| \sum_{j>s} Q_{m(j)} \left(\sum_i v_{i1}^s a_s \right) \right\| + \sum_s \sum_{j \le s} \left\| Q_{m(j)} \left(\sum_i v_{i1}^s a_s \right) \right\| \\
&\le \sum_s \|(1 - p_{m(s)}) y_s \mu_s a_s\| + \sum_s \sum_{j \le s} 0 \le \sum_s \frac{\varepsilon}{2^s} = \varepsilon.
\end{aligned}$$

Since the projections q and q' are selfadjoint, we obtain the second estimate as well.

Then, by Remark 7.4.6, $H \widetilde{\oplus} H'$ is the image of an adjointable embedding and, by Theorem 2.3.3, the decomposition $l_2(A) = H \widetilde{\oplus} H' \oplus (H^\perp \cap H'^\perp)$ holds. Denote by p and p' the projections onto H and H', respectively, which correspond to this decomposition, so that $pp' = p'p = 0$. Thus,

$$\|p - q\| < 3\varepsilon, \qquad \|p' - q'\| < 3\varepsilon, \qquad \|p\| < 1 + 3\varepsilon < 2, \qquad \|p'\| < 1 + 3\varepsilon < 2. \tag{7.19}$$

Indeed, let $x \in H \widetilde{\oplus} H'$, $\|x\| = 1$. Then $x = II^*Iy$ for some y and, by (7.9), $\|Iy\| \le 2(1 + \varepsilon)$. Therefore

$$\begin{aligned}
\|(p - q)x\| &= \|(p - q)(ii^*Iy + i'i'^*Iy)\| \\
&= \|(p - q)(q + q')Iy\| = \| - qq'Iy\| \le 2\varepsilon(1 + \varepsilon) < 3\varepsilon.
\end{aligned}$$

Besides, $\|p'Fp\| < 7\|F\|\varepsilon$. In fact,

$$\|p'Fp\| = \|(p' - q')Fp + q'Fp\| < 3\varepsilon\|F\| + \|q'Fp\|,$$

and, by (7.19), it is sufficient to prove that for $x \in H$, $\|x\| \leq 1$, the estimate $\|q'Fx\| < 2\varepsilon$ holds. Any such vector x can be presented as

$$\sum_s \sum_i v_{i1}^s a_s = \sum_s y_s \mu_s a_s, \qquad \|\sum_s a_s^* a_s\| \leq 1.$$

Then

$$\begin{aligned}\|q'Fx\| &= \left\| \sum_j Q_{m(j)} \sum_s \sum_i F v_{i1}^s a_s \right\| \\ &\leq \sum_s \sum_i \left\| \sum_{j>s} Q_{m(j)} F v_{i1}^s \right\| + \sum_s \sum_{j \leq s} \left\| Q_{m(j)} F \left(\sum_i v_{i1}^s a_s \right) \right\| \\ &\leq \sum_s \|(1 - p_{m(s)}) F y_s \mu_s a_s\| + \sum_s \sum_{j \leq s} \frac{\varepsilon}{2^j \cdot 2^s} \leq \sum_s \frac{\varepsilon}{2^s} + \varepsilon = 2\varepsilon.\end{aligned}$$

Note that one can obtain a similar statement not only for a single operator F (actually for two operators: F and Id), but also for a finite set (vertices of a simplicial complex) $F^{(1)}, \dots, F^{(N)}$. For this purpose, one has to argue for $F = F^{(1)}$ with a constant ε and to obtain projections P_1 and P_1'. After that one has to apply the algorithm to $P_1' F^{(2)} P_1$ and to obtain projections P_2' and P_2 such that

$$P_1' P_2' = P_2' P_1' = P_2', \quad P_1 P_2 = P_2 P_1 = P_2, \quad P_2 P_1 = P_1 P_2 = 0,$$

$$\|P_2' F^{(1)} P_2\| < \varepsilon, \quad \|P_2' F^{(2)} P_2\| < \varepsilon,$$

and so on. This completes the proof, since now one can apply the Neubauer homotopy. □

Let us complete this section with a discussion about the following example, which gives no hope that the class (K) is sufficiently large.

EXAMPLE 7.4.8. Under the conditions and notation of Example 2.5.6 the functional

$$f : l_2(A) \to A, \qquad f : (a_1, a_2, \dots) \mapsto \sum_i u_i a_i$$

does not have property (K).

REMARK 7.4.9. Characterization of topological spaces X such that $C(X)$ has property (K) was obtained in [**106**].

7.5. Dixmier-Douady Theorem for $l_2(A)$

Let us consider $l_2(A)$ as the completion of the algebraic tensor product $H \otimes A = L^2([0,1]) \otimes A$ with respect to the A-inner product $\langle f \otimes \gamma, g \otimes \beta \rangle = \langle f, g \rangle \gamma^* \beta$. We assume here that the inner product in $L^2([0,1])$ is linear in the second entry.

LEMMA 7.5.1 ([**33**, p. 250]). *There exist, for each $t \in [0,1]$, a closed linear subspace $H_t \subset H$ and, for each $t \in (0,1]$, a linear isometry $U_t : H_t \to H$ such that*

(i) *the map $t \mapsto P_t$, where P_t is the orthogonal projection onto H_t, is strongly continuous;*
(ii) *the maps $t \mapsto U_t P_t$ and $t \mapsto U_t^{-1}$ are strongly continuous;*
(iii) $H_1 = H, \quad H_0 = 0, \quad U_1 = 1.$

Recall that in [**33**] the subspaces H_t are defined in the following way:

$$H_t := \{f \in L^2([0,1]) \mid f(x) = 0 \quad \text{for} \quad x \geq t\}.$$

LEMMA 7.5.2. *If $F(t)$ is bounded and if $F_t \to F$ with respect to the strong topology on $B(H)$, as $t \to 0$, then $F_t \otimes \mathrm{Id}_A \to F \otimes \mathrm{Id}_A$ with respect to the left strict topology.*

PROOF. It is sufficient to prove that

$$\|(F_t \otimes \mathrm{Id}_A - F \otimes \mathrm{Id}_A)\theta_{x,y}\| \to 0 \qquad (t \to 0),$$

where

$$\theta_{x,y}(z) = x\langle y, z\rangle, \quad x = \sum_{i=1}^{N} h_i x_i \otimes \beta_i, \quad x_i \in \mathbf{C},\ \beta_i \in A, \quad \|x\| = \|y\| = 1,$$

and $\{h_i\}$ is an orthogonal basis of H. Then, for $z = \sum_i h_i z_i \otimes \mu_i$, one has

$$\|(F_t \otimes \mathrm{Id}_A - F \otimes \mathrm{Id}_A)\,\theta_{x,y}(z)\| = \left\| \sum_{i=1}^{N} (F_t - F)h_i x_i \otimes \beta_i \langle y, z\rangle \right\| < \varepsilon$$

if t is so close to 0 that

$$\|(F_t - F)h_i x_i\| \cdot \|\beta_i\| < \frac{1}{N}\,\varepsilon.$$

$\square$

LEMMA 7.5.3. *Let a set $\{G(t)\}$ be uniformly bounded (by a constant C) and let $G(t) \to G$ and $S(t) \to S$ $(t \to 0)$ with respect to the left strict topology. Then $G(t)S(t) \to GS$ $(t \to 0)$ with respect to the left strict topology.*

PROOF. Let $k \in \mathcal{K}_A$ be an arbitrary operator. Then $Sk \in \mathcal{K}_A$ and

$$\|S(t)k - Sk\| \to 0, \qquad \|(G(t) - G)(Sk)\| \to 0 \qquad (t \to 0).$$

Hence

$$\begin{aligned}\|G(t)S(t)k - GSk\| &\leq \|(G(t) - G)Sk + G(t)(S(t) - S)k\| \\ &\leq \|(G(t) - G)Sk\| + C\|(S(t) - S)k\| \to 0 \quad (t \to 0).\end{aligned}$$

$\square$

THEOREM 7.5.4. *The unitary group $\mathcal{U}$ of operators in $l_2(A)$ is contractible with respect to the left strict topology.*

PROOF. For any $\mathbf{U} \in \mathcal{U}$ and $t \in (0,1]$ we define

$$\Phi(\mathbf{U}, t) := (\mathrm{Id}_{l_2(A)} - P_t \otimes \mathrm{Id}_A) + (U_t^{-1} \otimes \mathrm{Id}_A)\,\mathbf{U}\,(U_t \otimes \mathrm{Id}_A)\,(P_t \otimes \mathrm{Id}_A)$$

and

$$\Phi(\mathbf{U}, 0) := \mathbf{U}.$$

The operator $\Phi(\mathbf{U}, t)$, $t \in (0,1]$, defines the identity mapping $H_t^{\perp} \otimes A$ and coincides with the restriction of the unitary map $(U_t^{-1} \otimes \mathrm{Id}_A)\,\mathbf{U}\,(U_t \otimes \mathrm{Id}_A)$ onto $H_t \otimes A$. Therefore

$$\Phi(\mathbf{U}, t) \in \mathcal{U}, \qquad \Phi(\mathbf{U}, 1) = \mathbf{U}.$$

Thus, since $U_t^{*} = U_t^{-1}$, all operators are adjointable.

It is clear from Lemma 7.5.3 that Φ is continuous in $t \in (0,1]$ and, similarly, in $(\mathbf{U},t)$. Indeed, let $(\mathbf{U}',t') \in \mathcal{U} \times (0,1]$ tend to $(\mathbf{U},t) \in \mathcal{U} \times (0,1]$. Then, for any $k \in \mathcal{K}_A$, one has

$$\begin{aligned}
&\|\Phi(\mathbf{U},t)k - \Phi(\mathbf{U}',t')k\| \\
&\quad = \|(\mathrm{Id}_{l_2(A)} - P_t \otimes \mathrm{Id}_A)k + (U_t^{-1} \otimes \mathrm{Id}_A)\, \mathbf{U}\, (U_t \otimes \mathrm{Id}_A)\, (P_t \otimes \mathrm{Id}_A)k \\
&\qquad - (\mathrm{Id}_{l_2(A)} - P_{t'} \otimes \mathrm{Id}_A)k - (U_{t'}^{-1} \otimes \mathrm{Id}_A)\, \mathbf{U}'\, (U_{t'} \otimes \mathrm{Id}_A)\, (P_{t'} \otimes \mathrm{Id}_A)k\| \\
&\quad \le \|(P_t \otimes \mathrm{Id}_A - P_{t'} \otimes \mathrm{Id}_A)k\| \\
&\qquad + \|[(U_t^{-1} \otimes \mathrm{Id}_A) - (U_{t'}^{-1} \otimes \mathrm{Id}_A)]\, \mathbf{U}\, (U_t \otimes \mathrm{Id}_A)\, (P_t \otimes \mathrm{Id}_A)k\| \\
&\qquad + \|(U_{t'}^{-1} \otimes \mathrm{Id}_A)\, [\mathbf{U} - \mathbf{U}']\, (U_t \otimes \mathrm{Id}_A)\, (P_t \otimes \mathrm{Id}_A)k\| \\
&\qquad + \|(U_{t'}^{-1} \otimes \mathrm{Id}_A)\, \mathbf{U}'\, [(U_t \otimes \mathrm{Id}_A) - (U_{t'} \otimes \mathrm{Id}_A)]\, (P_t \otimes \mathrm{Id}_A)k\| \\
&\qquad + \|(U_{t'}^{-1} \otimes \mathrm{Id}_A)\, \mathbf{U}'\, (U_{t'} \otimes \mathrm{Id}_A)\, [(P_t \otimes \mathrm{Id}_A) - (P_{t'} \otimes \mathrm{Id}_A)]k\| \\
&\quad \le \|(P_t \otimes \mathrm{Id}_A - P_{t'} \otimes \mathrm{Id}_A)k\| \\
&\qquad + \|[(U_t^{-1} \otimes \mathrm{Id}_A) - (U_{t'}^{-1} \otimes \mathrm{Id}_A)]\, k_1\| + \|[\mathbf{U} - \mathbf{U}']\, k_2\| \\
&\qquad + \|[(U_t \otimes \mathrm{Id}_A) - (U_{t'} \otimes \mathrm{Id}_A)]\, k_3\| + \|[(P_t \otimes \mathrm{Id}_A) - (P_{t'} \otimes \mathrm{Id}_A)]k\| \to 0
\end{aligned}$$

by Lemma 7.5.2. Here k_1, k_2 and k_3 are fixed operators from $\mathcal{K}_A$. Now let $(\mathbf{U}',t') \in \mathcal{U} \times (0,1]$ tend to $(\mathbf{U},0) \in \mathcal{U} \times [0,1]$. Then $P_{t'} \to 0$ with respect to the strong topology and $P_{t'} \otimes \mathrm{Id}_A \to 0$ with respect to the left strict topology by Lemma 7.5.2. Therefore, for any $k \in \mathcal{K}_A$,

$$\|(U_{t'}^{-1} \otimes \mathrm{Id}_A)\, \mathbf{U}\, (U_{t'} \otimes \mathrm{Id}_A)\, (P_{t'} \otimes \mathrm{Id}_A)\, k\| \le \|(P_{t'} \otimes \mathrm{Id}_A)\, k\| \to 0, \qquad \|\Phi(\mathbf{U},t)k\| \to 0.$$

□

7.6. Some generalizations

Note that in the proof of Theorem 7.5.4 we used only the boundedness of the set of invertible operators $\{\mathbf{U}\}$ but not its unitarity. Thus, we have actually proved the following statement.

THEOREM 7.6.1. *Every bounded set of invertible operators in a Hilbert space H is contractible with respect to the strong topology in the set of invertibles.*

Every bounded set of invertible operators from GL *(resp.* GL*) *is contractible in* GL *(resp.* GL**) with respect to the left strict topology.*

LEMMA 7.6.2. *Let S be a compact set and let*

$$f : S \to B(H), \qquad s \mapsto F_s,$$

be continuous in the strong topology. Then $\{\|F_s\|\}$ is bounded.

PROOF. Since S is compact, the set $\{\|F_s x\|\}$ is bounded by some $C(x)$ for any $x \in H$. Therefore, by the uniform boundedness principle [**35**, II.3.21], there exists a constant C such that

$$\|F_s x\| \le C, \qquad \forall\, s \in S, \quad x \in B_1(H).$$

Therefore $\|F_s\| \le C$. □

LEMMA 7.6.3. *Let S be a compact set and let*

$$f : S \to \mathrm{End}_A l_2(A) = \mathbf{LM}(\mathcal{K}_A), \qquad s \mapsto F_s,$$

be continuous in the left strict topology. Then $\{\|F_s\|\}$ is bounded.

PROOF. Let $x \in l_2(A)$ be an arbitrary element. Choose $k \in \mathcal{K}$ and z such that $x = kz$ (see Lemma 2.2.3). Then $s \mapsto F_s x$ is continuous: we apply the definition of the left strict topology to the inequality

$$\|F_s x - F_t x\| = \|F_s kz - F_t kz\| \leq \|F_s k - F_t k\| \, \|z\|.$$

The proof can be completed similarly to the proof of Lemma 7.6.2. □

Now, from Theorem 7.6.1 by Lemma 7.6.2 and Lemma 7.6.3, we obtain the following statement.

THEOREM 7.6.4. *The group $G(H)$ of invertible operators in a Hilbert space H is weakly contractible (i.e., the homotopy groups $\pi_i(G(H)) = 0$) with respect to the strong topology.*

The group GL *(resp.* GL^**) is weakly contractible with respect to the left strict topology.*

REMARK 7.6.5. We suppose that the results of this section in the part concerning Hilbert spaces were known earlier but we have not found them published anywhere.

7.7. Neubauer type homotopy

In this section we describe how to modify the homotopy from [**97**] for our purposes. Although we work with completely different objects, the construction in [**97**] is so universal that the proofs can be transferred almost without modifications.

LEMMA 7.7.1. *Let $\mathcal{M}$ be a Hilbert A-module, X a topological space, $T : X \to \mathcal{G} = \mathcal{G}(\mathcal{M})$ a continuous map and P and P' projections from $\mathcal{E} = \mathcal{E}(\mathcal{M})$ such that*

$$PP' = P'P = 0, \qquad H_0 : P'\mathcal{M} \cong P\mathcal{M},$$
$$H_0 P' \in \mathcal{E}, \quad H_0^{-1} P \in \mathcal{E}, \qquad P'T(x)P = 0 \quad \forall x \in X.$$

Then there is a homotopy $T \sim D$ in $\mathcal{G}$ such that

$$PD(x) = D(x)P = P \quad \forall x \in X.$$

PROOF. Put $Q := \mathrm{Id} - P$, $Q' := \mathrm{Id} - P'$,

$$\mathcal{P}(x) := T(x)PT(x)^{-1}Q', \qquad \mathcal{Q}(x) := Q' - \mathcal{P}(x).$$

Then $\mathcal{P}(x)$ is a projection on $T(x)P\mathcal{M}$ and there is a decomposition into projections, $\mathrm{Id} = \mathcal{Q}(x) + \mathcal{P}(x) + P'$, and the projections $\mathcal{Q}(x)$, $\mathcal{P}(x)$ and P' are such that the composition of any two of them vanishes for any x. Indeed,

$$\begin{aligned}
&Q'T(x)P = (\operatorname{Id} - P')T(x)P = T(x)P,\\
&PT(x)^{-1}Q'T(x)P = P, \quad T(x)P\mathcal{M} \subset \mathcal{P}(x)\mathcal{M} \subset T(x)P\mathcal{M},\\
&\mathcal{P}(x)\mathcal{P}(x) = T(x)PT(x)^{-1}(\operatorname{Id} - P')T(x)PT(x)^{-1}(\operatorname{Id} - P')\\
&\qquad = T(x)PT(x)^{-1}T(x)PT(x)^{-1}(\operatorname{Id} - P') = T(x)PT(x)^{-1}(\operatorname{Id} - P') = \mathcal{P}(x),\\
&\mathcal{Q}(x) + \mathcal{P}(x) + P' = Q' - \mathcal{P}(x) + P(x) + P' = \operatorname{Id},\\
&\mathcal{P}(x)P' = T(x)PT(x)^{-1}(\operatorname{Id} - P')P' = 0,\\
&P'\mathcal{P}(x) = P'T(x)PT(x)^{-1}(\operatorname{Id} - P') = 0.
\end{aligned}$$

So $\mathcal{P}(x) + P'$ is a projection, whence $\mathcal{Q}(x) = \operatorname{Id} - (\mathcal{P}(x) + P')$ is a projection too.

Define

$$H = -H_0P' + H_0^{-1}P.$$

Then, since $P'P = PP' = 0$, one has $P'H_0 = PH_0^{-1} = 0$ and

$$\begin{aligned}
H^2 &= (-H_0P' + H_0^{-1}P)(-H_0P' + H_0^{-1}P) = -(P' + P),\\
HP'H &= (-H_0P' + H_0^{-1}P)P'(-H_0P' + H_0^{-1}P)\\
&= -H_0P'H_0^{-1}P = -H_0H_0^{-1}P = -P,\\
Q'HP &= HP - P'HP = H_0^{-1}P - H_0^{-1}P = 0,\\
Q'HT(x)^{-1}\mathcal{P}(x) &= Q'HT(x)^{-1}T(x)PT(x)^{-1}Q' = 0,\\
\mathcal{P}(x)T(x)HP' &= T(x)PT(x)^{-1}Q'T(x)HP'\\
&= T(x)PT(x)^{-1}(1 - P')T(x)(-H_0P')\\
&= T(x)PT(x)^{-1}(1 - P')T(x)P(-H_0P')\\
&= T(x)PT(x)^{-1}T(x)P(-H_0P') = T(x)P(-H_0P') = T(x)HP'.
\end{aligned}$$

Assume that

$$G(x) := HT(x)^{-1}\mathcal{P}(x) + T(x)HP'.$$

Then

$$\begin{aligned}
G(x)^2 &= (HT(x)^{-1}\mathcal{P}(x) + T(x)HP')(HT(x)^{-1}\mathcal{P}(x) + T(x)HP')\\
&= HT(x)^{-1}\mathcal{P}(x)HT(x)^{-1}\mathcal{P}(x) + T(x)(-P)T(x)^{-1}\mathcal{P}(x)\\
&\quad + HT(x)^{-1}\mathcal{P}(x)T(x)HP' + T(x)HP'T(x)HP'\\
&= HT(x)^{-1}T(x)PT(x)^{-1}Q'HT(x)^{-1}\mathcal{P}(x)\\
&\quad + T(x)(-P)T(x)^{-1}T(x)PT(x)^{-1}Q'\\
&\quad + HT(x)^{-1}T(x)HP' + T(x)HP'T(x)HP'\\
&= 0 - T(x)PT(x)^{-1}Q' - P' + T(x)(-H_0P')T(x)(-H_0P')\\
&= 0 - \mathcal{P}(x) - P' + 0 = -(P' + \mathcal{P}(x)),\\
G(x)\mathcal{Q}(x) &= 0,
\end{aligned}$$

$$\begin{aligned}\mathcal{Q}(x)G(x) &= (Q' - \mathcal{P}(x))(HT(x)^{-1}\mathcal{P}(x) + T(x)HP') \\ &= Q'HT(x)^{-1}\mathcal{P}(x) + (\mathrm{Id} - P')T(x)(-PH_0P') \\ &\quad - \mathcal{P}(x)HT(x)^{-1}\mathcal{P}(x) - \mathcal{P}(x)T(x)HP' \\ &= 0 + T(x)HP' - (T(x)PT(x)^{-1}Q')(-H_0P' \\ &\quad + H_0^{-1}P)T(x)^{-1}(T(x)PT(x)^{-1}Q') - T(x)HP' \\ &= (T(x)PT(x)^{-1}Q'H_0[P'P]T(x)^{-1}Q') \\ &\quad - T(x)PT(x)^{-1}[Q'P']H_0^{-1}PT(x)^{-1}(T(x)PT(x)^{-1}Q') = 0.\end{aligned}$$

Hence, for

$$U(s,x) := \mathcal{Q}(x) + (1-s)(\mathcal{P}(x) + P') + sG(x),$$

we obtain that

$$U(s,x)^{-1} = \mathcal{Q}(x) + \frac{1}{s^2 + (1-s)^2}[(1-s)(\mathcal{P}(x) + P') - sG(x)].$$

Therefore $U(s,x)T(x)$ defines a homotopy in $\mathcal{G}$:

$$U(0,x)T(x) = \mathrm{Id} \circ T(x) \sim U(1,x) \circ T(x).$$

Thus, since $\mathcal{P}(x)T(x)P = T(x)P$, one has

$$\begin{aligned}U(1,x)T(x)P &= \mathcal{Q}(x)\mathcal{P}(x)T(x)P + G(x)\mathcal{P}(x)T(x)P \\ &= 0 + HT(x)^{-1}\mathcal{P}(x)T(x)P = HT(x)^{-1}T(x)P = HP.\end{aligned}$$

Since $H(P + P') = (P + P')H = H$, for

$$V(s) := QQ' + (1-s)(P + P') - sH,$$

we have

$$V(s)^{-1} = QQ' + \frac{1}{s^2 + (1-s)^2}[(1-s)(P + P') + sH].$$

Besides, $V(0) = QQ' + P + P' = \mathrm{Id}$. Therefore the following homotopy is defined:

$$R(x) := V(1)\,U(1,x)\,T(x) \sim U(1,x)\,T(x) \quad \text{in} \quad C(X, \mathcal{G}(\mathcal{M}))$$

and

$$\begin{aligned}R(x)P &= V(1)\,U(1,x)\,T(x)P \\ &= V(1)\,HP = QQ'HP - H^2P = 0 + (P + P')\,P = P.\end{aligned}$$

Put

$$R(s,x) := R(x) - sPR(x)Q.$$

For some $e \in \mathcal{M}$, let the equality $R(s,x)e = 0$ hold. Then

$$0 = R(s,x)e = R(x)(P + Q)e - sPR(x)Qe = Pe + R(x)Qe - sPR(x)Qe,$$

$$0 = QR(s,x)e = QR(x)Qe.$$

Let $f = PR(x)Qe$, whence $f = Pf$. Then

$$PR(x)(Qe - Pf) = f - Pf = 0, \qquad QR(x)Pf = 0.$$

Therefore $R(x)(Qe - Pf) = 0$, $Qe = Pf = f = 0$ and $PR(s,x)e = Pe = 0$, $e = 0$. One also has

$$R(x)\mathcal{M} = \mathcal{M}, \quad R(x)P = P, \quad QR(x)Q\mathcal{M} = QR(x)(1-P)\mathcal{M} = QR(x)\mathcal{M} = Q\mathcal{M}.$$

Therefore, with respect to the decomposition $\mathcal{M} = P\mathcal{M}\widetilde{\oplus}Q\mathcal{M}$ the operator $R(s,x)$ has the matrix

$$\begin{pmatrix} \mathrm{Id} & \star \\ 0 & QR(x)Q \end{pmatrix}, \qquad QR(x)Q\mathcal{M} = Q\mathcal{M}.$$

Hence, $R(s,x)$ is an epimorphism and $R(s,x)$ belongs to $\mathcal{G}(\mathcal{M})$ as an epimorphism without kernel. It remains to put $D(x) := R(1,x)$. □

LEMMA 7.7.2. *Let $\mathcal{M}$ be a Hilbert A-module, X a compact set, $T : X \to \mathcal{G}(\mathcal{M})$ a continuous map with $0 < \varepsilon < \min \|T(x)^{-1}\|^{-1}$ and P and P' projections from $\mathcal{E} = \mathcal{E}(\mathcal{M})$ such that*

$$\|P'T(x)P\| \leq \varepsilon \quad \forall x \in X.$$

Then there exists a homotopy $S(s,x)$ in $\mathcal{G}$ such that

$$S(0,x) = T(x), \qquad P'S(1,x)P = 0 \quad \forall x \in X.$$

PROOF. Put $S(s,x) := T(x) - sP'T(x)P$. Since

$$\|S(s,x) - T(x)\| \leq \varepsilon,$$

one has $S(s,x) \in \mathcal{G}(\mathcal{M})$. □

CHAPTER 8

Hilbert Modules and KK-Theory

In this chapter we formulate some definitions and theorems related to the Kasparov bivariant KK-functor. The details and the discussion of different applications can be found, e.g., in the original paper of Kasparov [**64**] and also in [**65, 119, 28, 55, 120, 3, 132, 24, 51, 114, 8, 9, 98**]. A more detailed bibliography is contained in [**39**]. Besides the original approach of Kasparov, we also discuss the Cuntz approach [**28, 29**], which is very useful in several situations. We briefly present some generalizations of Kapsarov A, B-bimodules (triples) and discuss some applications of this approach. In our presentation, we mainly follow [**65, 120, 55, 3**]. As before, we deal with complex C^*-algebras, although sometimes (cf. [**113**]) the real approach is more useful [**117**].

Let us emphasize that starting at some point we shall be forced to restrict ourselves to the case of separable C^*-algebras.

An application of the Kuiper type Theorem 7.2.2 to the construction of classifying spaces for $K(X;A)$ by means of A-Fredholm operators is discussed at the end of the chapter. We consider the simplest case. (For more general situation see [**126, 121**].) Relation to $KK(\mathbf{C}, C(X) \otimes A)$ is indicated (details can be found in [**30, 89, 134**]).

8.1. Tensor products

We need the following notions of tensor products of Hilbert C^*-modules.

DEFINITION 8.1.1. Let B_i be C^*-algebras and let $\mathcal{M}_i$ be Hilbert B_i-modules, $i = 1, 2$. Let $B_1 \otimes B_2$ be the C^*-tensor product (cf. Remark 5.6.3). The *exterior tensor product of modules* $\mathcal{M}_1 \otimes \mathcal{M}_2$ is the completion of the algebraic tensor product over $\mathbf{C}$ (which is a $B_1 \otimes B_2$-module) with respect to the norm defined by the $B_1 \otimes B_2$-valued inner product

$$\langle e_1 \otimes e_2,\, f_1 \otimes f_2 \rangle = \langle e_1,\, e_2 \rangle \otimes \langle f_1,\, f_2 \rangle, \quad e_i, f_i \in \mathcal{M}_i.$$

DEFINITION 8.1.2. Let A and B be two C^*-algebras, $\mathcal{M}_1$ and $\mathcal{M}_2$ Hilbert C^*-modules over A and B, respectively, and $\rho : A \to \operatorname{End}^*_B(\mathcal{M}_2)$ a *-homomorphism. Consider the degenerate B-valued inner product

$$\langle e_1 \otimes e_2,\, f_1 \otimes f_2 \rangle = \langle e_2,\, \rho(\langle e_1,\, f_1 \rangle)\, f_2 \rangle, \quad e_i, f_i \in \mathcal{M}_i. \tag{8.1}$$

Let N be the submodule of isotropic vectors in the algebraic tensor product $(\mathcal{M}_1 \otimes_{\mathrm{alg}} \mathcal{M}_2)$. Then the completion of the quotient space $(\mathcal{M}_1 \otimes_{\mathrm{alg}} \mathcal{M}_2)/N$ with respect to the product (8.1) equipped with the obvious Hilbert C^*-module structure is called the *interior tensor product* $\mathcal{M}_1 \otimes_\rho \mathcal{M}_2$. The action of B is defined by the formula $(e_1 \otimes e_2) \cdot b = e_1 \otimes e_2 \cdot b$; also, $e_1 \otimes \rho(a)\, e_2 = e_1 \cdot a \otimes e_2$.

8.2. Main definitions

REMARK 8.2.1. To explain the motivation for the conditions in the definition of KK-groups, note that this definition is close to the definition of Atiyah's functor *Ell* or, more precisely, to the idea of representing elements of K-homology by elliptic operators. An elliptic pseudodifferential operator acting in Sobolev spaces on a manifold can be distinguished among all bounded operators in some abstract Hilbert space by two features: it is Fredholm and it almost commutes with the operators of multiplication by functions. Evidently, to state this condition, the Hilbert space should be equipped with an appropriate representation of the algebra of functions on the manifold. Passing to a noncommutative algebra A instead of the algebra of functions and to a Hilbert B-module instead of a Hilbert space, we obtain the prototype for the notion of KK-groups and a motivation for the following definition.

DEFINITION 8.2.2. Let A and B be C^*-algebras. An A, B-*bimodule* is a pair $(\mathcal{M}, \pi)$, where $\mathcal{M}$ is a $\mathbf{Z}/2\mathbf{Z}$-graded Hilbert B-module equipped with an action of the C^*-algebra A given by a $*$-homomorphism $\pi : A \to \mathrm{End}^*(\mathcal{M})$, where the operator $\pi(a)$ is homogeneous of degree 0 for any $a \in A$: $\pi(A) \subset \mathrm{End}^*(\mathcal{M})^{(0)}$. Denote by $E(A, B)$ the set of all triples (Kasparov bimodules) $(\mathcal{M}, \pi, F)$, where $(\mathcal{M}, \pi)$ is an A, B-bimodule and the operator $F \in \mathrm{End}^*(\mathcal{M})$ is homogeneous of degree 0, such that the conditions

(i) $\pi(a)(F^2 - 1) \in \mathcal{K}(\mathcal{M})$,
(ii) the commutator $[\pi(a), F] \in \mathcal{K}(\mathcal{M})$

hold for each $a \in A$. A triple $(\mathcal{M}, \pi, F)$ is called *degenerate* if the conditions

(i) $\pi(a)(F^2 - 1) = 0$,
(ii) $[\pi(a), F] = 0$

hold for each $a \in A$. The set of all degenerate triples is denoted by $D(A, B)$. A *homotopy* in $E(A, B)$ is an element of $E(A, B[0, 1])$, where, as before, $B[0, 1]$ denotes the C^*-algebra of all continuous B-valued functions on the unit interval. Define addition on $E(A, B)$ by setting

$$(\mathcal{M}, \pi, F) + (\mathcal{M}', \pi', F') = (\mathcal{M} \oplus \mathcal{M}', \pi \oplus \pi', F \oplus F').$$

Define $KK(A, B)$ to be the set of homotopy classes of elements of $E(A, B)$.

This set turns out to be an Abelian group. It is easy to verify that any degenerate triple is homotopic to the triple $(0, 0, 0)$ and that the inverse element for $(\mathcal{M}, \pi, F)$ equals $(-\mathcal{M}, \pi, -F)$, where $-\mathcal{M}$ denotes $\mathcal{M}$ equipped with the inverse grading.

REMARK 8.2.3. A natural question arises: why do we consider operators from $\mathrm{End}^*(\mathcal{M})$, and not from $\mathrm{End}(\mathcal{M})$? The answer has two sides: on one hand, it is convenient to deal with adjointable operators, on the other hand, any pseudodifferential operator where the symbol is a homomorphism of bundles of projective modules over a "nice" C^*-algebra is adjointable. The details can be found in [**121**]. A "nice" C^*-algebra A means here that $\mathbf{LM}(A) = \mathbf{M}(A)$.

In the sequel we assume C^*-algebras to be *separable* and Hilbert C^*-modules to be *countably generated*. In some cases the first assumption, with a lot of difficulties, can be replaced by a weaker one, but this is not important for us here since only separable algebras are used in applications: the (reduced) group C^*-algebras of finitely presented groups (fundamental groups of manifolds) and commutative C^*-algebras of continuous functions on some "nice" spaces.

For a $*$-homomorphism $f : A_1 \to A_2$, define the map

$$f^* : E(A_2, B) \to E(A_1, B), \quad (\mathcal{M}, \pi, f) \mapsto (\mathcal{M}, \pi \circ f, F)$$

and, for a $*$-homomorphism $g : B_1 \to B_2$, define the map

$$g_* : E(A, B_1) \to E(A, B_2), \quad (\mathcal{M}, \pi, f) \mapsto (\mathcal{M} \otimes_g B_2, \pi \otimes 1, F \otimes 1),$$

where the tensor product is understood in the sense of Definition 8.1.2 and where $\pi \otimes 1$ is defined by the formulas

$$\pi \otimes 1 : A \to \operatorname{End}^*(\mathcal{M} \otimes_g B_2), \quad (\pi \otimes 1)(a)(e \otimes b) = \pi(a)e \otimes b.$$

By definition, this immediately implies the following statement.

THEOREM 8.2.4 ([**64**]). *The groups $KK(A, B)$ define a homotopy invariant bifunctor from the category of separable C^*-algebras to the category of Abelian groups, which is covariant in B and contravariant in A. One has $KK(\mathbf{C}, B) = K_0(B)$.*

To define the higher KK-groups, one can either use representations of Clifford algebras or one can set

$$KK_n(A, B) = KK(A, B(\mathbf{R}^n)).$$

We require the following definition in order to introduce the basic technical tool of KK-theory, namely, the Kasparov product.

DEFINITION 8.2.5. Define by $1_A \in KK(A, A)$ the class of the triple $(A, i_A, 0)$, where $A^{(1)} = A$, $A^{(0)} = 0$;

$$i_A : A \to \mathcal{K}(A) \subset \operatorname{End}^*(A), \quad i_A(a)b = ab, \quad a, b \in A.$$

Define also the map

$$\tau_D : KK(A,\, B) \to KK(A \otimes D,\, B \otimes D),$$
$$\tau_D(\operatorname{class}(\mathcal{M},\, \pi,\, F)) = \operatorname{class}(\mathcal{M} \otimes D,\, \pi \otimes 1_D,\, F \otimes 1).$$

THEOREM 8.2.6 ([**64**]). *There exists a well-defined pairing (Kasparov product)*

$$KK(A, D) \times KK(D, B) \to KK(A, B),$$

denoted by $(x, y) \mapsto x \otimes_D y$, which enjoys the following properties:

(i) *the pairing is covariant in B and contravariant in A;*

(ii) *if $f : D \to E$ is a $*$-homomorphism, then*

$$f_*(x) \otimes_E y = x \otimes_D f^*(y), \qquad x \in KK(A, D), \quad y \in KK(E, B)\,;$$

(iii) *the product is associative:*

$$(x \otimes_D y) \otimes_E z = x \otimes_D (y \otimes_E z)$$

$\forall\, x \in KK(A, D),\ y \in KK(D, E),\ z \in KK(E, B)$;

(iv) $\tau_E(x \otimes_B y) = \tau_E(x) \otimes_{B \otimes E} \tau_E(y) \quad \forall\, x \in KK(A, B), \quad \forall\, y \in KK(B, D)$;

(v) $x \otimes_B 1_B = 1_A \otimes x = x \qquad \forall\, x \in KK(A, B)$.

Due to these properties, the product can be generalized to a more general setting.

DEFINITION 8.2.7. For $x \in KK(A_1, B_1 \otimes D)$ and $y \in KK(D \otimes A_2, B_2)$ put

$$x \otimes_D y := \tau_{A_2}(x) \otimes_{B_1 \otimes D \otimes A_2} \tau_{B_1}(y) \in KK(A_1 \otimes A_2, B_1 \otimes B_2).$$

The original proof of this theorem and even the construction of the product itself is very complicated. That is why we will pass now to the presentation of the main aspects of the Cuntz approach to KK-theory [**28**]. This approach allows us to consider elements of $KK(A,B)$ as generalized homomorphisms from A to B and to define the product as their composition.

8.3. Cuntz's approach

First of all, let us note that for each class in $KK(A,B)$, one can choose a representing triple $(\mathcal{M},\pi,F)$ such that $F^2=1$. Indeed, for the opposite B-module $(-\mathcal{M})$, define the map

$$\pi' : A \to \mathcal{M}\oplus(-\mathcal{M}), \qquad \pi'(a)=\pi(a)\oplus 0,$$

and define $F'\in \mathrm{End}^*(\mathcal{M}\oplus(-\mathcal{M}))$ by the formula

$$F'(\xi,\zeta)=(F\xi+U^*\zeta, U(1-F^2)\xi-UFU^*\zeta),$$

$$F'=\begin{pmatrix} F & U^* \\ U(1-F^2) & -UFU^* \end{pmatrix},$$

where $U:\mathcal{M}\to(-\mathcal{M})$, $\deg U=1$ is the canonical identification. Then it is easy to check that the class of $(\mathcal{M}\oplus(-\mathcal{M}),\pi',F')$ coincides with the class of $(\mathcal{M},\pi,F)$ and that $F'^2=1$.

Moreover, for a class of $(\mathcal{M},\pi,F)$ with $F^2=1$ one can assume that $F=F^*$. Indeed, put $F'=F(FF^*)^{1/2}$. Since $FF^*=(F^*F)^{-1}$, one has $(FF^*)^{1/2}=(F^*F)^{-1/2}$ and the operator F' is invertible. Since

$$F'F'^*=F(FF^*)F^*=1,$$

the operator F' is unitary. Then, for any polynomial f, we have $Ff(F^*F)=f(FF^*)F$. Using approximation of the function $x\mapsto x^{-1/2}$ by polynomials on the spectrum of F^*F, we obtain the equality

$$F'=F(F^*F)^{-1/2}=(F^*F)^{-1/2}F=F'^{-1}=F'^*.$$

Under this assumption, $F_t=F(FF^*)^{t/2}$, $t\in[0,1]$, defines an *operator homotopy* between the triples $(\mathcal{M},\pi,F)$ and $(\mathcal{M},\pi,F')$. Namely, the bimodule in an operator homotopy is fixed and a family of operators F_t is norm continuous. This is a stronger equivalence relation on $E(A,B)$ than homotopy, but, nevertheless, one can define KK using operator homotopy.

So, we assume that for $(\mathcal{M},\pi,F)\in E(A,B)$, the equality $F=F^*=F^{-1}$ holds. In this case one can identify $\mathcal{M}^{(0)}$ and $\mathcal{M}^{(1)}$ using F. Thus, we arrive to a description of a representative of an element of $KK(A,B)$ as a trivially graded Hilbert B-module equipped with two representations, π^+ and π^-, of the C^*-algebra A:

$$\pi^+(a)=\pi(a)|_{\mathcal{M}^{(0)}}, \qquad \pi^-(a)=F\pi(a)F|_{\mathcal{M}^{(1)}}.$$

The Kasparov stabilization Theorem 1.4.2 allows us to restrict ourselves only to the module $H_B=l_2(B)$.

DEFINITION 8.3.1. Let A and B be C^*-algebras. A *quasi-homomorphism* from A to B is a couple of $*$-homomorphisms $\pi^\pm : A\to \mathrm{End}^*(H_B)$ such that

$$\pi^+(a)-\pi^-(a)\in B\otimes\mathcal{K}=\mathcal{K}(H_B)$$

holds for any $a\in A$, where $\mathcal{K}$ is the algebra of compact operators on a separable Hilbert space.

So, $KK(A,B)$ is the set of homotopy classes of quasi-homomorphisms from A to B [**27**].

For further constructions we need the C^*-algebra qA defined by using free products of C^*-algebras. Such products can be defined in two ways. We shall indicate one of them and present in detail another one. The first approach is the following. Let A and B be C^*-algebras and let $A *_{\mathrm{alg}} B$ be their algebraic free product. It is a $*$-algebra equipped with two canonical $*$-homomorphisms

$$j_A : A \to A *_{\mathrm{alg}} B, \qquad j_B : B \to A *_{\mathrm{alg}} B.$$

For any C^*-algebra D and for any pair (φ,ψ) of $*$-homomorphisms $\varphi : A \to D$, $\psi : B \to D$, there exists a unique $*$-homomorphism $(\varphi,\psi) : A *_{\mathrm{alg}} B \to D$ such that

$$(\varphi,\psi)\circ j_A = \varphi, \qquad (\varphi,\psi)\circ j_B = \psi.$$

Let

$$\|x\| = \sup_{D,\,(\varphi,\psi)} \|(\varphi,\psi)(x)\|, \qquad x \in A *_{\mathrm{alg}} B.$$

The corresponding completion of the algebraic free product is a C^*-algebra. This algebra is denoted by $A * B$ and is called the *C^*-free product* of A and B.

This free product can be defined alternatively in the following way (see [**55**, Appendix B] and [**21**]). For each $n = 1, 2, \ldots$, consider the following algebraic tensor products:

$$\begin{aligned}
A_n &= A \otimes_{\mathbf{C}} B \otimes_{\mathbf{C}} A \otimes_{\mathbf{C}} B \otimes_{C} A \otimes_{C} \cdots \otimes_{\mathbf{C}} B, && n \text{ tensor factors}, n \text{ even},\\
A_n &= A \otimes_{\mathbf{C}} B \otimes_{\mathbf{C}} A \otimes_{\mathbf{C}} B \otimes_{C} A \otimes_{C} \cdots \otimes_{\mathbf{C}} A, && n \text{ tensor factors}, n \text{ odd},\\
B_n &= B \otimes_{\mathbf{C}} A \otimes_{\mathbf{C}} B \otimes_{C} A \otimes_{\mathbf{C}} B \otimes_{C} \cdots \otimes_{\mathbf{C}} A, && n \text{ tensor factors}, n \text{ even},\\
B_n &= B \otimes_{\mathbf{C}} A \otimes_{\mathbf{C}} B \otimes_{C} A \otimes_{\mathbf{C}} B \otimes_{C} \cdots \otimes_{\mathbf{C}} B, && n \text{ tensor factors}, n \text{ odd}.
\end{aligned}$$

Consider the following direct sum of vector spaces:

$$C_0 := \bigoplus_{n=1}^{\infty} (A_n \oplus B_n).$$

For the purpose of obtaining multiplication in C_0, define the bilinear maps

$$\begin{array}{lll}
A_n \times B_m \to A_{n+m-1}, & n \text{ even}, & m \text{ even},\\
A_n \times B_m \to A_{n+m}, & n \text{ odd}, & m \text{ even},\\
B_n \times A_m \to B_{n+m-1}, & n \text{ even}, & m \text{ even},\\
B_n \times A_m \to B_{n+m}, & n \text{ odd}, & m \text{ even},\\
A_n \times B_m \to A_{n+m-1}, & n \text{ even}, & m \text{ odd},\\
A_n \times B_m \to A_{n+m}, & n \text{ odd}, & m \text{ odd},\\
B_n \times A_m \to B_{n+m-1}, & n \text{ even}, & m \text{ odd},\\
B_n \times A_m \to B_{n+m}, & n \text{ odd}, & m \text{ odd},\\
A_n \times A_m \to A_{n+m-1}, & n \text{ even}, & m \text{ any},\\
A_n \times A_m \to A_{n+m}, & n \text{ odd}, & m \text{ any},\\
B_n \times B_m \to B_{n+m-1}, & n \text{ even}, & m \text{ any},\\
B_n \times B_m \to B_{n+m}, & n \text{ odd}, & m \text{ any},
\end{array}$$

by setting, for the first row,

$$\begin{aligned}
&(a_1 \otimes b_2 \otimes a_3 \otimes \cdots \otimes b_n)\cdot(b_1 \otimes a_2 \otimes b_3 \otimes \cdots \otimes a_m)\\
&\qquad\qquad := (a_1 \otimes b_2 \otimes a_3 \otimes \cdots \otimes b_n b_1 \otimes a_2 \otimes b_3 \otimes \cdots \otimes a_m),
\end{aligned}$$

$$(a_1 \otimes b_2 \otimes a_3 \otimes \cdots \otimes a_n) \cdot (b_1 \otimes a_2 \otimes b_3 \otimes \cdots \otimes a_m)$$
$$:= (a_1 \otimes b_2 \otimes a_3 \otimes \cdots \otimes b_{n-1} \otimes a_n \otimes b_1 \otimes a_2 \otimes b_3 \otimes \cdots \otimes a_m).$$

The remaining maps are defined in a similar way. Define the bilinear map

$$\cdot : C_0 \times C_0 \to C_0$$

by the formula

$$(x_1, y_1, x_2, y_2, \dots) \cdot (x'_1, y'_1, x'_2, y'_2, \dots) := \sum_{i,j=1}^{n} (x_i \cdot x'_j + y_i \cdot x'_j + x_i \cdot y'_j + y_i \cdot y'_j),$$

where $x_i, x'_i \in A_i$, $y_i, y'_i \in B_i$. Then one can check that this bilinear map equips C_0 with the structure of an algebra. Define the involution $*$ on elementary tensors as the operation that transposes tensor factors in the inverse order and takes their adjoints in A or B. For example, $(a \otimes b)^* := b^* \otimes a^*$. Then, for an even n, one obtains maps $* : A_n \to B_n$, $* : B_n \to A_n$ and for an odd n, one obtains maps $* : A_n \to A_n$, $* : B_n \to B_n$. This turns C_0 into an involutive algebra with respect to the map

$$* : C_0 \to C_0,$$
$$(x_1, y_1, x_2, y_2, x_3, y_3, \dots)^* := (x_1^*, y_1^*, y_2^*, x_2^*, x_3^*, y_3^*, y_4^*, x_4^*, \dots),$$

where $x_i \in A_i$, $y_i \in B_i$.

Let $\varphi_A : A \to D$ and $\varphi_B : B \to D$ be two $*$-homomorphisms. Define the map $\varphi_n : A_n \to D$ on A_n for even n on elementary tensors by the formula

$$\varphi_n(a_1 \otimes b_1 \otimes a_2 \otimes b_2 \otimes \cdots \otimes a_{n/2} \otimes b_{n/2})$$
$$:= \varphi_A(a_1)\varphi_B(b_1)\varphi_A(a_2)\varphi_B(b_2) \cdots \varphi_A(a_{n/2})\varphi_B(b_{n/2}).$$

In the three remaining cases the map φ_n can be defined in a similar way. Then one can define the following $*$-homomorphism:

$$\overline{\varphi} : C_0 \to D, \quad \overline{\varphi}(x_1, y_1, x_2, y_2, \dots) := \sum_n (\varphi_n(x_n) + \varphi_n(y_n)), \quad x_n \in A_n,\ y_n \in B_n.$$

In particular, $*$-representations of C_0 on a Hilbert space H can be obtained in this way. Consider all representations $\pi : C_0 \to \mathcal{B}(H)$ (not necessarily of the mentioned form) and put

$$\|x\| := \sup_\pi \|\pi(x)\|, \qquad x \in C_0.$$

Each representation π generates $*$-representations of A and B:

$$\pi \circ i_A : A \to \mathcal{B}(H), \qquad i_A : A \cong A_1,$$
$$\pi \circ i_B : B \to \mathcal{B}(H), \qquad i_B : B \cong B_1.$$

Then one concludes that the norm $\|\cdot\|$ is finite on C_0 and that $\|i_A(a)\| = \|a\|$ and $\|i_B(b)\| = \|b\|$. Put $N := \{x \in C_0 \mid \|x\| = 0\}$. Then N is a two-sided $*$-ideal. Let C be a completion of the quotient algebra C_0/N with respect to the norm $\|\cdot\|$. One can easily verify that C is a C^*-algebra and, since i_A and i_B are isometries, the maps

$$j_A : A \to C, \quad a \mapsto i_A(a) + N, \qquad j_B : B \to C, \quad b \mapsto i_B(b) + N,$$

are isometric embeddings. Put $A * B := C$. The universal property, i.e., existence of the homomorphism (φ, ψ) (cf. the first approach to the definition), is now completely evident.

CONSTRUCTION 8.3.2. For a C^*-algebra A, let QA be the product $A * A$ and let $i^{\pm}$ denote the two canonical embeddings $A \to QA$, namely $i^+ := j_A$ and $i^- := j_B$ for $A = B$. Denote by qA the ideal of QA generated by all the elements of the form $\{q(a) := i^+(a) - i^-(a) | a \in A\}$. It is evident that

$$qA = \operatorname{Ker}\left\{\Delta = (\mathrm{Id}_A, \mathrm{Id}_A) : QA \to A\right\}$$

and we have the short exact sequence

$$0 \to qA \to QA \xrightarrow{\Delta} A \to 0,$$

where any of the two embeddings, i^+ and i^-, splits Δ. More precisely, this follows from the following statement.

LEMMA 8.3.3. *Elements of the form*

$$qa_1\, qa_2 \cdots qa_n \quad \text{and} \quad i^+(a_0)\, qa_1\, qa_2 \cdots qa_n, \qquad a_i \in A,$$

generate a dense involutive subalgebra C in qA and $qA = \operatorname{Ker}\Delta$.

PROOF. Since $(\mathrm{Id}_A, \mathrm{Id}_A)(qa) = 0$, one has $qA \subset \operatorname{Ker}\Delta$. It remains to show that C is dense in $\operatorname{Ker}\Delta$.

Denote by C_0 the linear span of elements of the same form as for C and also of elements of the form $i^+(a)$. Then C_0 is a dense involutive subalgebra in QA. Indeed,

$$\begin{aligned} q(ab) + q(a)q(b) &= i^+(ab) - i^-(ab) + (i^+(a) - i^-(a))(i^+(b) - i^-(b)) \\ &= i^+(ab) - i^-(ab) + i^+(ab) + i^-(ab) - i^+(a)i^-(b) - i^-(a)i^+(b) \\ &= 2i^+(ab) - i^+(a)i^-(b) - i^-(a)i^+(b), \end{aligned}$$

$$i^+(a)q(b) + q(a)i^+(b) = i^+(ab) - i^+(a)i^-(b) + i^+(ab) - i^-(a)i^+(b).$$

Hence,

$$q(a)i^+(b) = q(ab) + q(a)q(b) - i^+(a)q(b).$$

Since $i^- = i^+ - q$, we obtain, by induction, that $C_0 = A *_{\mathrm{alg}} A \subset QA$.

Suppose that $x \in \operatorname{Ker}\Delta \subset QA$ is an arbitrary element. Then $x = \lim_n x_n$, $x_n \in C_0$. Hence, there exists the limit

$$y = \lim_n \{x_n - i^+ \circ \Delta(x_n)\}, \qquad \{x_n - i^+ \circ \Delta(x_n)\} \in C,$$

$$\Delta\{x_n - i^+ \circ \Delta(x_n)\} = \Delta(x_n) - \Delta(x_n) = 0,$$

so $y \in \operatorname{Ker}\Delta$ and

$$0 = \Delta(x - y) = \lim_n \Delta \circ i^+ \circ \Delta(x_n) = \lim_n \Delta(x_n),$$

$$0 = \lim_n i^+ \circ \Delta(x_n), \qquad x = y.$$

□

Thus, any quasi-homomorphism $\pi^{\pm}$ from A to B defines a $*$-homomorphism $\pi = (\pi^+, \pi^-) : QA \to B$ satisfying the property $\pi(qA) \subset B \otimes \mathcal{K}$. Let $\varphi : qA \to B \otimes \mathcal{K}$ be the restriction of π. Conversely, using the Kasparov stabilization Theorem 1.4.2, one can show that each $*$-homomorphism $\varphi : qA \to B \otimes \mathcal{K}$ defines (up to a small deformation) a quasi-homomorphism from A to B. We obtain the following definition of the KK bifunctor due to Cuntz.

THEOREM 8.3.4 ([**27**]). *The set of homotopy classes of $*$-homomorphisms $qA \to B \otimes \mathcal{K}$ coincides with $KK(A, B)$.*

The details of the proof can also be found in [**55**, §5.2].

For an arbitrary C^*-algebra A denote by $p_A : qA \to A$ the restriction of $(\mathrm{Id}_A, 0) : QA \to A$ onto qA. Put $q^2A := q(qA)$.

Using the Pedersen derivation lifting theorem [**104**], one can prove the following result (see also [**55**, § 5.1]).

THEOREM 8.3.5 ([**28**]). *For any separable C^*-algebra A there exists a $*$-homomorphism $\varphi_A : qA \to M_2(q^2A)$ such that the operator $\varphi_A \circ p_{qA}$ is homotopic to the embedding j_2 of the algebra q^2A into $M_2(q^2A)$ in the class of $*$-homomorphisms and $(p_{qA} \otimes \mathrm{Id}_{M_2(\mathbf{C})}) \circ \varphi_A$ is homotopic to the embedding j_1 of the algebra qA into $M_2(qA)$ (for embeddings into two-by-two matrices the formula $x \mapsto \begin{pmatrix} x & 0 \\ 0 & 0 \end{pmatrix}$ is used).*

COROLLARY 8.3.6. *The map $p_B : qB \to B$ induces an isomorphism*

$$p_{B*} : KK(A, qB) \to KK(A, B).$$

PROOF. The inverse map for p_{B*} can be defined as follows. Let $\psi : qA \to B \otimes \mathcal{K}$ represent an element of $KK(A, B)$. It induces a $*$-homomorphism $q\psi : q^2A \to q(B \otimes \mathcal{K})$. Moreover, there exists a canonical $*$-homomorphism $\kappa : q(B \otimes \mathcal{K}) \to qB \otimes \mathcal{K}$. Then the desired element of $KK(A, B)$ is represented by the $*$-homomorphism

$$qA \xrightarrow{\varphi_A} M_2(q^2A) \xrightarrow{q\psi} M_2(q(B \otimes \mathcal{K})) \xrightarrow{\kappa} M_2(qB \otimes \mathcal{K}) \cong qB \otimes \mathcal{K}.$$

□

Since $\mathcal{K} \otimes \mathcal{K} \cong \mathcal{K}$, we obtain the following corollary.

COROLLARY 8.3.7 ([**28**]). *The set of homotopy classes of $*$-homomorphisms $qA \otimes \mathcal{K} \to qB \otimes \mathcal{K}$ coincides with $KK(A, B)$.*

After obtaining this identification, the Kasparov product can be defined to be the composition of $*$-homomorphisms.

8.4. Generalized Kasparov bimodules

In this section we shall briefly present some generalizations of the Kasparov A, B-bimodules (triples) and discuss some applications of this approach. We begin with a brief sketch of the construction of [**3**] (where the details can be found) based on unbounded operators in Hilbert C^*-modules and its application to an algebraic description of the Kasparov product.

DEFINITION 8.4.1. Let $\mathcal{E}$ be a $\mathbf{Z}/2\mathbf{Z}$-graded Hilbert C^*-module over a $\mathbf{Z}/2\mathbf{Z}$-graded C^*-algebra B and let $T : \mathrm{Dom}\, T \to \mathcal{E}$ be a densely defined operator on $\mathcal{E}$. The operator T is *regular* if

(i) T is closed;
(ii) the adjoint operator T^* is densely defined on $\mathcal{E}$;
(iii) $1 + T^*T$ has a dense image in $\mathcal{E}$.

Similarly to the case of bounded adjointable operators, T is B-linear in this case too. If $\mathcal{E} = B$, then regular operators on B are unbounded multipliers of B (see [**23**]). For any regular operator T, the operator $F := T(1+T^*T)^{-1/2}$ is bounded and adjointable and one has

$$T = FH^{-1}, \qquad T^* = F^*L^{-1}, \qquad F^* = T^*(1+TT^*)^{-1/2},$$

where $H := (1+T^*T)^{-1/2}$ and $L := (1+TT^*)^{-1/2}$.

DEFINITION 8.4.2. An *unbounded Kasparov A, B-bimodule* is a pair $(\mathcal{E}, D)$, where $\mathcal{E}$ is an A, B-bimodule (cf. Definition 8.2.2) and D is a regular operator of degree 1 satisfying

(i) $D = D^*$;
(ii) $\pi(a)(1+D^2)^{-1} \in \mathcal{K}(\mathcal{E})$ for any $a \in A$;
(iii) the set of all $a \in A$, such that the commutator $[D, a] = Da - (-1)^{\deg a} aD$ is defined on $\operatorname{Dom} D$ and can be extended up to a bounded adjointable operator, is dense in A.

Denote by $E_u(A, B)$ the set of all unbounded Kasparov A, B-bimodules.

With the help of a special approximate unit in $\mathcal{K}(\mathcal{E})$ the following proposition is proved in [**3**].

PROPOSITION 8.4.3. *Suppose that $(\mathcal{E}, D) \in E_u(A, B)$ and $F := D(1+D^2)^{-1/2}$. Then $(\mathcal{E}, F)$ is a Kasparov A, B-bimodule.*

The induced map $\beta : E_u(A, B) \to KK(A, B)$ is surjective.

Suppose that A_i is separable and B_i is σ-unital. One can verify that for $(\mathcal{E}_i, D_i) \in E_u(A_i, B_i)$, the operator $D_0 := D_1 \widehat{\otimes} 1 + 1 \widehat{\otimes} D_2$ admits a closure D, which is a regular operator on $\mathcal{E} := \mathcal{E}_1 \widehat{\otimes} \mathcal{E}_2$. Hence, a coupling

$$E_u(A_1, B_1) \times E_u(A_2, B_2) \to E_u(A_1 \widehat{\otimes} A_2, B_1 \widehat{\otimes} B_2)$$

is defined. It is proved in [**3**] that the diagram

$$\begin{array}{ccc} E_u(A_1, B_1) \times E_u(A_2, B_2) & \longrightarrow & E_u(A_1 \widehat{\otimes} A_2, B_1 \widehat{\otimes} B_2) \\ \Big\downarrow{\scriptstyle \beta \times \beta} & & \Big\downarrow{\scriptstyle \beta} \\ K(A_1, B_1) \times K(A_2, B_2) & \longrightarrow & K(A_1 \widehat{\otimes} A_2, B_1 \widehat{\otimes} B_2) \end{array}$$

commutes. This gives an algebraic description for the Kasparov product.

The approach based on unbounded modules turned out to be a useful tool and was used in a number of papers (see, e.g., [**51, 25, 26**]).

Among more recent results let us indicate the notion of a *Hilbert module with boundary* introduced by M. Hilsum, which allowed him to make progress in some topological problems [**52, 53**]. Unfortunately, these results (and other applications of the Hilbert C^*-module theory to elliptic theory) need an extended presentation of pseudodifferential calculus and other topics which are far from the main subject of our book.

8.5. Classifying spaces for some K- and KK-groups

Let A be a unital C^*-algebra. Let, as above, $\mathcal{P}(A)$ be the category of finitely generated projective A-modules. Recall the definition of the group $K(X; A)$ for a compact Hausdorff space X. The details and some generalizations can be found e.g. in [**121**].

DEFINITION 8.5.1. Consider the set of locally trivial bundles $E \to X$ with the fiber $P \in \mathcal{P}(A)$ and with the structure group of invertible continuous A-homomorphisms $P \to P$. We denote the symmetrization of the semigroup of the isomorphism classes of such bundles (with respect to the direct sum operation) by $K(X;A)$.

By H_A-bundles we mean locally trivial bundles with the fiber H_A and with the structure group $\mathrm{GL}^*(A)$.

LEMMA 8.5.2. *Each H_A-bundle over X is trivializable, i.e., isomorphic to $X \times H_A$.*

PROOF. This follows immediately from the contractibility of the structure group (Theorem 7.2.2). □

CONSTRUCTION 8.5.3. Denote by $\mathcal{F}_A^*$ the set of all adjointable A-Fredholm operators. Let

$$F \in C(X, \mathcal{F}_A^*(H_A)), \qquad F : x \mapsto F(x),$$

be a continuous family of Fredholm operators, $x \in X$. Then (cf. Theorem 2.7.6) we can choose an appropriate decomposition of H_A such that $\mathcal{M}_0^x = L_{n(x)}^\perp$, $\mathcal{N}_0^x = L_{n(x)}$. For each $x \in X$ there exists a neighborhood U_x such that for any $y \in U_x$, the map $F(y)|_{\mathcal{M}_0^x}$ is an isomorphism onto its image. Since X is compact, one can choose a finite subcovering $U_{x_1}, \dots, U_{x_s}$. Put $n := \min\{n(x_1), \dots, n(x_s)\}$, $\mathcal{N}_0 := L_n$ and $\mathcal{M}_0 := L_n^\perp$. Then $F(x)|_{\mathcal{M}_0}$ is an adjointable isomorphism onto its image for each x. Moreover, by Corollary 2.3.5, the module $F(x)(\mathcal{M}_0)$ has an orthogonal complement. This orthogonal complement (see the proof of Corollary 2.3.5) is defined by a continuous family of projections. Hence, it is locally trivial. Two subbundles arise in $X \times H_A$: $K_F := X \times \mathcal{N}_0$ and $C_F := (x, (F(x)\mathcal{M}_0)^\perp)$. Define the map

$$\operatorname{index} : C(X, \mathcal{F}_A^*(H_A)) \to K(X, A), \qquad F \mapsto [K_F] - [C_F],$$

where $[E]$ denotes the class of the bundle E in $K(X;A)$. Evidently, the result does not depend on the choice of n.

LEMMA 8.5.4. *The map* index *is invariant under small perturbations of F with respect to the uniform topology. Hence, it defines a map*

$$\operatorname{index} : [X, \mathcal{F}_A^*(H_A)] \to K(X, A).$$

PROOF. Indeed, since X is compact, for any F there exists some $\varepsilon > 0$ such that for any $G : X \to \mathcal{F}_A^*(H_A)$ satisfying the inequality $\|G - F\| < \varepsilon$, the submodule L_n still satisfies all restrictions and the projection onto $(x, (G(x)\mathcal{M}_0)^\perp)$ is close to the projection onto C_F. □

THEOREM 8.5.5 ([**126, 89, 30**]). *The map* index *defines an isomorphism*

$$\operatorname{index} : [X, \mathcal{F}_A^*(H_A)] \cong K(X, A).$$

PROOF. It is evident that this map is an epimorphism, since, for any bundles E and F, there exists a family of operators F_x making the following diagram commutative:

$$\begin{array}{ccc} (X \times H_A) \oplus E & \xrightarrow{\left(\begin{smallmatrix} 1 & 0 \\ 0 & 0 \end{smallmatrix}\right)} & (X \times H_A) \oplus G \\ \cong \uparrow & & \downarrow \cong \\ X \times H_A & \xrightarrow{\quad F_x \quad} & X \times H_A \end{array}$$

Here we have used Lemma 8.5.2. To prove injectivity, one has to construct a homotopy connecting an arbitrary map $F : X \to \mathcal{F}_0^*$ with $X \to \mathrm{Id}_{H_A}$, where $\mathcal{F}_0^*$ is the set of operators of index zero. For F, one has $[N_F] - [C_F] = 0$, i.e., after taking sums of these bundles and some trivial bundle $X \times A^k$, one obtains an isomorphism

$$J : N_F \oplus (X \times A^k) \cong C_F \oplus (X \times A^k),$$

where one can assume that $X \times A^k = F(X \times (L_{n+k} \ominus L_n))$ on the right-hand side. Hence, taking $n+k$ instead of n and writing J fiberwise as $J(x)$, define, for $t \in [0,1]$, the following family of maps:

$$F_t : X \to \mathrm{End}^*(H_A),$$

$$F_t(x)(h) := \begin{cases} F(x)(h) & \text{for } h \in (L_{n+k})^{\perp}, \\ [(1-t)F(x) + tJ(x)](h) & \text{for } h \in L_{n+k}. \end{cases}$$

Since the fiberwise perturbation is compact, the map takes values in $\mathcal{F}^*$ and the index is constant, hence, zero. For $t = 1$ one obtains a map to $\mathrm{GL}^*(A)$, which is homotopic (in the class of maps to $\mathrm{GL}^*(A)$) to a map into a single point, according to Theorem 7.2.2. □

REMARK 8.5.6. Using Theorem 7.2.2 it is proved in [**89, 30**] that

$$K(C(X) \otimes A) \cong [X, \mathcal{F}^*] \cong K(\mathbf{C}, C(X) \otimes A)$$

(the last identification in fact is by definition) Thus,

$$K(X; A) \cong K(C(X) \otimes A) \cong K(\mathbf{C}, C(X) \otimes A).$$

Bibliography

1. C. Akemann, G. Pedersen, and J. Tomiyama, *Multipliers of C^*-algebras*, J. Funct. Anal. **13** (1973), 277–301.
2. C. Anantharaman-Delaroche and J. F. Havet, *On approximate factorizations of completely positive maps*, J. Funct. Anal. **90** (1990), 411–428.
3. S. Baaj and P. Julg, *Théorie bivariante de Kasparov et opérateurs non bornés dans les C^*-modules hilbertiens*, C. R. Acad. Sci. Paris Sér. I **296** (1983), 875–878.
4. S. Baaj and G. Skandalis, *C^*-algèbres de Hopf et théorie de Kasparov équivariante*, K-Theory **2** (1989), 683–721.
5. M. Baillet, Y. Denizeau, and J.-F. Havet, *Indice d'une esperance conditionelle*, Compositio Math. **66** (1988), 199–236.
6. D. Bakić and B. Guljaš, *Extensions of Hilbert C^*-modules*, Houston J. Math. **30** (2004), 537–558.
7. A. O. Barut and R. Rączka, *Theory of group representations and applications*, PWN–Polish Scientific Publishers, Warszawa, 1977.
8. P. Baum and A. Connes, *Chern character for discrete groups*, A Fête of Topology, Academic Press, Boston, MA, 1988, pp. 163–232.
9. P. Baum, A. Connes, and N. Higson, *Classifying space for proper actions and K-theory of group C^*-algebras*, C^*-Algebras: 1943-1993, A Fifty Year Celebration, Contemporary Math., vol. 167, Amer. Math. Soc., Providence, RI, 1994, pp. 241–291.
10. B. Blackadar, *K-theory for operator algebras*, Springer-Verlag, New York, 1986.
11. D. P. Blecher, *A generalization of Hilbert modules*, J. Funct. Anal. **136** (1996), 365–421.
12. ———, *A new approach to Hilbert C^*-modules*, Math. Ann. **307** (1997), 253–290.
13. O. Bratteli and D. W. Robinson, *Operator algebras and quantum statistical mechanics.* I, Springer-Verlag, New York – Berlin – Heidelberg, 1981.
14. B. Brenken, *Representations and automorphisms of the irrational rotation algebra*, Pacific J. Math. **111** (1984), 257–282.
15. L. G. Brown, *Stable isomorphism of hereditary subalgebras of C^*-algebras*, Pacific J. Math. **71** (1977), 335–348.
16. ———, *Stable isomorphism of hereditary subalgebras of C^*-algebras*, Pacific J. Math. **71** (1977), 335–348.
17. ———, *Close hereditary C^*-subalgebras and the structure of quasi-multipliers*, Preprint MSRI 11211-85, 1985.
18. L. G. Brown, P. Green, and M. A. Rieffel, *Stable isomorphism and strong Morita equivalence of C^*-algebras*, Pacific J. Math. **71** (1977), 349–363.
19. L. G. Brown and G. K. Pedersen, *C^*-algebras of real rank zero*, J. Funct. Anal. **99** (1991), 131–149.
20. M.-D. Choi and G. A. Elliot, *Density of self-adjoint elements with finite spectrum in an irrational rotation C^*-algebra*, Math. Scand. **67** (1990), 73–86.
21. E. Christensen, E. Effros, and A. Sinclair, *Completely bounded multilinear maps and C^*-algebraic cohomology*, Invent. Math. **90** (1987), 279–296.
22. A. Connes, *C^*-algèbres et géométrie différentielle*, C. R. Acad. Sci. Paris Sér. I **290** (1980), 599–604.
23. ———, *An analogue of the Thom isomorphism for crossed products of a C^*-algebra by an action of* $\mathbf{R}$, Adv. in Math. **39** (1981), 31–55.
24. ———, *Noncommutative geometry*, Academic Press, San Diego, CA, 1994.

25. A. Connes and N. Higson, *Déformations, morphismes asymptotiques et K-théorie bivariante*, C. R. Acad. Sci. Paris Sér. I **311** (1990), 101–106.
26. A. Connes and H. Moscovici, *Cyclic cohomology, the Novikov conjecture and hyperbolic groups*, Topology **29** (1990), no. 3, 345–388.
27. J. Cuntz, *Generalized homomorphisms between C^*-algebras and KK-theory*, Dynamics and Processes, Lecture Notes in Math., vol. 1031, Springer-Verlag, 1983, pp. 31–45.
28. ———, *A new look at KK-theory*, K-Theory **1** (1987), 31–51.
29. ———, *A survey on some aspects of non-commutative geometry*, Jahresber. Deutsch. Math.-Verein. **95** (1993), 60–84.
30. J. Cuntz and N. Higson, *Kuiper's theorem for Hilbert modules*, Operator Algebras and Mathematical Physics, Contemporary Math., vol. 62, Amer. Math. Soc., Providence, RI, 1987, pp. 429–435.
31. J. Dixmier, *Les C^*-algèbres et leurs représentations*, Gauthier-Villars, Paris, 1967.
32. ———, *Les algèbres d'operateurs dans l'espace Hilbertien*, Gauthier-Villars, Paris, 1969.
33. J. Dixmier and A. Douady, *Champs continus d'espaces Hilbertiens*, Bull. Soc. Math. France **91** (1963), 227–284.
34. R. G. Douglas and C. Pearcy, *On the spectral theorem for normal operators*, Proc. Cambridge Philos. Soc. **68** (1970), 393–400.
35. N. Dunford and J. T. Schwartz, *Linear operators. I. General theory*, Interscience, New York, 1958.
36. M. J. Dupré and P. A. Fillmore, *Triviality theorems for Hilbert modules*, Topics in Modern Operator Theory, 5th International Conference on Operator Theory (Timisoara and Herculane (Romania), June 2–12, 1980), Birkhäuser Verlag, Basel – Boston – Stuttgart, 1981, pp. 71–79.
37. G. A. Elliott, T. Natsume, and R. Nest, *The Heisenberg group and K-theory*, K-Theory **7** (1993), 409–428.
38. G. A. Elliott, K. Saitô, and J. D. M. Wright, *Embedding AW^*-algebras as double commutants in type I algebras*, J. London Math. Soc. **28** (1983), 376–384.
39. S. C. Ferry, A. Ranicki, and J. Rosenberg (eds.), *Novikov conjectures, index theorems and rigidity*, Vols. 1, 2, London Math. Soc. Lecture Note Ser., vols. 226, 227, Cambridge Univ. Press, 1995.
40. O. G. Filippov, *On C^*-algebras A over which the Hilbert module $l_2(A)$ is self-dual*, Vestnik Moskov. Univ. Ser. I Mat. Mekh. (1987), no. 4, 74–76; English transl., Moscow Univ. Math. Bull. **42** (1987), no. 4, 87–90.
41. M. Frank, *A set of maps from K to $\mathrm{End}_a(l_2(a))$ isomorphic to $\mathrm{End}_{a(k)}(l_2(a(k)))$. Applications*, Ann. Global Anal. Geom. **3** (1985), no. 2, 155–171.
42. ———, *Self-duality and C^*-reflexivity of Hilbert C^*-modules*, Z. Anal. Anwendungen **9** (1990), 165–176.
43. ———, *Hilbert C^*-modules and related subjects — a guided reference overview*, Leipzig University. ZHS-NTZ preprint No 13, 1996, `funct-anbabbage.sissa.it`, preprint # 9605003.
44. ———, *Geometrical aspects of Hilbert C^*-modules*, Positivity **3** (1999), 215–243.
45. M. Frank and E. Kirchberg, *On conditional expectations of finite index*, J. Operator Theory **40** (1998), no. 1, 87–111.
46. M. Frank and V. M. Manuilov, *Diagonalizing "compact" operators on Hilbert W^*-modules*, Z. Anal. Anwendungen **14** (1995), 33–41.
47. M. Frank, V. M. Manuilov, and E. V. Troitsky, *On conditional expectations arising from group actions*, Z. Anal. Anwendungen **16** (1997), 831–850.
48. M. Frank and E. V. Troitsky, *Lefschetz numbers and geometry of operators in W^*-modules*, Funktsional Anal. i Prilozhen. **30** (1996), no. 4, 45–57; English transl., Functional Anal. Appl. **30** (1996), 257–266.
49. P. Griffiths and J. Harris, *Principles of algebraic geometry*. Vol. 1, Wiley-Interscience, New York, 1978.
50. K. Grove and G. K. Pedersen, *Diagonalizing matrices over $C(X)$*, J. Funct. Anal. **59** (1984), 64–89.
51. M. Hilsum, *Signature operator on Lipschitz manifolds and unbounded Kasparov bimodules*, Operator Algebras and their Connections with Topology and Ergodic Theory (H. Araki, C. C. Moore, Ş. Stratila, and D. Voiculescu, eds.), Lecture Notes in Math., vol. 1132, Springer–Verlag, 1985, pp. 254–288.

52. ———, *Index classes of Hilbert modules with boundary*, Preprint 2001-03 Institut de mathématiques de Jussieu, 2001.
53. ———, *Hilbert modules of foliated manifolds with boundary*, Foliations: Geometry and Dynamics (Walczak, Pawel et al., eds.), Proceedings of the Euroworkshop (Warsaw, Poland, May 29-June 9, 2000), World Scientific, Singapore, 2002, pp. 315-332.
54. A. Irmatov, *On a new topology in the space of Fredholm operators*, Ann. Global Anal. Geom. **7** (1989), no. 2, 93–106.
55. K. K. Jensen and K. Thomsen, *Elements of KK-theory*, Mathematics: Theory and Applications, Birkhäuser, Boston, MA, 1991.
56. B. E. Johnson, *Centralisers and operators reduced by maximal ideals*, J. London Math. Soc. **43** (1968), 231–233.
57. R. V. Kadison, *Diagonalizing matrices*, Amer. J. Math. **106** (1984), 1451–1468.
58. R. V. Kadison and J. R. Ringrose, *Fundamentals of the theory of opeator algebras*. Vols. 1–2, Graduate Studies in Math., Amer. Math. Soc., Providence, RI, 1997.
59. I. Kaplansky, *Modules over operator algebras*, Amer. J. Math. **75** (1953), 839–858.
60. M. Karoubi, *K-theory. An introduction*, Grundlehren Math. Wiss., vol. 226, Springer-Verlag, Berlin – Heidelberg – New York, 1978.
61. V. A. Kasimov, *Homotopy properties of the general linear group of the Hilbert module $l_2(A)$*, Mat. Sbornik **119** (1982), 376–386; English transl., Math. USSR-Sb. **47** (1984), 365–376.
62. G. G. Kasparov, *Topological invariants of elliptic operators, I: K-homology*, Izv. Akad. Nauk SSSR Ser. Mat. **39** (1975), 796–838; English transl., Math. USSR-Izv. **9** (1975), 751–792.
63. ———, *Hilbert C^*-modules: theorems of Stinespring and Voiculescu*, J. Operator Theory **4** (1980), 133–150.
64. ———, *The operator K-functor and extensions of C^*-algebras*, Izv. Akad. Nauk SSSR Ser. Mat. **44** (1980), 571–636; English transl., Math. USSR-Izv. **16** (1981), 513–572.
65. ———, *Novikov's conjecture on higher signatures: the operator K-theory approach*, Representation Theory of Groups and Algebras (J. Adams, R. Herb, S. Kudla, J.-S. Li, R. Lipsman, and J. Rosenberg, eds.), Contemporary Math., vol. 145, Amer. Math. Soc., Providence, RI, 1993, pp. 79–99.
66. A. Ya. Khelemskij, *Banach and locally convex algebras*, Clarendon Press, Oxford, 1993.
67. D. Kucerovsky, *Finite rank operators and functional calculus on Hilbert modules over abelian C^*-algebras*, Canad. Math. Bull. **40** (1997), no. 2, 193–197.
68. N. Kuiper, *The homotopy type of the unitary group of the Hilbert space*, Topology **3** (1965), 19–30.
69. A. Kumjian, *On equivariant sheaf cohomology and elementary C^*-bundles*, J. Operator Theory **20** (1988), 207–240.
70. E. C. Lance, *On nuclear C^*-algebras*, J. Funct. Anal. **12** (1973), 157–176.
71. ———, *Hilbert C^*-modules - a toolkit for operator algebraists*, London Math. Soc. Lecture Note Ser., vol. 210, Cambridge Univ. Press, England, 1995.
72. N. P. Landsman, *Rieffel induction as generalized quantum Marsden–Weinstein reduction*, J. Geom. Phys. **15** (1995), 285–319.
73. H. Lin, *Bounded module maps and pure completely positive maps*, J. Operator Theory **26** (1991), 121–138.
74. ———, *Injective Hilbert C^*-modules*, Pacific J. Math. **154** (1992), 131–164.
75. N. N. Lusin, *Theory of functions of real variable*, Uchpedgiz, Moscow, 1948. (Russian)
76. A. S. Lyskova, *On Schrödinger operator in magnetic field*, Uspekhi Mat. Nauk **36** (1981), 189–190; English transl., Russian Math. Surveys **36** (1981), no. 2, 181–182.
77. B. Magajna, *Hilbert C^*-modules in which all closed submodules are complemented*, Proc. Amer. Math. Soc. **125** (1997), 849–852.
78. V. M. Manuilov, *On K_0-group of continuous family of algebras A_θ*, Uspekhi Mat. Nauk **44** (1989), no. 3, 163–164; English transl., Russian Math. Surveys **44** (1989), no. 3, 206–207.
79. ———, *Diagonalization of compact operators in Hilbert modules over W^*-algebras of finite type*, Uspekhi Mat. Nauk **49** (1994), no. 2, 159–160; English transl., Russian Math. Surveys **49** (1994), 166–167.
80. ———, *On eigenvalues of perturbed Schrödinger operators in magnetic field with irrational magnetic flow*, Funktsional Anal. i Prilozhen. **28** (1994), 57–60; English transl., Functional Anal. Appl. **28** (1994), no. 2, 120–122.

81. ______, *Diagonalization of compact operators in Hilbert modules over finite W^*-algebras*, Ann. Global Anal. Geom. **13** (1995), 207–226.
82. ______, *Adjointability of operators on Hilbert C^*-modules*, Acta Math. Univ. Comenian. **65** (1996), 161–169.
83. ______, *Diagonalization of operators over continuous fields of C^*-algebras*, Mat. Sb. **188** (1997), no. 6, 99–118; English transl., Sb. Math. **188** (1997), no. 6, 893–911.
84. ______, *On almost commuting operators*, Funktsional Anal. i Prilozhen. **31** (1997), no. 3, 80–82; English transl., Functional Anal. Appl. **31** (1997), no. 3, 212–213.
85. ______, *An example of non-completed Hilbert W^*-module*, Vestnik Moskov. Univ. Ser. I Mat. Mekh. (2000), no. 5, 59–60; English transl., Moscow Univ. Math. Bull. **55** (2000), no.5, 38–39.
86. J. Milnor, *On spaces having the homotopy type of a CW-complex*, Trans. Amer. Math. Soc. **90** (1959), 272–280.
87. J. Mingo and W. Phillips, *Equivariant triviality theorems for Hilbert C^*-modules*, Proc. Amer. Math. Soc. **91** (1984), 225–230.
88. J. A. Mingo, *On the contractibility of the general linear group of Hilbert space over a C^*-algebra*, J. Integral Equations Operator Theory **5** (1982), 888–891.
89. ______, *K-theory and multipliers of stable C^*-algebras*, Trans. Amer. Math. Soc. **299** (1987), 397–411.
90. A. S. Mishchenko, *Banach algebras, pseudodifferential operators and their applications to K-theory*, Uspekhi Mat. Nauk **34** (1979), no. 6, 67–79; English transl., Russian Math. Surveys **34** (1979), no. 6, 77-91.
91. ______, *Representations of compact groups on Hilbert modules over C^*-algebras*, Trudy Mat. Inst. Steklov. **166** (1984), 161–176; English transl., Proc. Steklov Inst. Math. **166** (1986), 179–195.
92. A. S. Mishchenko and A. T. Fomenko, *The index of elliptic operators over C^*-algebras*, Izv. Akad. Nauk SSSR Ser. Mat. **43** (1979), 831–859; English transl., Math. USSR-Izv. **15** (1980) 87–112.
93. G. D. Mostow, *Cohomology of topological groups and solvmanifolds*, Ann. of Math. **73** (1961), 20–48.
94. G. J. Murphy, *C^*-algebras and operator theory*, Academic Press, San Diego, 1990.
95. ______, *Diagonality in C^*-algebras*, Math. Z. **199** (1990), 279–284.
96. F. J. Murray and J. von Neumann, *On rings of operators*, Ann. of Math. **37** (1936), 116–229.
97. G. Neubauer, *Der Homotopietyp der Automorphismegruppe in der Räumen l_p und c_0*, Math. Ann. **174** (1967), 33–40.
98. V. Nistor, *A bivariant Chern-Connes character*, Ann. of Math. **138** (1993), 555–590.
99. S. P. Novikov, *Two-dimensional Schrödinger operators in periodic fields*, Modern Problems of Math., vol. 23, VINITI, Moscow, 1983, pp. 3–32.
100. W. L. Paschke, *Inner product modules over B^*-algebras*, Trans. Amer. Math. Soc. **182** (1973), 443–468.
101. ______, *The double B-dual of an inner product module over a C^*-algebra*, Canad. J. Math. **26** (1974), 1272–1280.
102. ______, *Integrable group actions on von Neumann algebras*, Math. Scand. **40** (1977), 234–248.
103. A. A. Pavlov, *Algebras of multipliers and spaces of quasimultipliers*, Vestnik Moskov. Univ. Ser. I Mat. Mekh. (1998), no. 6, 14–18; English transl., Moscow Univ. Math. Bull. **53** (1998), 13–16.
104. G. K. Pedersen, *C^*-algebras and their automorphism groups*, Academic Press, London – New York – San Francisco, 1979.
105. P. Popov and A. Buchina, *Quasi-orthogonalization of functionals on $l_2(A)$*, Acta. Appl. Math. **68** (2001), 123–135.
106. P. S. Popov, *A topological criterion for almost orthocomplementation of all functionals on $l_2(C(X))$*, Mat. Zametki **65** (1999), no. 4, 636–640; English transl., Math. Notes **65** (1999), no. 3-4, 532–536.
107. M. Reed and B. Simon, *Methods of modern mathematical physics. V. 4. Analysis of operators*, Academic Press, New York – London, 1978.

108. M. A. Rieffel, *Induced representations of C^*-algebras*, Adv. in Math. **13** (1974), 176–257.
109. ———, *Morita equivalence for C^*-algebras and W^*-algebras*, J. Pure Appl. Algebra **5** (1974), 51–96.
110. ———, *Strong Morita equivalence of certain transformation group C^*-algebras*, Math. Ann. **222** (1976), 7–22.
111. ———, *C^*-algebras associated with irrational rotations*, Pacific J. Math. **93** (1981), 415–429.
112. ———, *Continuous fields of C^*-algebras coming from group cocycles and actions*, Math. Ann. **283** (1989), 631–643.
113. J. Rosenberg, *C^*-algebras, positive scalar curvature, and the Novikov conjecture*, Publ. Math. I.H.E.S. **58** (1983), 197–212.
114. ———, *K and KK: Topology and operator algebras*, Proc. Sympos. Pure Math. **51-1** (1990), 445–480.
115. W. Rudin, *Functional analysis*, McGraw-Hill, New York, 1973.
116. S. Sakai, *C^*-algebras and W^*-algebras*, Springer-Verlag, Berlin – New York, 1971.
117. H. Schröder, *K-theory for real C^*-algebras and applications*, Pitman Res. Notes Math. Ser., Longman Sci. Tech., Harlow, 1993.
118. V. V. Seregin, *Reflexivity of C^*-Hilbert modules obtained by the actions of a group*, Vestnik Moskov. Univ. Ser. I Mat. Mekh. (2003), no. 1, 40–45; English transl., Moscow Univ. Math. Bull. **58** (2003), no. 1, 44–48.
119. G. Skandalis, *Some remarks on Kasparov's theory*, J. Funct. Anal. **56** (1984), 337–347.
120. ———, *Kasparov's bivariant K-theory and applications*, Expositiones Math. **9** (1991), 193–250.
121. Yu. P. Solovyov and E. V. Troitsky, *C^*-algebras and elliptic operators in differential topology*, Amer. Math. Soc., Providence, RI, 2001.
122. V. S. Sunder and K. Thomsen, *Unitary orbits of selfadjoints in some C^*-algebras*, Houston J. Math. **18** (1992), 127–137.
123. M. Takesaki, *Theory of operator algebras, 1*, Springer-Verlag, New York – Heidelberg – Berlin, 1979.
124. V. A. Trofimov, *Reflexivity of Hilbert modules over the algebra of compact operators with adjoint identity*, Vestnik Moskov. Univ. Ser. I Mat. Mekh. (1986), no. 5, 60–64; English transl., Moscow Univ. Math. Bull. **41** (1986), no.5, 51–55.
125. ———, *Reflexive and self-dual Hilbert modules over some C^*-algebras*, Uspekhi Mat. Nauk **42** (1987), no. 2, 247–248; English transl., Russian Math. Surveys **42** (1987), 303–304.
126. E. V. Troitsky, *The classifying space of a K-functor related to a C^*-algebra*, Vestnik Moskov. Univ. Ser. I Mat. Mekh. (1985), no. 1, 96–98; English transl., Moscow Univ. Math. Bull. **40** (1985), 111–115.
127. ———, *Contractibility of the full general linear group of the C^*-Hilbert module $l_2(A)$*, Funktsional Anal. i Prilozhen. **20** (1986), no. 4, 50–64; English transl., Functional Anal. Appl. **20** (1986), 301–307.
128. ———, *The eqivariant index of elliptic operators over C^*-algebras*, Izv. Akad. Nauk SSSR Ser. Mat. **50** (1986), no. 4, 849–865; English transl., Math. USSR-Izv. **29** (1987), 207–224.
129. ———, *Orthogonal complements and endomorphisms of Hilbert modules and C^*-elliptic complexes*, Novikov Conjectures, Index Theorems and Rigidity, Vol. 2 (S. C. Ferry, A. Ranicki, and J. Rosenberg, eds.), London Math. Soc. Lecture Note Ser., vol. 227, Cambridge Univ. Press, 1995, pp. 309–331.
130. ———, *Functionals on $l_2(A)$, Kuiper and Dixmier-Douady type theorems for C^*-Hilbert modules*, Trudy Mat. Inst. Steklov. **225** (1999), 362–380; English transl., Proc. Steklov Inst. Math. **225** (1999), 344–362.
131. ———, *Discrete group actions and corresponding modules*, Proc. Amer. Math. Soc. **131** (2003), 3411–3422.
132. A. Valette, *Les fibrés en théorie de Kasparov*, Mem. Acad. Roy. Belg. Cl. Sci. **XLV** (1988), no. 6.
133. Y. Watatani, *Index for C^*-subalgebras*, Mem. Amer. Math. Soc., vol. 424, Amer. Math. Soc., Providence, RI, 1990.
134. N. E. Wegge-Olsen, *K-theory and C^*-algebras*, Oxford Univ. Press, Oxford, 1993.

135. S. L. Woronowicz, *Unbounded elements affiliated with C^*-algebras and non-compact quantum groups*, Comm. Math. Phys. **136** (1991), 399–432.
136. S. L. Woronowicz and K. Napiórkowski, *Operator theory in the C^*-algebra framework*, Rep. Math. Phys. **31** (1992), 353–371.
137. S. Zhang, *Diagonalizing projections in multiplier algebras and in matrices over a C^*-algebra*, Pacific J. Math. **145** (1990), 181–200.

Notation Index

Index

Titles in This Series

226 **V. M. Manuilov and E. V. Troitsky,** Hilbert C^*-modules, 2005

225 **S. M. Natanzon,** Moduli of Riemann surfaces, real algebraic curves, and their superanaloges, 2004

224 **Ichiro Shigekawa,** Stochastic analysis, 2004

223 **Masatoshi Noumi,** Painlevé equations through symmetry, 2004

222 **G. G. Magaril-Il'yaev and V. M. Tikhomirov,** Convex analysis: Theory and applications, 2003

221 **Katsuei Kenmotsu,** Surfaces with constant mean curvature, 2003

220 **I. M. Gelfand, S. G. Gindikin, and M. I. Graev,** Selected topics in integral geometry, 2003

219 **S. V. Kerov,** Asymptotic representation theory of the symmetric group and its applications to analysis, 2003

218 **Kenji Ueno,** Algebraic geometry 3: Further study of schemes, 2003

217 **Masaki Kashiwara,** D-modules and microlocal calculus, 2003

216 **G. V. Badalyan,** Quasipower series and quasianalytic classes of functions, 2002

215 **Tatsuo Kimura,** Introduction to prehomogeneous vector spaces, 2003

214 **L. Š. Grinblat,** Algebras of sets and combinatorics, 2002

213 **V. N. Sachkov and V. E. Tarakanov,** Combinatorics of nonnegative matrices, 2002

212 **A. V. Mel′nikov, S. N. Volkov, and M. L. Nechaev,** Mathematics of financial obligations, 2002

211 **Takeo Ohsawa,** Analysis of several complex variables, 2002

210 **Toshitake Kohno,** Conformal field theory and topology, 2002

209 **Yasumasa Nishiura,** Far-from-equilibrium dynamics, 2002

208 **Yukio Matsumoto,** An introduction to Morse theory, 2002

207 **Ken'ichi Ohshika,** Discrete groups, 2002

206 **Yuji Shimizu and Kenji Ueno,** Advances in moduli theory, 2002

205 **Seiki Nishikawa,** Variational problems in geometry, 2001

204 **A. M. Vinogradov,** Cohomological analysis of partial differential equations and Secondary Calculus, 2001

203 **Te Sun Han and Kingo Kobayashi,** Mathematics of information and coding, 2002

202 **V. P. Maslov and G. A. Omel′yanov,** Geometric asymptotics for nonlinear PDE. I, 2001

201 **Shigeyuki Morita,** Geometry of differential forms, 2001

200 **V. V. Prasolov and V. M. Tikhomirov,** Geometry, 2001

199 **Shigeyuki Morita,** Geometry of characteristic classes, 2001

198 **V. A. Smirnov,** Simplicial and operad methods in algebraic topology, 2001

197 **Kenji Ueno,** Algebraic geometry 2: Sheaves and cohomology, 2001

196 **Yu. N. Lin′kov,** Asymptotic statistical methods for stochastic processes, 2001

195 **Minoru Wakimoto,** Infinite-dimensional Lie algebras, 2001

194 **Valery B. Nevzorov,** Records: Mathematical theory, 2001

193 **Toshio Nishino,** Function theory in several complex variables, 2001

192 **Yu. P. Solovyov and E. V. Troitsky,** C^*-algebras and elliptic operators in differential topology, 2001

191 **Shun-ichi Amari and Hiroshi Nagaoka,** Methods of information geometry, 2000

190 **Alexander N. Starkov,** Dynamical systems on homogeneous spaces, 2000

189 **Mitsuru Ikawa,** Hyperbolic partial differential equations and wave phenomena, 2000

188 **V. V. Buldygin and Yu. V. Kozachenko,** Metric characterization of random variables and random processes, 2000

187 **A. V. Fursikov,** Optimal control of distributed systems. Theory and applications, 2000

186 **Kazuya Kato, Nobushige Kurokawa, and Takeshi Saito,** Number theory 1: Fermat's dream, 2000

185 **Kenji Ueno,** Algebraic Geometry 1: From algebraic varieties to schemes, 1999

184 **A. V. Mel′nikov,** Financial markets, 1999

183 **Hajime Sato,** Algebraic topology: an intuitive approach, 1999

182 **I. S. Krasil′shchik and A. M. Vinogradov, Editors,** Symmetries and conservation laws for differential equations of mathematical physics, 1999

181 **Ya. G. Berkovich and E. M. Zhmud′,** Characters of finite groups. Part 2, 1999

180 **A. A. Milyutin and N. P. Osmolovskii,** Calculus of variations and optimal control, 1998

179 **V. E. Voskresenskiĭ,** Algebraic groups and their birational invariants, 1998

178 **Mitsuo Morimoto,** Analytic functionals on the sphere, 1998

177 **Satoru Igari,** Real analysis—with an introduction to wavelet theory, 1998

176 **L. M. Lerman and Ya. L. Umanskiy,** Four-dimensional integrable Hamiltonian systems with simple singular points (topological aspects), 1998

175 **S. K. Godunov,** Modern aspects of linear algebra, 1998

174 **Ya-Zhe Chen and Lan-Cheng Wu,** Second order elliptic equations and elliptic systems, 1998

173 **Yu. A. Davydov, M. A. Lifshits, and N. V. Smorodina,** Local properties of distributions of stochastic functionals, 1998

172 **Ya. G. Berkovich and E. M. Zhmud′,** Characters of finite groups. Part 1, 1998

171 **E. M. Landis,** Second order equations of elliptic and parabolic type, 1998

170 **Viktor Prasolov and Yuri Solovyev,** Elliptic functions and elliptic integrals, 1997

169 **S. K. Godunov,** Ordinary differential equations with constant coefficient, 1997

168 **Junjiro Noguchi,** Introduction to complex analysis, 1998

167 **Masaya Yamaguti, Masayoshi Hata, and Jun Kigami,** Mathematics of fractals, 1997

166 **Kenji Ueno,** An introduction to algebraic geometry, 1997

165 **V. V. Ishkhanov, B. B. Lur′e, and D. K. Faddeev,** The embedding problem in Galois theory, 1997

164 **E. I. Gordon,** Nonstandard methods in commutative harmonic analysis, 1997

163 **A. Ya. Dorogovtsev, D. S. Silvestrov, A. V. Skorokhod, and M. I. Yadrenko,** Probability theory: Collection of problems, 1997

162 **M. V. Boldin, G. I. Simonova, and Yu. N. Tyurin,** Sign-based methods in linear statistical models, 1997

161 **Michael Blank,** Discreteness and continuity in problems of chaotic dynamics, 1997

160 **V. G. Osmolovskiĭ,** Linear and nonlinear perturbations of the operator div, 1997

159 **S. Ya. Khavinson,** Best approximation by linear superpositions (approximate nomography), 1997